Statistical Signal Processing for Neuroscience and Neurotechnology

Statistical Signal Processing for Neuroscience and Neurotechnology

Edited by
Karim G. Oweiss

AMSTERDAM • BOSTON • HEIDELBERG • LONDON
NEW YORK • OXFORD • PARIS • SAN DIEGO
SAN FRANCISCO • SINGAPORE • SYDNEY • TOKYO
Academic Press is an imprint of Elsevier

Academic Press is an imprint of Elsevier
30 Corporate Drive, Suite 400, Burlington, MA 01803, USA
The Boulevard, Langford Lane, Kidlington, Oxford, OX5 1GB, UK

Library of Congress Cataloging-in-Publication Data
Statistical signal processing for neuroscience and neurotechnology / edited by Karim G. Oweiss.
    p. ; cm.
    Includes bibliographical references and index.
    ISBN 978-0-12-375027-3
1. Electroencephalography. 2. Signal processing–Digital techniques. 3. Brain mapping. I. Oweiss, Karim G.
    [DNLM: 1. Brain–physiology. 2. Signal Processing, Computer-Assisted. 3. Brain Mapping–instrumentation.
4. Man-Machine Systems. 5. Prostheses and Implants. 6. User-Computer Interface. WL 26.5 S7976 2011]
    QP376.5.S73 2011
    616.8'047547–dc22

                                                                    2010012521

British Library Cataloguing-in-Publication Data
A catalogue record for this book is available from the British Library.

For information on all Academic Press publications,
visit our Web site at *www.elsevierdirect.com*

Typeset by: diacriTech, India

Printed in the United States of America
10 11 12 13 14 15   9 8 7 6 5 4 3 2 1

Contents

Preface

Neuroscience, like many other areas of biology, has experienced a data explosion in the last few years. Many remarkable findings have been triggered since and have undeniably improved our understanding of many aspects of the nervous system. However, the study of the brain and its inner workings is an extraordinarily complex endeavor; undoubtedly it requires the collective effort of many experts across multiple disciplines to accelerate the science of information management that neuroscience data have invigorated.

Signal processing and statistics have a long history of mutual interaction with neuroscience, going back to the Hudgkin-Huxley circuit models of membrane potential and leading up to statistical inference about stimulus–response relationships using Bayes' rule. Classical signal processing, however, quickly erodes in the face of dynamic, multivariate neuroscience and behavioral data. Realizing the immediate need for a rigorous theoretical and practical analysis of such data, this book was written by experts in the field to provide a comprehensive reference summarizing state-of-the-art solutions to some fundamental problems in neuroscience and neurotechnology. It is intended for theorists and experimentalists at the graduate and postdoctoral level in electrical, computer, and biomedical engineering, as well as for mathematicians, statisticians, and computational and systems neuroscientists. A secondary audience may include neurobiologists, neurophysiologists, and clinicians involved in basic and translational neuroscience research. The book can also be used as a supplement in a quantitative course in neurobiology or as a textbook for instruction on neural signal processing and its applications. Industry partners who want to explore research in and clinical applications of these concepts in the context of neural interface technology will certainly find useful content in a number of chapters.

This book was written with the following goals in mind:

- To serve as an introductory text for readers interested in learning how traditional and modern principles, theories, and techniques in statistical signal processing, information theory, and machine learning can be applied to neuroscience problems.

- To serve as a reference for researchers and clinicians working at the interface between neuroscience and engineering.

- To serve as a textbook for graduate-level courses within an electrical engineering, biomedical engineering, or computational and systems neuroscience curriculum.

Readers are expected to have had an introductory course in probability and statistics at the undergraduate level as part of an engineering or neuroscience course of study. The book is organized so that it progressively addresses specific problems arising in a neurophysiological experiment or in the design of a brain-machine interface system. An ultimate goal is to encourage the next generation of scientists to develop advanced statistical signal processing theory and techniques in their pursuit of answers to the numerous intricate questions that arise in the course of these experiments.

I owe a considerable amount of gratitude to my fellow contributors and their willingness to spare a significant amount of time to write, edit, and help review every chapter in this book since its inception. I am also very thankful to a number of colleagues who helped me in the review of early versions of the manuscript.

Developing the tools needed to understand brain function and its relation to behavior from the collected data is no less important than developing the devices used to collect these data, or developing the biological techniques to decipher the brain's anatomical structure. I believe this book is a first step towards fulfilling an urgent need to hasten the development of these tools and techniques that will certainly thrive in the years to come.

Karim G. Oweiss
East Lansing, Michigan

About the Editor

Karim G. Oweiss received his B.S. (1993) and M.S. (1996) degrees with honors in electrical engineering from the University of Alexandria, Egypt, and his Ph.D. (2002) in electrical engineering and computer science from the University of Michigan, Ann Arbor. In that year he also completed postdoctoral training with the Department of Biomedical Engineering at the University of Michigan. In 2003, he joined the Department of Electrical and Computer Engineering and the Neuroscience Program at Michigan State University, where he is currently an associate professor and director of the Neural Systems Engineering Laboratory. His research interests are in statistical signal processing, information theory, machine learning, and control theory, with direct applications to studies of neuroplasticity, neural integration and coordination in sensorimotor systems, neurostimulation and neuromodulation in brain-machine interfaces, and computational neuroscience.

Professor Oweiss is a member of the IEEE and the Society for Neuroscience. He served as a member of the board of directors of the IEEE Signal Processing Society on Brain-Machine Interfaces and is currently an active member of the technical and editorial committees of the IEEE Biomedical Circuits and Systems Society, the IEEE Life Sciences Society, and the IEEE Engineering in Medicine and Biology Society. He is also associate editor of *IEEE Signal Processing Letters*, *Journal of Computational Intelligence and Neuroscience*, and *EURASIP Journal on Advances in Signal Processing*. He currently serves on an NIH Federal Advisory Committee for the Emerging Technologies and Training in Neurosciences. In 2001, Professor Oweiss received the Excellence in Neural Engineering Award from the National Science Foundation.

About the Contributors

Chapter 2

Karim Oweiss (see About the Editor)

Mehdi Aghagolzadeh received his B.Sc. degree in electrical engineering with honors from University of Tabriz, Iran, in 2003 and his M.Sc. degree in electrical engineering (with a specialty in signal processing and communication systems) from the University of Tehran in 2006. Since September 2007, he has been a research assistant in the Neural Systems Engineering Lab (directed by Professor Karim Oweiss) working toward a Ph.D. degree in the Department of Electrical and Computer Engineering at Michigan State University, East Lansing. His research interests include statistical signal processing, neural engineering, information theory, and system identification.

Chapter 3

Don H. Johnson received his S.B. and S.M. degrees in 1970, his E.E. degree in 1971, and his Ph.D. degree in 1974, all in electrical engineering, from the Massachusetts Institute of Technology. After spending three years at MIT's Lincoln Laboratory working on digital speech systems, he joined the Department of Electrical and Computer Engineering at Rice University, where he is currently J.S. Abercrombie Professor Emeritus. Professor Johnson is a fellow of the IEEE. He is also a recipient of the IEEE Millennium Medal and the Signal Processing Society's Meritorious Service Award. He is former a president of the Signal Processing Society and a former chair of the IEEE Kilby Medal Committee. Currently, he chairs the program committee of the Computational Neuroscience Meeting. Professor Johnson co-authored the graduate-level textbook *Array Signal Processing*, published in 1993. He holds four patents.

Chapter 4

Dong Song received his B.S. degree in biophysics from the University of Science and Technology of China, Hefei, in 1994 and his doctorate in biomedical engineering from the University of Southern California (USC), Los Angeles, in 2003. From 2004 to 2006, he worked as a postdoctoral research associate at the Center for Neural Engineering at USC. He is currently a research assistant professor in the USC's Department of Biomedical Engineering. Professor Song's main research interests are nonlinear systems analysis of the nervous system, cortical neural prostheses, electrophysiology of the hippocampus, long-term and short-term synaptic plasticity, and the development of modeling methods incorporating both parametric (mechanistic) and nonparametric (input-output) modeling techniques. He is a member of the Biomedical Engineering Society, the American Statistical Association, and the Society for Neuroscience.

Theodore W. Berger holds the David Packard Professorship in Engineering at USC, where he is also a professor of biomedical engineering and neuroscience, and director of the Center for Neural Engineering. He received his Ph.D. from Harvard University in 1976. In 1979 he joined the departments of Neuroscience and Psychiatry at the University of Pittsburgh and became a full professor in 1987. Since 1992, Dr. Berger has been a professor of biomedical engineering and neurobiology at USC and was awarded the David Packard Chair of Engineering in 2003. In 1997, he became director of the Center for Neural Engineering, an organization that helps to unite USC faculty with cross-disciplinary interests in neuroscience, engineering, and medicine.

Dr. Berger's research interests are in the development of biologically realistic, experimentally based mathematical models of higher brain (hippocampus) function; the application of biologically realistic neural network models to real-world signal processing problems; VLSI-based implementations of biologically realistic models of higher brain function; neuron–silicon interfaces for bi-directional communication between the brain and VLSI systems; and next-generation brain-implantable, biomimetic signal processing devices for neural prosthetic replacement and/or enhancement of brain function.

Chapter 5

Seif Eldawlatly received hisB.Sc. and M.Sc. degrees in electrical engineering from the Computer and Systems Engineering Department, Ain Shams University, Cairo, in 2003 and 2006, respectively. Since August 2006, he has been a research assistant in the Neural Systems Engineering Lab (directed by Professor Karim Oweiss) and working toward his Ph.D. in the Department of Electrical and Computer Engineering, Michigan State University, East Lansing. His research focuses on developing signal processing and machine learning techniques to understand brain connectivity at the neuronal level.

Karim G. Oweiss (see About the Editor)

Chapter 6

Zhe Chen received his Ph.D. degree in electrical and computer engineering in 2005 from McMaster University in Hamilton, Ontario. From 2001 to 2004, he worked there as a research assistant in the Adaptive Systems Laboratory (directed by Professor Simon Haykin). During the summer of 2002, Dr. Chen was a summer intern at Bell Laboratories, Lucent Technologies, in Murray Hill, New Jersey. Upon completion of his doctorate, he joined the RIKEN Brain Science Institute, Saitama, Japan, in June 2005 and worked as a research scientist in the Laboratory of Advanced Brain Signal Processing (headed by Professor Shun-ichi Amari and Professor Andrzej Cichocki). Since March 2007 Dr. Chen has worked in the Neuroscience Statistics Research Laboratory (directed by Professor Emery Brown) at Massachusetts General Hospital and the Harvard Medical School, and in the Department of Brain and Cognitive Sciences at MIT. His main research interests are neural signal processing, neural computation and machine learning, Bayesian modeling, and computational neuroscience.

Riccardo Barbieri received his M.S. degree in electrical engineering from the University of Rome "La Sapienza" in 1992, and his Ph.D. in biomedical engineering from Boston University in 1998. Currently he is an assistant professor of anaesthesia at Harvard Medical School and Massachusetts General Hospital, and a research affiliate at MIT. His main research interests are in the development of signal processing algorithms for analysis of biological systems, and he is currently focusing his studies on the application of multivariate and statistical models to characterize heart rate and heart rate variability as related to cardiovascular control dynamics. He is also studying computational modeling of neural information encoding.

Emery N. Brown received his B.A. degree in applied mathematics from Harvard College, his M.D. from Harvard Medical School, and his A.M and Ph.D. degrees in statistics from Harvard University. He is a professor of computational neuroscience and health sciences and technology at MIT and the Warren M. Zapol Professor of Anaesthesia at Harvard Medical School and Massachusetts General Hospital. In his experimental research, Dr. Brown uses functional imaging techniques to study how anesthetic drugs act in the human brain to create the state of general anesthesia. In his statistics research, he develops signal processing algorithms and data statistical methods to analyze a broad range of neuroscience data.

Professor Brown is a fellow of the American Statistical Association (ASA), a fellow of the American Association for the Advancement of Science (AAAS), a fellow of the IEEE, a fellow of the American Institute of

Biomedical Engineering, and a member of the Institute of Medicine. In 2007 he received an NIH Director's Pioneer Award.

Chapter 7

Byron M. Yu received his B.S. degree in electrical engineering and computer sciences from the University of California, Berkeley, in 2001 and his M.S. and Ph.D. degrees in electrical engineering in 2003 and 2007, respectively, from Stanford University. From 2007 to 2009, he was a postdoctoral fellow jointly in electrical engineering and neuroscience at Stanford and at the Gatsby Computational Neuroscience Unit, University College London. He then joined the faculty of Carnegie Mellon University, Pittsburgh, where he is an assistant professor in electrical and computer engineering and biomedical engineering.

Professor Yu is a recipient of a National Defense Science and Engineering Graduate Fellowship and a National Science Foundation Graduate Fellowship. His research interests focus on statistical techniques for characterizing multi-channel neural data and their application to the study of neural processing, and on decoding algorithms for neural prosthetic systems.

Gopal Santhanam received his B.S. degree in electrical engineering and computer science and his B.A. degree in physics from the University of California, Berkeley, and his M.S. and Ph.D. degrees in electrical engineering from Stanford University in 2002 and 2006, respectively. His research involves neural prosthetics system design, neural signal processing, and embedded neural recording systems. He also has extensive industry experience in high technology, especially embedded systems, through various positions and consulting projects.

Dr. Santhanam is the recipient of notable awards, including a National Defense Science and Engineering Graduate fellowship and a National Science Foundation Graduate Fellowship.

Maneesh Sahani attended the California Institute of Technology (Caltech), where he earned a B.S. in Physics in 1993 and a Ph.D. in Computation and Neural Systems in 1999. Between 1999 and 2002, he was a senior research fellow at the Gatsby Computational Neuroscience Unit, University College, London (UCL), and then moved to the University of California, San Francisco, for further postdoctoral research. He returned to UCL to join the faculty of the Gatsby Unit in 2004 and is now a reader in theoretical neuroscience and machine learning. He is also a visiting member of the faculty in the Department of Electrical Engineering at Stanford University. Professor Sahani's research interests include probabilistic methods as applied to neural data analysis, behavioral modeling, and machine perception.

Krishna V. Shenoy received his B.S. degree in electrical engineering from the University of California, Irvine, in 1990, and his S.M. and Ph.D. degrees in electrical engineering from MIT in 1992 and 1995, respectively. He was a neurobiology postdoctoral fellow at Caltech from 1995 to 2001. Presently Dr. Shenoy is a member of the faculty at Stanford University, where he is an associate professor in the departments of Electrical Engineering and Bioengineering, and in the Neurosciences Program. His research interests include computational motor neurophysiology and neural prosthetic system design.

Dr. Shenoy received the 1996 Hertz Foundation Doctoral Thesis Prize, a Burroughs Wellcome Fund Career Award in the Biomedical Sciences, an Alfred P. Sloan Research Fellowship, a McKnight Endowment Fund in Technological Innovations in Neurosciences Award, and a 2009 NIH Director's Pioneer Award.

Chapter 8

Antonio R. C. Paiva received his *licenciatura* degree in electronics and telecommunications engineering from the University of Aveiro, Portugal, in 2003, and his M.S. and Ph.D. degrees in electrical and computer engineering from the University of Florida, Gainesville, in 2005 and 2008, respectively. After completing his undergraduate studies, he joined the Institute for Electronics and Telematics of Aveiro University, working with Dr. A. Pinho in image compression. In the fall of 2004, he joined the Computational Neuro-Engineering Laboratory at the University of Florida, where he worked under the supervision of Dr. Jose C. Principe. His doctoral dissertation focused on the development of a reproducing kernel Hilbert spaces framework for analysis and processing of point processes, with applications to single-unit neural spike trains.

Currently Dr. Paiva is a postdoctoral fellow in the Scientific Computing and Imaging Institute at the University of Utah, working on neural circuitry reconstruction and on the application of machine learning to image analysis and natural vision systems. His research interests are signal and image processing, with special emphasis on biomedical and biological applications.

Il Park, also known as Memming, received his BS degree in computer science from Korea Advanced Institute of Science and Technology (KAIST) and his MS degree in electrical engineering from the University of Florida in 2007. He is currently a doctoral candidate in biomedical engineering at the University of Florida. His current research interests include statistical modeling and analysis of spike trains, information-theoretic learning, and computational neuroscience.

Jose C. Principe is Distinguished Professor of Electrical and Biomedical Engineering at the University of Florida, Gainesville, where he teaches advanced signal processing and machine learning. He received his BE in electrical engineering from the University of Porto (Bachelor), Portugal, in 1972, and his M.Sc and Ph.D. from the University of Florida in 1974 and 1979, respectively. He is BellSouth Professor and founder and director of the University of Florida Computational NeuroEngineering Laboratory (CNEL), where he conducts research in biomedical signal processing, in particular brain–machine interfaces and the modeling and applications of cognitive systems. He has authored four books and more than 160 publications in refereed journals, and over 350 conference papers. He has also directed over 60 Ph.D. dissertations and 61 master's theses.

Dr. Principe is a fellow of both the IEEE and the AIMBE and is a recipient of the IEEE Engineering in Medicine and Biology Society Career Achievement Award. He is also a former member of the scientific board of the Food and Drug Administration and a member of the advisory board of the McKnight Brain Institute at the University of Florida. He is editor-in-chief of *IEEE Reviews on Biomedical Engineering*, past editor-in-chief of *IEEE Transactions on Biomedical Engineering*, past president of the International Neural Network Society, and former secretary of the Technical Committee on Neural Networks of the IEEE Signal Processing Society.

Chapter 9

Paul Sajda is associate professor of biomedical engineering and radiology (physics) at Columbia University, New York, where he is also the director of the Laboratory for Intelligent Imaging and Neural Computing (LIINC). He received his B.S. degree in Electrical Engineering from MIT and his M.S.E. and Ph.D. degrees in bioengineering from the University of Pennsylvania in 1992 and 1994, respectively.

Robin I. Goldman is an associate research scientist at the Laboratory for Intelligent Imaging and Neural Computing (LIINC) at Columbia University. She received her B.S. degree in physics from the University of California, Santa Cruz, and her Ph.D. in biomedical physics from UCLA in 1994 and 2002, respectively.

Mads Dyrholm is a postoctoral fellow in the Center for Visual Cognition at the University of Copenhagen and was previously a postdoctoral fellow in the Laboratory for Intelligent Imaging and Neural Computing (LIINC) at Columbia University. He received his Ph.D. from the Technical University of Denmark in 2006.

Truman R. Brown is both Percy Kay and Vida L. W. Hudson Professor of Biomedical Engineering and Radiology (Physics) and director of the Hatch MR Center at Columbia University. He received his B.S. and Ph.D. degrees from MIT in 1967 and 1970, respectively.

Chapter 10

Rajesh P. N. Rao is an associate professor in the Computer Science and Engineering department and head of the Neural Systems Laboratory, both at the University of Washington, Seattle. He received his PhD from the University of Rochester in 1998 and was an Alfred P. Sloan postdoctoral fellow at the Salk Institute for Biological Studies in San Diego. Professor Rao is a recipient of an NSF CAREER award, an ONR Young Investigator Award, a Sloan Faculty Fellowship, and a David and Lucile Packard Fellowship for Science and Engineering. He has co-edited two books, *Probabilisitic Models of the Brain* and *Bayesian Brain*, both published by MIT Press. His research spans computational neuroscience, robotics, and brain−machine interfaces.

Reinhold Scherer is a postdoctoral research fellow in the Department of Computer Science and Engineering at the University of Washington, Seattle. In 2008 he received his doctorate in computer science from Graz University of Technology, Graz, Austria. Professor Scherer is a member of the Neural Systems Laboratory and the Neurobotics Laboratory at the University of Washington, and a member of the Institute for Neurological Rehabilitation and Research, Judendorf-Strassengel Clinic, Austria. His research interests include brain−machine interfaces based on EEG and ECoG signals, statistical and adaptive signal processing, and robotics-mediated rehabilitation.

Chapter 11

Emily R. Oby received her B.S. degree in physics from Denison University, Granville, Ohio, in 2002, and her M.A. degree in physics and astronomy from Johns Hopkins University, Baltimore, in 2004. She is currently a doctoral student in the Northwestern University Interdepartmental Neuroscience Program. Her primary research interests are the study of the plasticity and adaptation of cortical motor representations and the development of a cortically controlled functional electrical stimulation system.

Matthew J. Bauman graduated from Goshen College, Goshen, Indiana, in 2006 with a B.A. degree in physics and mathematics. He is currently a lab technician in the physiology department at Northwestern University, Chicago, in the Limb Motor Control Laboratory of Dr. Lee Miller. He plans to pursue a Ph.D. in biomedical engineering.

Christian Ethier received his B.Eng. degree in electrical engineering and his Ph.D. degree in neurobiology from Laval University, Quebec City, in 2002 and 2008, respectively. He is a postdoctoral fellow in the Department of Physiology at Northwestern University, where he is developing a brain–machine interface for the restoration of grasp function in a primate model of spinal cord injury. His research interests include the organization and function of the motor cortex and the development of neuroprostheses.

Jason H. Ko received his B.A. degree in economics with minors in philosophy and chemistry from Duke University, Durham, North Carolina, in 1999, and his M.D. degree from the Duke University School of Medicine in 2004. He is currently a resident in the Integrated Plastic and Reconstructive Surgery program at Northwestern University. His research interests include targeted reinnervation, peripheral nerve surgery, and neuroma physiology and treatment.

Eric J. Perreault received his B.Eng (1989) and M. Eng (1992) degrees in electrical engineering from McGill University, Montreal, and his Ph.D. (2000) in biomedical engineering from Case Western Reserve University, Cleveland. He also completed a postdoctoral fellowship in physiology at Northwestern University. Currently, he is an associate professor at Northwestern University, with appointments in the departments of Biomedical Engineering and Physical Medicine and Rehabilitation. He also is a member of the Sensory Motor Performance Program at the Rehabilitation Institute of Chicago. Professor Perreault's research focuses on the neural and biomechanical factors involved in the normal control of movement and posture, and their modification following stroke and spinal cord injury. The goal of his research is to provide a scientific basis to guide rehabilitative strategies and to develop user interfaces for restoring function to individuals with motor deficits.

Lee E. Miller received his B.A. degree in physics from Goshen College, Goshen, Indiana, in 1980 and his M.S. degree in biomedical engineering and Ph.D. in physiology from Northwestern University in 1983 and 1989, respectively. He also completed two years of postdoctoral training in the Department of Medical Physics, University of Nijmegen, The Netherlands. Currently a professor in the departments of Physiology, Physical Medicine and Rehabilitation, and Biomedical Engineering at Northwestern, Professor Miller's primary research interests are in the cortical control of muscle activity and limb movement, and in the development of brain–machine interfaces that attempt to mimic normal physiological systems.

Introduction

Karim G. Oweiss

1.1 BACKGROUND

Have you ever wondered what makes the human brain so unique compared to nonhuman brains with similar building blocks when it comes to the many complex functions it can undertake, such as instantaneously recognizing faces, reading a piece of text, or playing a piano piece with seemingly very little effort? Although this long-standing question governs the founding principles of many areas of neuroscience, the last two decades have witnessed a paradigm shift in the way we seek to answer it, along with many others.

For over a hundred years, it was believed that profound understanding of the neurophysiological mechanisms underlying behavior required monitoring the activity of the brain's basic computational unit, the neuron. Since the late 1950s, techniques for intra- and extracellular recording of single-unit activity have dominated the analysis of brain function because of perfection in isolating and characterizing individual neurons' physiological and anatomical characteristics (Gesteland et al., 1959; Giacobini et al., 1963; Evarts, 1968; Fetz and Finocchio, 1971). Many remarkable findings have fundamentally rested on the success of these techniques and have largely contributed to the foundation of many areas of neuroscience, such as computational and systems neuroscience.

Monitoring the activity of a single unit while subjects perform certain tasks, however, hardly permits gaining insight into the dynamics of the underlying neural system. It is widely accepted that such insight is contingent upon the scrutiny of the collective and coordinated activity of many neural elements, ranging from single units to small voxels of neural tissue, many of which may not be locally observed. Not surprisingly, there has been a persistent need to *simultaneously* monitor the coordinated activity of these elements—*within* and *across* multiple areas—to gain a better understanding

Statistical Signal Processing for Neuroscience and Neurotechnology. DOI: 10.1016/B978-0-12-375027-3.00001-6

of the mechanisms underlying complex functions such as perception, learning, and motor processing.

Fulfilling such a need has turned out to be an intricate task given the *space-time trade-off*. As Figure 1.1 illustrates, the activities of neural elements can be measured at a variety of temporal and spatial scales. Action potentials (APs) elicited by individual neurons—or *spikes*—are typically 1–2 ms in duration and are recorded intra- or extracellularly with penetrating microelectrodes (McNaughton et al., 1983; Gray et al., 1995; Nicolelis, 1999) or through voltage-sensitive dye-based calcium imaging at shallow depths (Koester and Sakmann, 2000; Stosiek et al., 2003). These spike trains are all-or-none representations of each neuron's individual discharge pattern. Local field potentials (LFPs) can also be recorded with penetrating microelectrodes and are known to represent the synchronized activity of large populations of neurons (Mitzdorf, 1985). Recent studies suggest that they reflect aggregated and sustained local population activity within $\sim 250\,\mu$m of the recording electrode (Katzner et al., 2009), typical of somato-dendritic currents (Csicsvari et al., 2003). Electrocorticograms (ECoGs)—also known as intracranial EEG (iEEG)—are intermediately recorded using subdural electrode grids implanted through skull penetration and therefore are considered semi-invasive as they do not penetrate the blood-brain barrier. Originally discovered in the 1950s (Penfield and Jasper, 1954), they are believed to reflect synchronized post-synaptic potentials aggregated over a few tens of

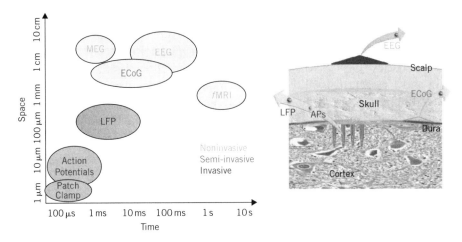

FIGURE 1.1

Spatial and temporal characteristics of neural signals recorded from the brain. Please see this figure in color at the companion web site: www.elsevierdirect.com/companions/9780123750273

millimeters. Noninvasive techniques include functional magnetic resonance imaging (fMRI), magnetoencephalography (MEG), and electroencephalography (EEG). The latter signals are recorded with surface electrodes patched to the scalp and are thought to represent localized activity in a few cubic centimeters of brain tissue, but they are typically smeared out because of the skull's low conductivity effect (Mitzdorf, 1985).

Despite the immense number of computing elements and the inherent difficulty in acquiring reliable, sustained recordings over prolonged periods of time (Edell et al., 1992; Turner et al., 1999; Kralik et al., 2001), information processing at the individual and population levels remains vital to understand the complex adaptation mechanisms inherent in many brain functions that can only be studied at the microscale, particularly those related to synaptic plasticity associated with learning and memory formation (Ahissar et al., 1992). Notwithstanding the existence of microwire bundles since the 1980s, the 1990s have witnessed some striking advances in solid-state technology allowing microfabrication of high-density microelectrode arrays (HDMEAs) on single substrates to be streamlined (Drake et al., 1998; Normann et al., 1999). These HDMEAs have significantly increased experimental efficiency and neural yield (see for example Nicolelis (1999) for a review of this technology). As a result, a paradigm shift in neural recording techniques has been witnessed in the last 15 years that has paved the way for this technology to become a building block in a number of emerging clinical applications.

As depicted in Figure 1.2, substantial progress in invasive brain surgery, in parallel with the revolutionary progress in engineering the devices just discussed, has fueled a brisk evolution of brain-machine interfaces (BMIs). Broadly defined, a BMI system provides a direct communication pathway between the brain and a man-made device. The latter can be a simple electrode, an active circuit on a silicon chip, or even a network of computers. The overarching theme of BMIs is the restoration/repair of damaged sensory, cognitive, and motor functions such as hearing, sight, memory, and movement via direct interactions between an artificial device and the nervous system. It may even go beyond simple restoration to conceivably augmenting these functions, previously imagined only in the realm of science fiction.

1.2 **MOTIVATION**

The remarkable advances in neurophysiology techniques, engineering devices, and impending clinical applications have outstripped the progress in statistical signal processing theory and algorithms specifically tailored to: (1) performing large-scale analysis of the immense volumes of collected

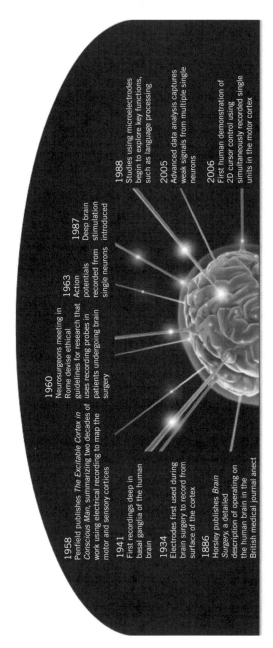

1886
Horsley publishes *Brain Surgery*, a detailed description of operating on the human brain in the British medical journal anect

1934
Electrodes first used during brain surgery to record from surface of the cortex

1941
First recordings deep in basal ganglia of the human brain

1958
Penfield publishes *The Excitable Cortex in Conscious Man*, summarizing two decades of work using electrical recording to map the motor and sensory cortices

1960
Neurosurgeons meeting in Rome devise ethical guidelines for research that uses recording probes in patients undergoing brain surgery

1963
Action potentials recorded from single neurons

1987
Deep brain stimulation introduced

1988
Studies using microelectrodes begin to explore key functions, such as language processing

2005
Advanced data analysis captures weak signals from multiple single neurons

2006
First human demonstration of 2D cursor control using simultaneously recorded single units in the motor cortex

FIGURE 1.2

Timeline of neural recording and stimulation during invasive human brain surgery.
Source: Modified from Abbott (2009).

neural data; (2) explaining many aspects of *natural* signal processing that characterize the complex interplay between the central and peripheral nervous systems; and (3) designing software and hardware architectures for practical implementation in clinically viable neuroprosthetic and BMI systems.

Despite the existence of a growing body of recent literature on these topics, there does not exist a comprehensive reference that provides a unifying theme among them. This book is intended to exactly fulfill this need and is therefore exclusively focused on the most fundamental statistical signal processing issues that are often encountered in the analysis of neural data for basic and translational neuroscience research. It was written with the following objectives in mind:

1. To apply classical and modern statistical signal processing theory and techniques to fundamental problems in neural data analysis.
2. To present the latest methods that have been developed to improve our understanding of natural signal processing in the central and peripheral nervous systems.
3. To demonstrate how the combined knowledge from the first two objectives can help in practical applications of neurotechnology.

1.3 OVERVIEW AND ROADMAP

A genuine attempt was made to make this book comprehensive, with special emphasis on signal processing and machine learning techniques applied to the analysis of neural data, and less emphasis on modeling complex brain functions. (Readers interested in the latter topic should refer to the many excellent texts on it.[1]) The sequence of chapters was structured to mimic the process that researchers typically follow in the course of an experiment. First comes data acquisition and preconditioning, followed by information extraction and analysis. Statistical models are then built to fit experimental data, and goodness-of-fit is assessed. Finally, the models are used to design and build actual systems that may provide therapeutic benefits or augmentative capabilities to subjects.

In Chapter 2, Oweiss and Aghagolzadeh focus on the joint problem of detection, estimation, and classification of neuronal action potentials in noisy microelectrode recordings—often referred to as spike detection and sorting.

---

[1] See for example, Dayan and Abbott, *Theoretical Neuroscience: Computational and Mathematical Modeling of Neural Systems*; Koch and Segev, *Methods in Neuronal Modeling: From Ions to Networks*; Izhikevich, *Dynamical Systems in Neuroscience: The Geometry of Excitability and Bursting*; and Rieke, Warland, van Steveninck and Bialek: *Spikes: Exploring the Neural Code*.

The importance of this problem stems from the fact that its outcome affects virtually *all* subsequent analysis. In the absence of a clear consensus in the community on what constitutes the best method, spike detection and sorting have been and will continue to be a subject of intense research because techniques for multiunit recordings have started to emerge. The chapter provides an in-depth presentation of the fundamentals of detection and estimation theory as applied to this problem. It then offers an overview of traditional and novel methods that revolve around the theory, in particular contrasting the differences—and potential benefits—that arise when detecting and sorting spikes with a single-channel versus a multi-channel recording device. The authors further link multiple aspects of classic and modern signal processing techniques to the unique challenges encountered in the extracellular neural recording environment. Finally, they provide a practical way to perform this task using a computationally efficient, hardware-optimized platform suitable for real-time implementation in neuroprosthetic devices and BMI applications.

In Chapter 3, Johnson provides an overview of classical information theory, rooted in the 1940s pioneering work of Claude Shannon (Shannon, 1948), as applied to the analysis of neural coding once spike trains are extracted from the recorded data. He offers an in-depth analysis of how to quantify the degree to which neurons can individually or collectively process information about external stimuli and can encode this information in their output spike trains. Johnson points out that extreme care must be exercised when analyzing neural systems with classical information-theoretic quantities such as *entropy*. This is because little is known about how non-Poisson communication channels that often describe neuronal discharge patterns provide optimal performance bounds on stimulus coding. The limits of classical information theory as applied to information processing by spiking neurons are discussed as well. Finally, Johnson provides some interesting thoughts on post-Shannon information theory to address the more puzzling questions of how stimulus processing by some parts of the brain results in *conveying* useful information to other parts, thereby triggering *meaningful* actions rather than just *communicating* signals between the input and the output of a communication system—the hallmark of Shannon's pioneering work.

In Chapter 4, Song and Berger focus on the system identification problem—that is, determining the input-output relationship in the case of a multi-input, multi-output (MIMO) system of spiking neurons. They first describe a nonlinear multiple-input, single-output (MISO) neuron model consisting of a number of components that transform the neuron's input spike trains into synaptic potentials and then feed back the neuron's output spikes through nonlinearities to generate spike-triggered after-potential contaminated by noise to capture system uncertainty. Next they describe how this MISO model is combined with similar models to predict the MIMO transformation taking place in the hippocampus CA3-CA1 regions, both known to play a central role in declarative memory formation. Song and Berger suggest that such a model can serve as a computational framework for the development of memory prostheses which replaces damaged hippocampus circuitry. As a means to bypass a damaged region, they demonstrate the utility of their MIMO model to predict output spike trains from the CA1 region using input spike trains to the CA3 region. They point out that the use of hidden variables to represent the internal states of the system allows simultaneous estimation of all model parameters directly from the input and output spike trains. They conclude by suggesting extensions of their work to include nonstationary input-output transformations that are known to take place as a result of cortical plasticity—for example during learning new tasks—a feature not captured by their current approach.

In Chapter 5, Eldawlatly and Oweiss take a more general approach to identifying distributed neural circuits by inferring connectivity between neurons locally within and globally across multiple brain areas, distinguishing between two types of connectivity: *functional* and *effective*. They first review techniques that have been classically used to infer connectivity between various brain regions from continuous-time signals such as fMRI and EEG data. Given that inferring connectivity among neurons is more challenging because of the stochastic and discrete nature of their spike trains and the large dimensionality of the neural space, the authors provide an in-depth focus on this problem using graphical techniques deeply rooted in statistics and machine learning. They demonstrate that graphical models offer a number of advantages over other techniques, for example by distinguishing between mono-synaptic and poly-synaptic connections and in inferring inhibitory connections among other features that existing methods cannot capture. The authors demonstrate the application of their method in the analysis of neural activity in the medial prefrontal cortex (mPFC) of an awake, behaving rat performing a working memory task. Their results demonstrate that the networks inferred for similar behaviors are strongly consistent and exhibit graded transition between their dynamic states during the recall process, thereby providing additional evidence in support of the long withstanding Hebbian cell assembly hypothesis (Hebb, 1949).

In Chapter 6, Chen, Barbieri, and Brown discuss the application of a subclass of graphical models—the state-space models—to the analysis of neural spike trains and behavioral learning data. They first formulate point process and binary observation models for the data and review the framework of filtering and smoothing under the state space paradigm. They provide a number of examples where this approach can be useful: ensemble spike train decoding, analysis of neural receptive field plasticity, analysis of individual and population behavioral learning, and estimation of cortical UP/DOWN states. The authors demonstrate that the spiking activity of 20 to 100 hippocampal place cells can be used to decode the position of an animal foraging in an open circular environment. They also demonstrate that the evolution of the 1D and 2D place fields of hippocampus CA1 neurons reminiscent of experience-dependent plasticity can be tracked with steepest descent and particle filter algorithms. They further show how the state space approach can be used to measure learning progress in behavioral experiments over a sequence of repeated trials. Finally, the authors demonstrate the utility of this approach in estimating neuronal UP and DOWN states—the periodic fluctuations between increased and decreased spiking activity of a neuronal population—in the primary somatosensory cortex of a rat. They derive an expectation maximization algorithm to estimate the number of transitions between these states to compensate for the partially observed state variables. In contrast to multiple hypothesis testing, their approach dynamically assesses population learning curves in four operantly conditioned animal groups.

In Chapter 7, Yu, Santhanam, Sahani, and Shenoy discuss the problem of decoding spike trains in the context of motor and communication neuroprosthetic systems. They examine closely two broad classes of decoders, continuous and discrete, depending on the behavioral correlate believed to be represented by the observed neural activity. For continuous decoding, the goal is to capture the moment-by-moment statistical regularities of movements exemplified by the movement trajectory; for discrete decoding, a fast and accurate classifier of the desired movement target suffices. The authors point to the existence of a trade-off between accuracy in decoding and computational complexity, which is an important design constraint for real-time implementation. To get the best compromise, they describe a "probabilistic mixture of trajectory models" (MTM) decoder and demonstrate its use in analyzing ensemble neural activity recorded in premotor cortex in macaque monkeys as they plan and execute goal-directed arm reach movements. The authors compare their MTM approach to a number of decoders reported in the literature and demonstrate substantial improvements. In conclusion, they suggest the need for new methods for investigating the dynamics of plan and movement activity that do not smear useful information over large temporal intervals by assuming an explicit relationship between neural activity and arm movement. This becomes very useful when designing decoding algorithms for closed-loop settings where subjects continuously adapt the neuronal firing properties to attain the desired goal.

In Chapter 8, Paiva, Park, and Principe develop a novel theory of inner products for representation and learning in the spike train domain. They motivate their work by considering the effect of binning spike trains—a commonly used method to reduce a signal's dimensionality and variance—on the information loss in precise spike times. This lossy binning is seen as a transformation of the *randomness in time* typically observed in spike trains into the *randomness in amplitude* of firing rate functions that may be correlated with a stochastic, time-varying stimulus. The authors build on the earlier framework of Houghton and colleagues (Houghton and Sen, 2008) to develop a functional representation of the spike trains using inner product operators for spike train outputs of a biophysical leaky integrate and to fire neural models. They show that many classes of operations on spike trains such as convolution, projection for clustering purposes, and linear discriminant analysis for classification all involve some form of an inner product operation. Because inner products are deeply rooted in reproducing kernel Hilbert spaces (RKHS) theory (Aronszajn, 1950), the authors highlight the potential benefit of using this representation in the development of a multitude of analysis and machine learning algorithms. For example, they demonstrate how known algorithms such as principal component analysis (PCA) and Fisher linear discriminant (FLD) analysis can be derived from this representation. They include some results of this approach applied to synthetic spike trains generated from a number of homogeneous renewal point processes, and show that spike trains can be analyzed in an unsupervised manner while discerning between intrinsic spike train variability and extrinsic variability observed under distinct signal models. They conclude with some suggestions on how these inner products can be useful in the development of efficient machine learning algorithms for unsupervised and supervised learning.

The next two chapters emphasize noninvasive neural signal processing. In Chapter 9, Sajda, Goldman, Dyrholm, and Brown present an elegant approach to investigating the relationship—and potentially fusing the information—between simultaneously acquired EEG and fMRI signals. The motivation is to combine the superior temporal resolution of EEG and the spatial resolution of fMRI in a multimodal neuroimaging technique. This may provide more insight into the relationship between local neural activity exemplified by EEG signals and hemodynamic changes exemplified by fMRI signals to permit capturing trial-to-trial variations in these signals that are hypothesized to underlie many aspects of perception and cognition. In this realm, the authors emphasize a number of issues related to hardware design, such as MR-induced artifacts in the EEG and lead-induced artifacts in the MR images. They also address issues related to the extraction of task-relevant discriminant components using supervised learning. Their results for an auditory detection paradigm show how single-trial analysis of the EEG/fMRI data provides evidence of novel activations reminiscent of latent brain states such as attention modulation. The latter cannot be measured by traditional event-related regressors in standard fMRI analyses. The chapter concludes with a discussion of the challenges and future directions of EEG/fMRI, particularly those related to direct modulation of EEG activity (for example using transcranial magnetic stimulation (TMS)) to infer causal changes between EEG patterns and BOLD signals.

In Chapter 10, Rao and Scherer provide an overview of the utility of noninvasive or semi-invasive signals in brain-computer interface (BCI) applications. The motivation is to compare ECoGs and EEGs; in BCI control, both have similar temporal resolution but their spatial resolution differs significantly (ECoGs are $\sim$100 times larger than EEG signal amplitudes). Techniques for enhancing the signal-to-noise ratio of EEGs are therefore of paramount importance, as are those for feature extraction and classification. All are reviewed and their performance is evaluated in the context of BCI applications. As would be expected, nonstationarity and the inherent variability of EEG and ECoG signals are by far the major challenges to overcome.

Rao and Scherer also discuss the problem of detecting event-related potentials (ERPs) and event-related desynchronization/synchronization (ERD/ERS) potentials occurring with different types of motor imagery. These form the foundations of the two major paradigms in EEG/ECoG BCI: *cue-guided* and *self-paced*. While cue-guided BCIs are easier to train because of pattern robustness and repeatability, self-paced BCIs require analysis and interpretation of the signals throughout the entire period of BCI operation. The authors point out that signal processing in these applications is subject-specific, meaning that optimization of the system for each subject is essential. Therefore, the trade-off between cost, practicality, and reliability becomes a governing factor when optimizing variables such as the number of electrodes, the level of sophistication of the signal processing algorithm used, and the amount of time needed to train the systems.

In Chapter 11, Oby, Ethier, Bauman, Perreault, Ko, and Miller provide an application-driven framework for employing some of the concepts introduced in earlier chapters to provide therapeutic benefit to spinal cord-injured patients through a BMI system. In contrast to predicting movement intent from plan activity, as Yu and colleagues demonstrated in Chapter 7, their objective is to develop a neuroprosthesis that predicts certain movement attributes such as grip force, joint torque, and muscle activity (EMG signals) from primary motor cortex (M1) activity in real time to provide voluntary control of hand grasping following a peripheral nerve block (temporary paralysis). The rationale is the wide range of grasping behaviors that can be represented in a low-dimensional space, permitting only a few principal components to account for ~80% of the variance in hand shape. Voluntary control of hand grasp is achieved by direct functional electrical stimulation (FES) of flexor muscles of the forearm and hand using the predicted EMG signals from M1 activity.

Electrical stimulation, however, has important limitations. One is that muscle fibers are activated synchronously, opposite to the normal order, so the large, fatiguing fibers are recruited first. Despite this, Oby and colleagues' use of simple, linear decoders of M1 activity has been quite successful, in part because it allows them to harness cortical plasticity to allow the monkey to adapt its control commands to accommodate the unnatural mapping to muscle activation. They suggest future investigation of this adaptation process to determine whether adaptive algorithms may also improve control, particularly when more complex and naturalistic limb movements with multiple degrees of freedom are desired. They also intend to study how task dependence in the relationship between motor cortex activity and muscle activation patterns may be minimized by identifying a "muscle space" in the cortical neural space and by selectively activating groups of muscle synergies (as opposed to individual muscles in the current approach). This may lead to simpler yet more robust control. Techniques for blind source separation are very relevant in this context.

Despite the diversity of problems treated throughout this book, all chapters revolve in whole or in part around the following fundamental issues: pattern recognition, system identification, and information processing.

- *Pattern recognition* refers to the clustering and classification of multiple patterns, whether in the input or in the output of a given system.

- *System identification* refers to the establishment of an input-output relationship through full or partial knowledge of system structure.

- *Information processing* refers to the system's ability to filter out "irrelevant" information to guide desired transformations/actions.

In each of the problems addressed, some important questions are, What constitutes a pattern? Given a pattern, how can its recognition be carried out? If this is feasible, how can the system be identified for a better understanding of the natural signal and information processing taking place in it, and what is its role in shaping an entire behavior? The answers to these questions, among many others, will fundamentally rest on quantum advances in methods for analyzing neural and behavioral data.

The pace of research in this area is so rapid that new theories and techniques continue to swiftly emerge. Nevertheless, this book should provide a good starting point for readers to get a first grip on the many different aspects of statistics, signal processing, machine learning, and information theory as applied to neural and behavioral data processing and analysis, as well as on how these aspects may highlight important considerations in designing actual systems for improving human welfare.

References

Abbott, A., 2009. Neuroscience: opening up brain surgery. Nature 461 (7266), 866–868.

Ahissar, E., Vaadia, E., Ahissar, M., Bergman, H., Arieli, A., Abeles, M., 1992. Dependence of cortical plasticity on correlated activity of single neurons and on behavioral context. Science 257 (5075), 1412.

Aronszajn, N., 1950. Theory of reproducing kernels. Trans. Am. Math. Soc. 68 (May), 337–404.

Csicsvari, J., Henze, D.A., Jamieson, B., Harris, K.D., Sirota, A., Barthó, P., et al., 2003. Massively parallel recording of unit and local field potentials with silicon-based electrodes. Am. Physiol. Soc. 90, 1314—1323.

Drake, K.L., Wise, K.D., Farraye, J., Anderson, D.J., BeMent, S.L., 1988. Performance of planar multisite microprobes in recording extracellular single-unit intracortical activity. IEEE Trans. Biomed. Eng. 35 (9), 719–732.

Edell, D.J., Toi, V.V., McNeil, V.M., Clark, L.D., 1992. Factors influencing the biocompatibility of insertable silicon microshafts in cerebral cortex. IEEE Trans. Biomed. Eng. 39 (6), 635–643.

Evarts, E.V., 1968. A technique for recording activity of subcortical neurons in moving animals. Electroencephalogr. Clin. Neurophysiol. 24 (1), 83–86.

Fetz, E.E., Finocchio, D.V., 1971. Operant conditioning of specific patterns of neural and muscular activity. Science 174 (4007), 431–435

Gesteland, R.C., Howland, B., Lettvin, J.Y., Pitts, W.H., 1959. Comments on microelectrodes. Proc. IRE 47 (1856), 1862.

Giacobini, E., Handelman, E., Terzuolo, C.A., 1963. Isolated neuron preparation for studies of metabolic events at rest and during impulse activity. Science 140 (356), 74–75.

Gray, C.M., Maldonado, P.E., Wilson, M., McNaughton, B., 1995. Tetrodes markedly improve the reliability and yield of multiple single-unit isolation from multi-unit recordings in cat striate cortex. J. Neurosci. Methods 63 (1–2), 43–54.

Hebb, D.O., 1949. The Organization of Behavior; a Neuropsychological Theory. Wiley, New York.

Houghton, C., Sen, K., 2008. A new multineuron spike train metric. Neural Comput. 20 (6), 1495–1511.

Katzner, S., Nauhaus, I., Benucci, A., Bonin, V., Ringach, D.L., Carandini, M., 2009. Local origin of field potentials in visual cortex. Neuron 61 (1), 35–41.

Koester, H.J., Sakmann, B., 2000. Calcium dynamics associated with action potentials in single nerve terminals of pyramidal cells in layer 2/3 of the young rat neocortex. J. Physiol. 529 Pt 3, 625–646.

Kralik, J.D., Dimitrov, D.F., Krupa, D.J., Katz, D.B., Cohen, D., Nicolelis, M.A., 2001. Techniques for long-term multisite neuronal ensemble recordings in behaving animals. Methods 25 (2), 121—150.

McNaughton, B.L., O'Keefe, J., Barnes, C.A., 1983. The stereotrode: a new technique for simultaneous isolation of several single units in the central nervous system from multiple unit records. J. Neurosci. Methods 8 (4), 391–397.

Mitzdorf, U., 1985. Current source-density method and application in cat cerebral cortex: investigation of evoked potentials and EEG phenomena. Physiol. Rev. 65 (1), 37–100.

Nicolelis, M.A.L., 1999. Methods for Neural Ensemble Recordings. CRC Press, Boca Raton, FL.

Normann, R.A., Maynard, E.M., Rousche, P.J., Warren, D.J., 1999. A neural interface for a cortical vision prosthesis. Vision Res. 39 (15), 2577–2587.

Penfield, W., Jasper, H., 1954. Epilepsy and the functional anatomy of the human brain. Brain 77 (4), 639–641.

Shannon, C.E., 1948. A mathematical theory of communication. Bell Syst. Tech. J. 27, 379–423, 623–656.

Stosiek, C., Garaschuk, O., Holthoff, K., Konnerth, A., 2003. In vivo two-photon calcium imaging of neuronal networks. Proc. Natl. Acad. Sci. 100 (12), 7319.

Turner, J.N., Shain, W., Szarowski, D.H., Andersen, M., Martins, S., Isaacson, M., et al., 1999. Cerebral astrocyte response to micromachined silicon implants. Exp. Neurol. 156 (1), 33–49.

Detection and Classification of Extracellular Action Potential Recordings

2

Karim Oweiss*†, Mehdi Aghagolzadeh*

*Electrical and Computer Engineering Department, †Neuroscience Program, Michigan State University, East Lansing, Michigan

2.1 INTRODUCTION

The theories of detection and estimation play a crucial role in processing neural signals, largely because of the highly stochastic nature of these signals and the direct impact this processing has on any subsequent information extraction. Detection theory is rooted in statistical hypothesis testing, in which one needs to decide which generative model, or hypothesis, among many possible ones, may have generated the observed signals (Neyman and Pearson, 1933). The degree of complexity of the detection task can be viewed as directly proportional to the degree of "closeness" of the candidate generative models (i.e., the detection task becomes more complex as the models get closer together in a geometrical sense). Estimation theory can then be used to estimate the values of the parameters underlying each generative model (Kay, 1993). Alternatively, one may estimate the parameters first and then, based on the outcome, make a decision on which model best fits the observed data. Therefore, detection and estimation are often carried out simultaneously.

In this chapter, we focus exclusively on the discrete-time detection and classification of action potentials (APs)—or *spikes*—in extracellular neural recordings. These spikes are usually recorded over a finite number of time instants in the presence of noise. At each time instant, we assign a hypothesis to each signal value. We may then build an estimate of individual spikes *sample by sample*, testing the presence of each. Estimation can thus be viewed as a limiting case of detection theory in which an estimate of the spike is built from a continuum of possible hypotheses (Poor, 1988). Alternatively, we may restrict our estimation to a subset of spike parameters—or features—for the

Statistical Signal Processing for Neuroscience and Neurotechnology. DOI: 10.1016/B978-0-12-375027-3.00002-8

purpose of *classifying* spikes elicited by multiple neurons—a process often termed as *spike sorting*—because the primary information in the spikes resides in the *timing* of their individual occurrences and not in their actual shapes, making estimation of their exact shapes of secondary importance relative to their actual labels.

We focus exclusively on spikes for a number of reasons. First, their origins—the neurons—are the fundamental computational units of the brain. Second, spikes fired by a given neuron tend to be highly stereotypical, at least over short intervals, permitting to model their shapes with probability distributions that have relatively few parameters and thus facilitating their classification. Third, spike trains carry an affluent amount of information about the neurons' encoding of external covariates—that is, their *tuning* characteristics to stimuli. This permits modeling the dynamics of brain function at very high temporal and spatial resolutions, as well as *decoding* the spike trains, for example to drive a neuroprosthetic device (e.g., Serruya et al., 2002). Fourth, spike occurrence from multiple neurons in a certain order is known to mediate plastic changes in synaptic connectivity between them—a widely accepted mechanism that underlies many of the brain's complex functions such as learning and memory formation.

2.2 SPIKE DETECTION

Spikes typically fall within the category of transient signals. A transient signal is known to be nonzero only for a subset of samples, L, within an observation interval of length N. Detecting a transient presence involves estimating three variables: the arrival time, k_0, the support (or time scale), L, and the waveform shape, s, as illustrated in the neural trace of Figure 2.1. The most important parameter is the spike arrival time, k_0, as it is believed to carry all of the information needed by neurons to encode external covariates. Therefore, k_0 needs to be estimated with high reliability. The other two parameters are believed to play no role in encoding these covariates and are mainly used to *sort* the spikes when multiple neurons are recorded simultaneously.

Knowledge of the transient parameters is important to formulate a test statistic, T, from the observations, y. T is used to solve a binary hypothesis testing problem in which the goal is to determine which of the following two models is consistent with the data:

$$\begin{aligned} \text{Null hypothesis:} &\quad H_0 : y = n \\ \text{Alternative hypothesis:} &\quad H_1 : y = s + n \end{aligned} \tag{2.1}$$

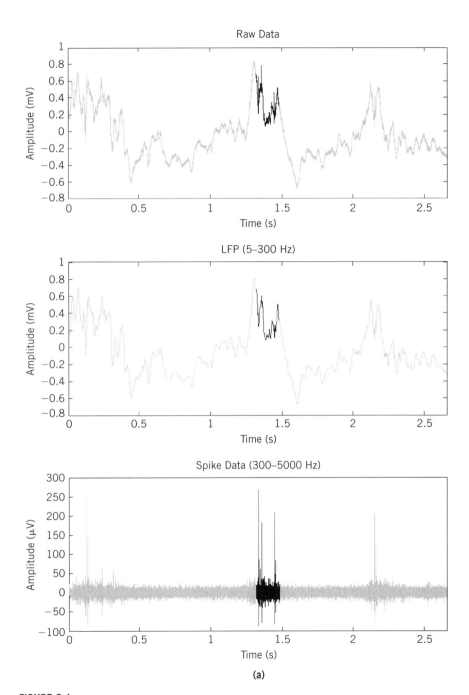

FIGURE 2.1

(*Continued*)

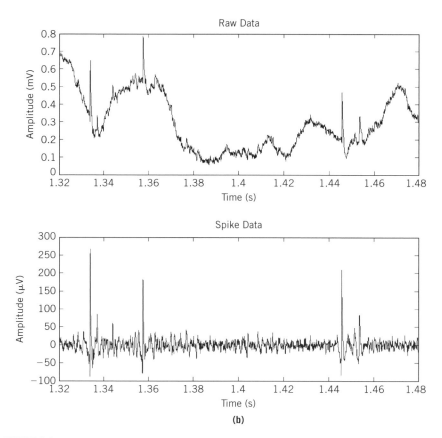

FIGURE 2.1

Sample neural trace recorded with a penetrating microelectrode in the rat somatosensory cortex with different time and magnitude scaling. (a) Top panel: 2.6 seconds of raw data; middle panel: local field potential (LFP) signal after passing through a fifth-order Butterworth filter with frequency range 5–300 Hz; bottom panel: spike data obtained after filtering the raw data in the top panel with the same Butterworth filter with frequency range 300–5000 Hz. (b) Expanded time and amplitude scale display of the data segment in (a): top: LFP signal; bottom: spike data showing a number of spikes (highly transient) with distinct waveforms. Each spike presumably represents the extracellular action potential from a nearby neuron. Please see this figure in color at the companion web site: www.elsevierdirect.com/companions/9780123750273

Here s denotes the L-dimensional spike to be detected and n is an additive noise term. The observations y can be a single sample or a vector (block detection) of length N. The models H_0 and H_1 are called simple hypotheses when the distribution of y can be determined under perfectly known signal parameters.

In this case, Bayesian analysis yields the optimal test: the likelihood ratio test (LRT), which is of the form (Poor, 1988)

$$T(y) \equiv \ln \frac{p(y|H_1)}{p(y|H_0)} \underset{H_0}{\overset{H_1}{\underset{<}{>}}} \ln \frac{p(H_0)(c_{10} - c_{00})}{p(H_1)(c_{01} - c_{11})} \equiv \gamma \qquad (2.2)$$

where c_{ij} is the cost of deciding model i when model j is in effect (typically $c_{ij} > c_{ii}$ for $i \neq j$), and $p(H_i)$ is the probability of model H_i. These quantities can be lumped in one threshold value, γ, chosen by the user to fulfill a constant probability of false positives, P_F (otherwise known as the power of the test). The probabilities of the models, $p(H_0)$ and $p(H_1)$, may be known a priori from knowledge of the neuron's firing characteristics—for example, tuning to a specific input stimulus.

The assumptions we make about the contaminating noise, n, play a crucial role in assessing how far apart the two models are. A firm understanding of the noise process in extracellular recordings reveals the following categorization of its possible sources:

- Thermal and electrical noise due to electronic circuitry in the electrode data acquisition system. This includes amplifiers and filters in the headstage as well as quantization noise introduced by the analog-to-digital conversion (ADC). This noise type can be modeled as a stationary, zero-mean, Gaussian white noise process, $n \sim \mathcal{N}(0, \sigma^2 I)$, where I is the identity matrix. The only unknown parameter in this model is the noise variance σ^2.

- Background neural noise caused by activity from sources far from the recording electrode. This noise type is typically colored and may often be regarded as the context that can inherently drive a neuron to fire (or not fire) depending on the intensity of the population's driving input. This in turn depends on the behavioral context under which the neural data was collected. Spike detection (and consequently sorting) can be greatly complicated when this noise component is high.

In the sections that follow, we discuss a few possible scenarios and formulate the detector in each case.

2.2.1 Known Spike

Here we assume that the spike waveform is fully known; in other words, we have complete knowledge of the spike support L and the shape s. This may be the case when spike templates are available from training data. The goal is to determine k_0, the time of arrival of the waveform. Assume that the noise is colored and Gaussian-distributed with zero mean and covariance Σ (i.e., $n \sim \mathcal{N}(0, \Sigma)$). Although the Gaussian assumption may not be realistic, it facilitates the mathematical tractability of the problem to a large extent and is a reasonable assumption for sufficiently large observation intervals. Under the alternative hypothesis, H_1, the distribution of the observations is

$$p_{Y|H_1,s}(y|H_1,s) = \frac{1}{\sqrt{2\pi |\Sigma|}} \exp\left(-\frac{1}{2}(y-s)^T \Sigma^{-1}(y-s)\right) \qquad (2.3)$$

where $|.|$ denotes the matrix determinant, $(.)^T$ denotes a transpose, and $(.)^{-1}$ denotes the matrix inverse. The optimal test in this case is

$$s^T \Sigma^{-1} y \underset{H_0}{\overset{H_1}{\underset{<}{\gtrless}}} \gamma \qquad (2.4)$$

This is a block LRT test because it computes the test statistic using the N observations to detect the presence of the spike. Note here that this is not a real-time detector because it has to wait for N samples to be collected before it computes the test statistic. One way to interpret this detector is to spectrally factor the kernel matrix Σ using singular value decomposition (SVD) to yield the form $\Sigma = UDU^T$ (Klema and Laub, 1980), where U is an eigenvector matrix and D is a diagonal matrix of eigenvalues arranged in decreasing order of magnitude. The detector in this case takes the form

$$\left(D^{-\frac{1}{2}}U^{-1}s\right)^T \left(D^{-\frac{1}{2}}U^{-1}y\right) \underset{H_0}{\overset{H_1}{\underset{<}{\gtrless}}} \gamma \qquad (2.5)$$

Since Σ is a square positive definite covariance matrix, the eigenvector matrix inverse U^{-1} is simply its transpose U^T. This factorization illustrates the role of the noise kernel Σ^{-1}. The transformation $D^{-\frac{1}{2}}U^{-1}$ is lower triangular and has the effect of whitening the observation vector with respect to the noise. Because it is assumed that s is known a priori, the detector becomes a two-stage system, where the first stage is a whitening filter that tries to minimize the effect of the noise and the second is a matched filter that measures the degree of correlation of the observations with the known spike. Clearly, a key factor in this detector is the ability to obtain a good estimate of the spike

s and the noise covariance Σ, which can diminish performance if the noise assumptions are violated or if the spike waveform becomes nonstationary.

The trade-off in selecting the detection threshold γ is clear: When γ is set too high, the probability of false positives, P_F, decreases at the cost of increasing the probability of missed events, P_M. On the other hand, low thresholds increase P_F but simultaneously decrease P_M. Typically, γ is set by the user to satisfy a constant P_F. Figure 2.2 demonstrates the performance of this detector under various SNRs as a function of the number of samples N. In the first graph, the observations are used in the test of Eq. (2.4), assuming the inner product kernel to be the identity matrix. This corresponds to the white noise case. In the second graph, the observations have been decorrelated using the whitening filter $D^{-\frac{1}{2}}U^{-1}$. It can be clearly seen that when the noise is colored, the whitening approach leads to substantial performance gain compared to the case when the noise is assumed white but is not. For example, by fixing $N = 10$ samples, an increase in P_D of more than 20% can be seen at SNR $= 4$ dB.

We can assess the performance of the detector in Eq. (2.4) by noting that the test statistic is a linear function of a Gaussian variable and hence is also Gaussian. We can determine the distribution of the test statistic T for the two hypotheses in terms of the means and variances:

Under H_0 : $E[T] = E\left[s^T \Sigma^{-1} y\right] = 0$

$$E\left[(T - \overline{T})^2\right] = E\left[(s^T \Sigma^{-1} y)^T (s^T \Sigma^{-1} y)\right] = s^T \Sigma^{-1} s$$

Under H_1 : $E[T] = E\left[s^T \Sigma^{-1} y\right] = s^T \Sigma^{-1} s$ (2.6)

$$E\left[(T - \overline{T})^2\right] = s^T \Sigma^{-1} s$$

The probabilities of detection P_D and false alarm P_F can be calculated using the tail probability of a unit normal Q(.) (Van Trees, 2002):

$$P_D = P_{H_1}(T > \gamma) = \frac{1}{2}\left[1 - Q\left(\frac{\gamma - s^T \Sigma^{-1} s}{s^T \Sigma^{-1} s \sqrt{2}}\right)\right]$$
$$P_F = P_{H_0}(T > \gamma) = \frac{1}{2}\left[1 - Q\left(\frac{\gamma}{s^T \Sigma^{-1} s \sqrt{2}}\right)\right]$$ (2.7)

Hence, for a Gaussian-distributed test statistic, the distance between the two hypotheses is simply $s^T \Sigma^{-1} s$. It is straightforward to show that this distance is maximized when the noise is uncorrelated with the signal (Poor, 1988).

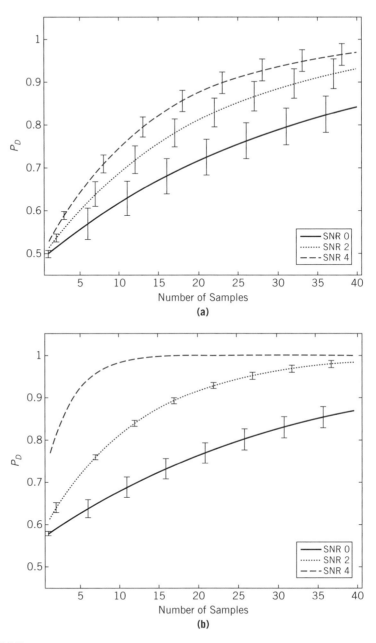

FIGURE 2.2

Performance of the spike detector in the known signal case in the presence of colored noise. (a) Observations are used in the test statistic without whitening. (b) Observations are used in the test statistic with whitening using the eigendecomposition of the inner product kernel Σ^{-1}.

It is easy to interpret the distance term in the case where the observation interval contains only a single sample ($N = 1$). In such a case, we obtain a detector that can be implemented in real time. The expressions for P_D and P_F reduce to

$$P_D = P_{H_1}(T > \gamma) = \frac{1}{2}\left[1 - Q\left(\frac{\gamma - \frac{s_k^2}{\sigma_k^2}}{\frac{s_k^2}{\sigma_k^2}\sqrt{2}}\right)\right]$$

$$P_F = P_{H_0}(T > \gamma) = \frac{1}{2}\left[1 - Q\left(\frac{\gamma}{\frac{s_k^2}{\sigma_k^2}\sqrt{2}}\right)\right]$$

(2.8)

where σ_k^2 is the noise variance. In this case the distance between the two hypotheses becomes $\frac{s_k^2}{\sigma_k^2}$, which is the energy of the signal divided by the variance of the noise, or the signal-to-noise ratio (SNR). Figure 2.3 illustrates the receiver operating characteristics (ROC) under different SNRs.

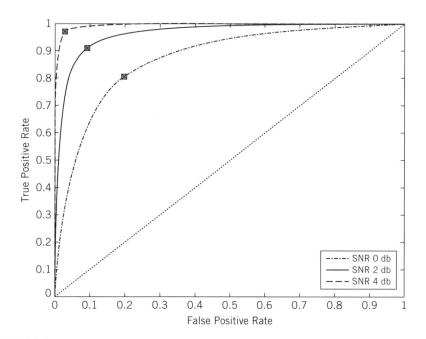

FIGURE 2.3

Receiver operating characteristics of the known signal detector (matched filter) as a function of SNR.

When the observations contain more than one sample—for example, when we have an N-sample time interval from a single electrode, or a single-sample interval from an array of N electrodes, we obtain a block detector that will result in better performance over the sequential test if the following condition is satisfied:

$$s^T \Sigma^{-1} s > \frac{s_k^2}{\sigma_k^2}$$

When the noise samples are independent and identically distributed (iid) (in either time or space), the covariance Σ is diagonal. The term $s^T \Sigma^{-1} s$ reduces to $d^2 = \sum_{k=1}^{N} \frac{s_k^2}{\sigma_k^2}$, or the SNR average power multiplied by the number of samples N. In this case, the block test performs better because the sum of the individual SNRs is larger than any one SNR. In the more general case, the effect of noise correlation on performance will depend on the sign of the noise correlation coefficient.

2.2.2 Unknown Spike

Prior knowledge of the deterministic spike shape, s, or its parameters if stochastic, greatly simplifies the analysis. In practice, however, this information may not be fully available. Spikes may also be nonstationary over short time intervals, for example during bursting periods (Harris et al., 2000; Csicsvari et al., 2003). The shapes can also change over prolonged periods, for example as a result of electrode drift or cell migration during chronic recordings with implanted electrodes (Buzsaki et al., 2004; Gold et al., 2006). In this case, the models H_0 and H_1 may involve some nuisance (unknown) parameters for which performance depends on how close a chosen signal model is to the real spike waveform. The optimal detector here selects the maximum value of the distribution $p_{Y|H_1,s}(y|H_1,s)$ with respect to the unknown parameters. This happens when $y = s$, or

$$\max_s p_{Y|H_1,s}(y|H_1,s) = \frac{1}{\sqrt{2\pi |\Sigma|}} \tag{2.9}$$

The other hypothesis does not depend on the signal, and the LRT becomes a generalized LRT (GLRT), often termed the square law or incoherent energy detector; it takes the form

$$y^t \Sigma^{-1} y \underset{H_0}{\overset{H_1}{\underset{<}{\overset{>}{\gtrless}}}} \gamma \tag{2.10}$$

Similar interpretation of the noise kernel applies, but now the product involves a whitened observation vector, $y' = D^{-\frac{1}{2}} U^{-1} y$. The test statistic in this case becomes a blind energy detector, where the observations are whitened, squared, and then compared to a threshold.

Performance of this detector can be easily derived by noting that the test statistic becomes the sum of the squares of statistically independent Gaussian variables and is chi-square–distributed (Papoulis et al., 1965). Similar to the analysis in Eq. (2.7), it can be shown that this distribution has the following form under the null hypothesis:

$$p_{H_0}(T,0) = \frac{T^{\frac{N}{2}-1} e^{-\frac{T}{2}}}{\Gamma\left(\frac{N}{2}\right) 2^{\frac{N}{2}}} \tag{2.11}$$

Where $\Gamma(.)$ is the gamma function. Under H_1, the test statistic has the noncentrality parameter $\alpha = s^T \Sigma^{-1} s$ (Urkowitz, 1967).

$$p_{H_1}\left(T, s^T \Sigma^{-1} s\right) = \frac{T^{\frac{N}{2}-1} e^{-\frac{T+\alpha}{2}}}{2^{\frac{N}{2}}} \sum_{k=0}^{\infty} \frac{(\alpha T)^k}{2^{2k} k! \, \Gamma\left(k + \frac{N}{2}\right)} \tag{2.12}$$

If we denote by $F\left(\frac{T}{N}, \alpha\right)$ the cumulative distribution function of a noncentral chi-square–distributed random variable T with N degrees of freedom and noncentrality parameter α, these results can help us calculate the performance of the detector in terms of P_D and P_F as follows:

$$P_D = P_{H_1}(T > \gamma) = 1 - F_{H_1}(\gamma) = 1 - F_{H_1}\left(\gamma | N, s^T \Sigma^{-1} s\right)$$
$$P_F = P_{H_0}(T > \gamma) = 1 - F_{H_0}(\gamma) = 1 - F_{H_0}(\gamma | N, 0) \tag{2.13}$$

Figure 2.4 illustrates the performance of this detector. As in the known signal case, the performance improves when the number of samples N increases and/or the ratio of the average spike power to the average noise power increases.

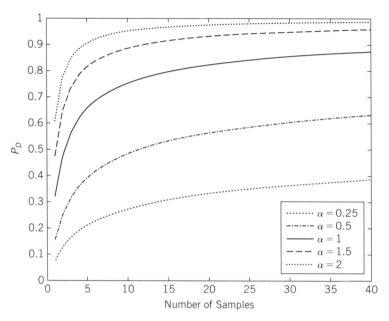

FIGURE 2.4

Performance of the GLRT energy detector versus the number of samples for different values of the noncentrality parameter α.

2.2.3 Practical Limitations

We have summarized how detecting a spike waveform should be carried out within the framework of statistical hypothesis testing. In most experimental neural recording applications, however, a much simpler, suboptimal but robust, approach is used. The raw data is first bandpass-filtered to eliminate low-frequency signals such as local field potentials (LFPs) that are much larger in magnitude than the spikes depicted in Figure 2.1. Spike detection follows by comparing the filtered data to a voltage threshold, usually computed as two to three times the standard deviation of the noise (Gozani and Miller, 1994). A pair of thresholds may also be used to detect the rising and falling edges of the spike (Guillory and Normann, 1999). Once the observations surpass this threshold, a spike vector is extracted around the detection point. The length of that vector is empirically set to be approximately 0.5 ms before the detection point and approximately 1 ms after that point (Obeid et al., 2004). This approach assumes that the unknown support of the spike, L, is typically around 1.5 ms to match that of many extracellular AP waveforms observed in mammalian nervous systems (Kandel et al., 2000).

Many studies suggest, however, that the scale parameter (i.e., the support of the spike) varies considerably depending on cell-specific intrinsic properties, such as the morphology of the cell's dendritic tree and the ion channels' distribution (e.g., fast spiking interneurons versus regular spiking pyramidal neurons (Buzsaki et al., 2004)), in addition to some extrinsic properties, such as the degree of inhomogeneity of the extracellular space and the distance relative to the recording electrode (Gold et al., 2006). A poor choice of the scale parameter can either discard useful information from the spike or introduce spurious samples in the extracted spike vector. Both can significantly degrade the performance of the spike sorting step that follows. The spike shape may also be affected by the preprocessing bandpass filters (Quiroga, 2009). These considerations have led some recent studies to argue against the use of spike waveform shape in spike detection and sorting (Ventura, 2009).

The simple threshold-crossing method is appealing because it is very simple to implement in hardware and therefore enables spike detection to take place in real time (Bankman, 1993; Obeid et al., 2004). This is particularly useful in cortically controlled neuroprosthetic and brain–machine interface (BMI) applications (Isaacs et al., 2000; Hochberg et al., 2006; Oweiss, 2006), where spike detection and sorting needs to be implemented with (or close to) real-time performance to permit instantaneous decoding of the neural signals. However, as one would expect, this method relies on very strong simplifying assumptions about the signal and noise processes which are often violated, such as the absence of correlation between the signal and noise and the noise being white and Gaussian-distributed. A spike alignment step is usually needed to center the detected waveforms around a common reference point, usually the peak (McGill and Dorfman, 1984; Wheeler and Smith, 1988). This step increases the reliability of the threshold-crossing method to a certain extent, as it uses some interpolation techniques (e.g., cubic splines) to minimize errors from asynchronous sampling of the spikes. The spike alignment task is computationally intense (Jung et al., 2006), and multiple alternative methods have been proposed to mitigate this effect (Hulata et al., 2002; Bokil et al., 2006; Kim and McNames, 2007).

Detection thresholds of time domain detectors vary from being too conservative, with a high probability of misses, to being too relaxed, with high a probability of false positives. As stated earlier, changes to waveform shapes have been consistently observed across multiple sessions of chronic recordings (Kralik et al., 2001; Buzsaki et al., 2004) and even within the same recording session (Wolf and Burdick, 2009). Systematic variations in signal amplitude and time scale of these variations are consistent with many physiological cycles of the subject (Chestek et al., 2007). This inherent non-stationarity mandates that spike detection be minimally supervised because unrealistic assumptions about signal and noise characteristics can make the detection susceptible to changes in experimental conditions. Assuming prior knowledge of spike parameters, such as duration and shape, limits detection performance to experimental conditions in which these assumptions are valid. This may be temporary, and the trade-off between supervision and blind detection becomes dependent on the anticipated neural yield—the number of detected cells per electrode–for the intended application.

2.2.4 Transform-Based Methods

As stated before, when the signal parameters are unknown, usually no Uniformly Most Powerful (UMP) test exists and the estimates depend on the chosen signal model. If the model is not appropriate, a transform domain representation may be adequate. Transforms that yield the most compact representation of the signal (i.e., those that are inherently localized) are the most desirable because of the transient nature of spikes. Most of the early work on transient detection schemes used transforms that are well localized in time and frequency (Frisch and Messer, 1992; Del Marco et al., 1997; Kulkarni and Bellary, 2000; Liu and Fraser-Smith, 2000). The most popular by far are wavelet-based techniques (Letelier and Weber, 2000; Oweiss and Anderson, 2000, 2001a,b, 2002a,b; Hulata et al., 2002; Oweiss, 2002c; Kim and Kim, 2003; Quiroga et al., 2004; Nenadic and Burdick, 2005; Oweiss, 2006). We briefly review wavelet theory and then discuss wavelets' application to spike detection and later spike sorting. We focus on the discrete case for practical implementation purposes.

2.2.4.1 *Theoretical Foundation*

The wavelet transform is an efficient tool for sparse representation of signals on an orthonormal basis $\{\varphi_n\}$; that is,

$$s = \sum_{n=1}^{N} c_n \varphi_n \tag{2.14}$$

The sparsity condition implies that $k \ll N$ of the coefficients c_n are nonzero, which enables the transform to compress the spike energy into very few coefficients. In matrix form,

$$s = \Phi c \tag{2.15}$$

where Φ is the $N \times N$ matrix of bases (stacked as column vectors) and c is a k-sparse vector. Computing the transform coefficients amounts to computing

$$c = \Phi^T s \qquad (2.16)$$

or, in the noisy case,

$$c = \Phi^T y = \Phi^T (s+n) = \Phi^T s + \Phi^T n = y_s + y_n \qquad (2.17)$$

The detection in the transform domain becomes

$$\begin{aligned} H_0 &: c = y_n \\ H_1 &: c = y_s + y_n \end{aligned} \qquad (2.18)$$

The sparsity of the transformed signal implies that few large coefficients are typically in the neighborhood of sharp signal transitions, which usually occur near the spike's zero crossing. This feature permits denoising by discarding coefficients below a predetermined threshold (Donoho, 1995) because the noise becomes a low-level "signal" scattered across all of the coefficients. The resulting above-threshold coefficients are used for spike detection and possibly reconstruction. Mathematically, the thresholded observations can be expressed as

$$\bar{y} = c.I\,(|c| > TH) \qquad (2.19)$$

$I(.)$ is an indicator function, and the threshold is typically estimated as $TH = \sigma \sqrt{2 \log N}$, where σ^2 is the noise variance (in each subband) calculated as

$$\sigma^2 = MAD(c)/0.6745 \qquad (2.20)$$

MAD denotes median absolute deviation. Noise coefficients typically spread out across many scales, making them relatively easier to isolate (Oweiss and Anderson, 2000). The GLRT energy detector in this case takes the form

$$\bar{y}^T \bar{y} \underset{H_0}{\overset{H_1}{\underset{<}{\gtrless}}} \gamma \qquad (2.21)$$

Because the basis is fixed, this unsupervised detection scheme falls under the category of "unknown signal" for which the test in Eq. (2.21) is optimal, depending on how well the basis sparsifies the observations (Oweiss and Anderson, 2002a,b).

2.2.4.2 *Performance*

Here we demonstrate some performance examples of transform-based spike detection and compare them to detection schemes discussed earlier. Figure 2.5 illustrates a qualitative example of the denoising ability of the transform as well as its ability to preserve the spikes faithfully for a sample neural trace.

Figure 2.6 illustrates quantitatively the performance of the transform-based detector over the time domain threshold-crossing detector under a moderately high SNR for a sufficiently large data set. It can be seen that in both the known and unknown cases, the transform-based detector performs significantly better than the time domain detector.

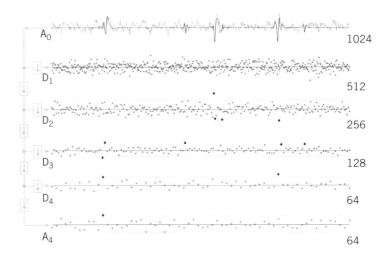

FIGURE 2.5

Sample spike data (upper trace) and its sparse discrete wavelet transform representation (using the *symlet* 4 basis (Coifman and Wickerhauser, 1992)) across four decomposition levels. Approximation coefficients are denoted by 'A', while details are denoted by 'D' in each level. The reconstructed waveform is obtained using the above threshold coefficients, preserving most of the details of the spikes while substantially reducing the contaminating noise. Please see this figure in color at the companion web site: www.elsevierdirect.com/companions/9780123750273

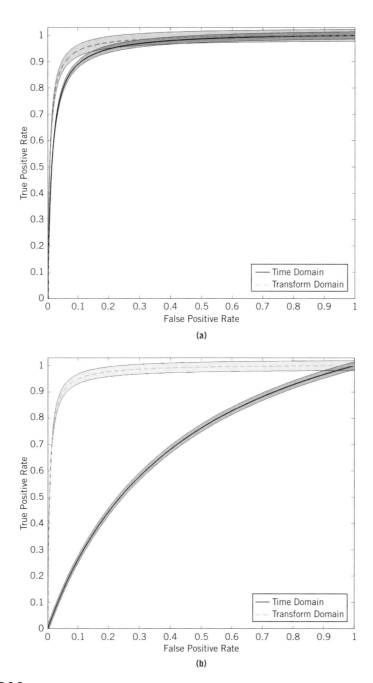

FIGURE 2.6

Performance of the transform domain and time domain detectors in the known (a) and unknown (b) spike cases SNR = 3 dB.

2.3 SPIKE SORTING

The process of segregating the activity of each recorded neuron from the observed mixture is often referred to as spike sorting. The objective is to characterize the ability of each neuron to encode, in isolation or as part of a population, one or more of the external covariates. Spike sorting has been one of the most extensively studied problems since techniques for extracellular recordings began to emerge (McNaughton et al., 1983; Abeles and Gerstein, 1988; Gray et al., 1995). The need for more sophisticated spike sorting techniques continues to grow apace with the ability to simultaneously sample large populations of neurons, particularly with high-density microelectrode arrays being regularly implanted in awake, behaving subjects (Normann et al., 1999; Wise et al., 2004).

We distinguish here between two approaches that have been proposed to solve this problem: (1) the pattern recognition approach, which operates on the individual spikes extracted after the spike detection step; (2) the blind source separation (BSS) approach, which operates on the raw data *without* the need for a priori spike detection. Each approach has a number of advantages and disadvantages, as will be shown next.

2.3.1 Pattern Recognition Approach

Historically, spike sorting has been categorized as a pattern recognition problem (Jain et al., 2000). This is the most common approach and it relies on the sequence of steps illustrated in Figure 2.7.

After spikes are detected and extracted, they are aligned relative to a common reference point. This step is important because spikes may be translated and potentially scaled from previous occurrences relative to the detection point, especially when the extraction window length N is empirically chosen by the user. The feature extraction step extracts the most *discriminative d* features from each spike waveform and uses them to represent the spike as a single point (vector) in a d-dimensional space. The criterion for choosing the features should allow the corresponding pattern vectors from different spike classes to occupy compact and disjoint regions in this d-dimensional space.

The clustering step attempts to find these regions by partitioning the feature space into distinct clusters, as shown in the 2D projection of the feature space in Figure 2.7. This is done by identifying statistical decision boundaries based on perceived similarity among the patterns. Because clusters can have very different shapes and their number is unknown beforehand, this is the most complex step. The final step is labeling the waveforms in each cluster as belonging to a single neuron (or unit). These steps constitute the training

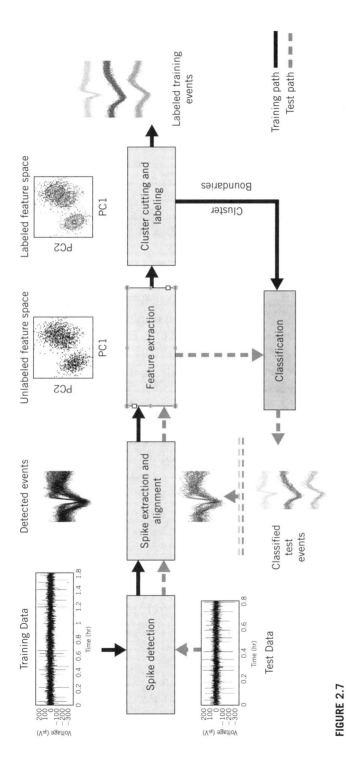

FIGURE 2.7

Pattern recognition approach to spike sorting. Spike events are extracted after detection and discriminative features are computed and clustered in a 2D or 3D feature space. Please see this figure in color at the companion web site: www.elsevierdirect.com/companions/
9780123750273

mode of the pattern classifier. Supervision by the user allows optimizing the preprocessing and feature extraction/selection strategies (e.g., determining the number of clusters). Similar preprocessing and feature extraction take place in the test mode for the incoming unlabeled spike. A decision-making step compares the features of the unlabeled spike to the set of stored features and assigns it to the closest pattern class using some distance metric.

The effectiveness of the representation space (feature set) is determined by how well patterns from different classes can be separated, whereas original details about spike waveform shapes become less important once these patterns are reasonably separable. For this reason, most of the work in spike sorting has focused primarily on the feature extraction and clustering steps. Initial work relied on extracting simple features of the waveform shape such as *amplitude* and *duration* (Dinning and Sanderson, 1981). Because these are highly variable features, cluster separability was often poor unless the SNR was considerably high and the waveforms were stationary. *Template matching*, on the other hand, relies on determining the degree of similarity between stored template spike waveforms and incoming spikes (Wheeler and Heetderks, 1982). The templates are obtained by averaging many occurrences of labeled spikes during the training mode. This method is computationally demanding because the comparison needs to take place for all possible translations and rotations of the waveforms; it is also vulnerable to spike nonstationarity unless the stored templates are regularly updated.

Principal component analysis (PCA) is the most popular method of pattern recognition because it attempts to alleviate these limitations. If the detected spike waveforms are stacked as rows of the matrix S, each spike can be represented as a weighted sum of orthonormal eigenvectors of S:

$$s = \sum_{n=1}^{N} \lambda_n u_n \tag{2.22}$$

Typically, the PCA feature space uses $d = 2$ features (sometimes $d = 3$ depending on how separable the clusters are) to represent each spike. The features, or PCA *scores*, are computed as (Fee et al., 1996)

$$s_i = u_i^T s, \quad i = 1, 2, \ldots, d \tag{2.23}$$

These features are subsequently clustered using algorithms such as expectation-maximization (EM) (Dempster et al., 1977), fuzzy c-means (Pal and Bezdek, 1995), or superparamagnetic clustering (Blatt et al., 1996). Figure 2.8 illustrates an example PCA feature space with two scores

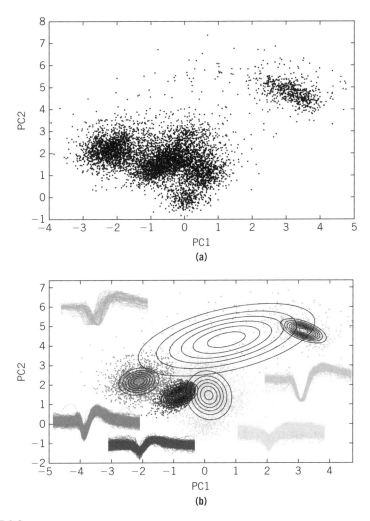

FIGURE 2.8

PCA for feature extraction and EM for unsupervised cluster cutting of 6469 spike events.
(a) Unlabeled spikes projected onto the two dominant eigenvectors of the data matrix.
(b) Five distinct units can be separated using this approach as indicated by the EM contour
plots. More units could be separated with user supervision—for example, by breaking the
lightest gray cluster into two or three smaller clusters. Please see this figure in color at the
companion web site: www.elsevierdirect.com/companions/9780123750273

representing each spike. In this case, we have a number of clusters with various
degrees of separability. An unsupervised EM algorithm was able to separate
five distinct spike shapes that indicate the presence of five putative single units
in the observed ensemble activity.

Despite being a reasonable approach to spike sorting, pattern recognition has some limitations. One is the uncertainty in the number of neurons being simultaneously recorded. The clustering step has to determine the number of *clearly isolated* clusters; this depends on how *compact* these clusters are, which often requires user supervision. A clear example is the large cluster centered around the (0,1) coordinate in the feature space illustrated in Figure 2.8. This cluster could be broken down to two or three smaller clusters, depending on the anticipated cluster variance. The compactness of the clusters relies to some extent on how well spikes were detected and extracted in the previous step as this can significantly minimize the number of outliers in the data.

The pattern recognition approach relies entirely on the shape of s to discriminate spikes. There are a number of cases where ambiguity might occur, such as when a large cell far away from the electrode tip yields a spike waveform similar to that of a smaller nearby cell. Nonstationarity of spike waveforms during chronic recording can also make clusters gradually *overlap* over time because cell migration, electrode drift, and/or tissue encapsulation can significantly alter waveform shapes (Fee et al., 1996; Lewicki, 1998; Gerstein and Kirkland, 2001; Wood et al., 2004).

When spike clusters from different neurons have disjoint, oval-like shapes in the feature space, reminiscent of multivariate Gaussian distributions of their underlying pattern, they are often attributed to *clearly isolated* multiple single-unit activity (MSUA) and can be efficiently sorted using likelihood-based methods such as the EM algorithm (Lewicki, 1998). Even if these cluster shapes are not typical of Gaussian distributions (for example, when they are heavy-tailed and thus more indicative of multivariate t-distributions (Shoham et al., 2003)), cluster cutting can be achieved with a variety of clustering algorithms. In a typical extracellular recording experiment, however, MSUAs often constitute only 20–40% of the detectable signals (Gerstein, 2000). The rest are either complex overlapping spikes resulting from a few of the clearly isolated cells firing *simultaneously* in the vicinity of the electrode tip or they are superimposed spikes from numerous cells that are not adjacent to the electrode tip, often referred to as *multi-unit* activity (MUA).

Studying the spike overlap case is important because it indicates the degree of synchrony in local populations that has been widely hypothesized to underlie certain aspects of stimulus coding (Perkel et al., 1967; Eggermont, 1993; Gerstein and Kirkland, 2001; Goldberg et al., 2002; Jackson et al., 2003; Dzakpasu and Zochowski, 2005; Sakamoto et al., 2008; Yu et al., 2008). It is also useful when studying the synaptic plasticity that is widely believed to accompany learning and memory formation (Caporale and Dan, 2008; Feldman, 2009). To date, there is no consensus on a reliable method to resolve complex spikes. Nevertheless, under certain conditions, this problem can be reasonably solved with the BSS approach discussed next.

2.3.2 Blind Source Separation Approach

The blind source separation (BSS) approach assumes that P signal sources (neurons) *impinge* on an array of M sensors (electrodes) to form an observed mixture Y (snippet), as illustrated in Figure 2.9. The signal mixture can be expressed as

$$X = \sum_{p=1}^{P} A_p^T S_p \qquad (2.24)$$

where A_p represents the columns of the $M \times P$ mixing matrix A, and S_p denotes the $1 \times N$ signal vector from neuron p within the observation interval. Note here that this signal vector may comprise *any* number of spike waveforms s_p. For P neurons, the observation model is

$$Y = AS + Z = X + Z \qquad (2.25)$$

where the signals from the P neurons are stacked as rows of the $P \times N$ matrix S, and Z denotes an additive noise $M \times N$ matrix with arbitrary temporal and

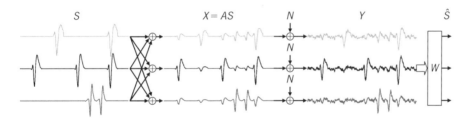

FIGURE 2.9

BSS approach to spike sorting. Spikes from different units form the signal matrix S and are linearly mixed using the matrix A to produce the noise-free observation matrix X. The data matrix Y is obtained by adding noise to the X matrix. When the number of sources $P <$ the number of channels M, the BSS estimates the matrix S directly from Y by finding the matrix W that optimizes an objective function (for example it maximizes the independence between the rows of S in the ICA approach). When $P > M$, sparsifying the S matrix first through a transform-based representation (e.g., a discrete wavelet transform with basis Φ) helps reduce the dimensionality of the signal subspace along the direction of selected bases to permit reconstruction of S. These bases are determined by their ability to preserve the *array manifold*—the space spanned by the columns of the mixing matrix A. Please see this figure in color at the companion web site: www.elsevierdirect.com/companions/9780123750273

spatial covariance. The mixing matrix is sometimes called *the array manifold* (Comon, 1994).

2.3.2.1 *Fewer Sources Than Mixtures*

Source separation algorithms attempt to solve the cocktail party problem— that is, to recover the signal matrix S from Y. Therefore, they can, in principle, recover the entire spike train of each neuron from the observed mixture. When knowledge of the mixing matrix A is available, a least squares solution exists in the form of a minimum-variance best linear unbiased estimate (MVBLUE) (Henderson, 1975). This estimate is obtained when the mean square error (MSE) between the signal matrix S and its estimate $\hat{S}$ is minimized. In such a case we obtain

$$\hat{S} = \left(A^T R_Z^{-1} A \right)^{-1} A^T R_Z^{-1} Y \tag{2.26}$$

where R_Z^{-1} is the inverse of the noise spatial covariance. When the noise is uncorrelated across the electrodes, R_Z becomes a diagonal matrix with individual noise variances along the diagonal elements. The MVBLUE solution requires knowledge of the signal and noise subspaces (i.e., the mixing matrix A and the noise covariance R_Z), and is therefore not considered "blind."

BSS algorithms, on the other hand, attempt to recover S from Y when knowledge of the mixing matrix A is unavailable. Many BSS algorithms have been proposed to estimate S (Cardoso, 1998). A popular approach is independent component analysis (ICA) (Comon, 1994; Bell and Sejnowski, 1995; Hyvärinen and Oja, 2000). In the absence of noise, ICA can recover S by iteratively finding a linear transformation W such that

$$\hat{S} = WX = WAS \tag{2.27}$$

The matrix W is selected to "optimize the independence" between the variables $\hat{S}_p$ by maximizing the mutual information between the input X and the output $\hat{S}$. In the absence of noise and for generalized Gaussian distributions (GGDs) of the form $p(x) = K\exp(-\alpha|x|^\gamma)$ with $\gamma \neq 2$, ICA or any of its variants (Hyvärinen, 1999; Hyvärinen and Oja, 2000) can be used to exactly recover S when W asymptotically approaches the pseudo-inverse of A up to a permutation and a scale factor. An example application of BSS-ICA to spike sorting is illustrated in Figure 2.10 (Oweiss and Anderson, 2002a,b).

Figure 2.11 illustrates another example of a slow wave with synchronized spike trains A and B to it but not to each other. The coincident firing was resolved efficiently with MVBLUE and ICA (with some artifacts).

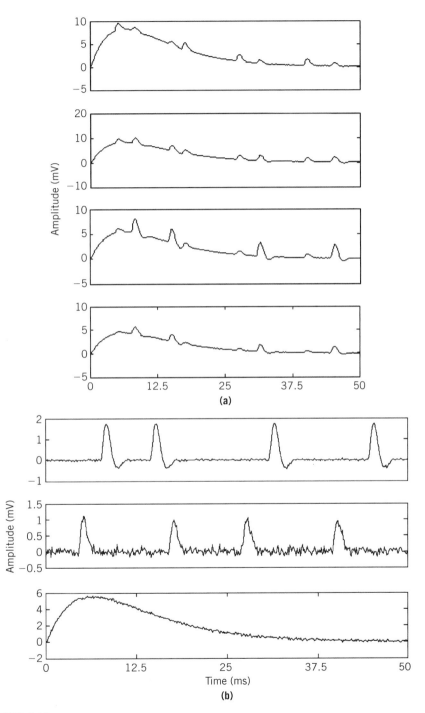

(a)

(b)

FIGURE 2.10

(*Continued*)

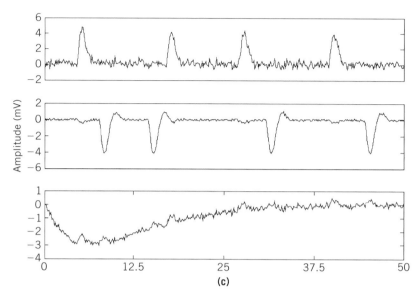

FIGURE 2.10

Comparison of ICA and MVBLUE estimates. (a) Four traces of synthetic neural data simulated with two units and a slow wave represented by different weights across an array of four electrodes. (b) Three components extracted by MVBLUE. (c) The three components obtained by ICA. Except for the sign in the bottom two components, the distributions across the array are very similar to the original. In the ICA case, the information was recovered with no prior knowledge of mixing matrix A.

Source: Adapted from Oweiss and Anderson (2002b).

Both methods demonstrate that source signals can be effectively recovered from the mixture even when spike overlap occurs in the data. MVBLUE estimates are less noisy but require knowledge of the signal and noise subspaces. ICA does not require this information.

ICA, however, has a number of limitations: First, the number of sources P has to be less than or equal to the number of electrodes M; typically, a dimensionality reduction algorithm such as PCA is applied before ICA. Second, the signals (rows of S) have to be independent so that the matrix X is full-rank, and there are situations where this requirement may not be fulfilled. For example, X would be rank-deficient when S is rank-deficient, and this might happen when a subset of the signals is perfectly coherent (i.e., at least one of the signals is just a scaled and delayed version of another). This might happen when two cells with similar spike waveforms are perfectly synchronized, possibly with a time lag. In this case, one of the signals will not be recovered. The matrix X will also be rank-deficient when A is

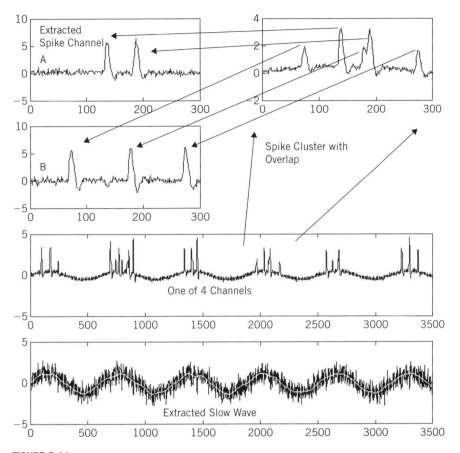

FIGURE 2.11

Two spike trains are synchronized to the slow wave but not to each other. For a more strenuous evaluation, the simulation includes some spikes that overlap in time (top right). The sine wave simulates a low-frequency LFP. In the top left, a cluster of spikes has been separated from the slow wave and shown on an expanded time base. The two plots in the top left demonstrate that the two extracted spike channels can be separated from each other even when the spikes overlap (the third spike in the top right is two spikes that overlap in time), producing the complex spike waveforms. The bottom plot is the extracted slow wave. The white waveform inside the noisy one is the wavelet-denoised version of the slow wave.

rank-deficient, yielding an unmixing matrix W that is almost singular. This might happen when spike waveforms from two (or more) cells impinge on the array with approximately equal power. Then the corresponding columns of A will be linearly dependent, making it rank-deficient. Although these situations are rarely encountered in practice, inhomogeneities of the extracellular medium may affect spike amplitude roll-off, and consequently spike power, as a function of the distance from the electrode (Plonsey, 1974). ICA also requires at most one of the sources to be normally distributed and only allows very small noise levels for faithful signal recovery (Comon, 1994).

2.3.2.2 *More Sources Than Mixtures*

In practice, the number of *theoretically recordable* neurons is larger than the number of *actually recorded* ones, and the numbers of both are typically larger than the number of electrodes (Buzsaki et al., 2004). In the latter case, separating the spike train of each neuron from the observed mixture becomes an undetermined system of equations in the BSS model. More important, examination of the observed signals reveals that the contaminating noise is predominantly neural with considerable temporal and spatial correlation, even for spontaneous activity. An example can be seen in Figure 2.12 in which a snippet of spontaneous activity recorded using an array of four electrodes in the dorsal cochlear nucleus (DCN) of an anesthetized guinea pig is illustrated (Figure 2.12(a)). Figure 2.12(b) demonstrates that channels with an apparent lack of detectable spiking activity (e.g., channel 1) exhibit power spectra that are typical of colored noise processes and that significantly overlap with the signal spectrum (e.g., channel 4). In Figure 2.12(c), evidence of spatial noise correlation among channels can be clearly seen for both spontaneous and stimulus-driven activity.

In general, it is impossible to recover a P-dimensional vector from an M-dimensional mixture when $M < P$. However, it is interesting to study the case when each of the spike trains S_p is k_p-sparse. As before, the spike train of neuron p can be expressed in terms of basis vectors as

$$S_p = \sum_{k=1}^{K} c_{kp} \varphi_k \tag{2.28}$$

where now the φ_k's are elements of an *overcomplete* dictionary of size $K > N$. In matrix form,

$$S = C\Phi \tag{2.29}$$

C is the $P \times K$ coefficient matrix. We note here that the size of the dictionary K is much greater than N or P but that Φ is selected to make C as sparse as possible.

An estimate of the noise-free observations can be expressed (using Eqs. (2.19) and (2.20)) as follows:

$$\overline{Y} = \hat{X} = AC\Phi \tag{2.30}$$

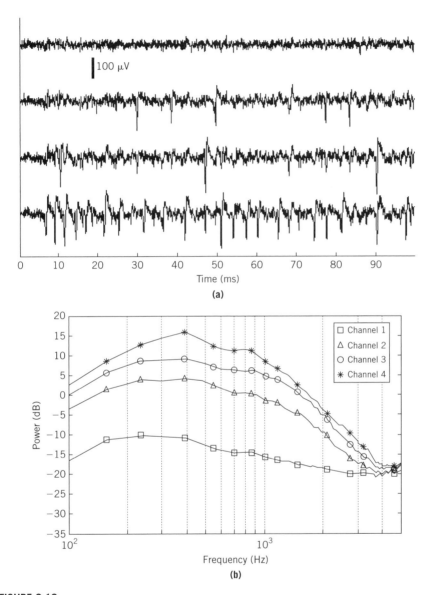

FIGURE 2.12

(*Continued*)

It is worth examining the $M \times K$ product matrix AC in the case where C is sparse. From the formulation in Eq. (2.29), each of the P rows of C will have only $k_p \ll K$ entries that are nonzero. This observation can be used to *sequentially* estimate the signal subspace spanned by the columns of the mixing matrix A. More precisely, assume $\{k_p\}$ to denote the set of nonzero

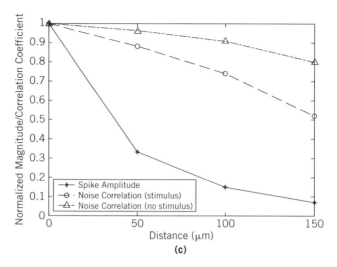

FIGURE 2.12

(a) Sample 100-ms trace of spontaneous 4-electrode recordings in the dorsal cochlear nucleus of an anesthetized guinea pig. (b) Estimated power spectrum of the data in (a) using Welch's averaged periodogram method. (c) Spike amplitude roll-off and noise correlation along the array for the data in (a). Stimulus-driven data were collected under white noise stimuli.

Source: Adapted with modifications from Oweiss (2006).

entries of the k_p-sparse vector c_p. For each entry in this set, the eigenvectors of $\overline{Y}$ can be calculated using SVD:

$$\overline{Y}_k = U_k D_k V_k^T = \sum_{i=1}^{M} \lambda_i^k u_i^k v_i^{k^T} \quad k \in \{k_p\} \tag{2.31}$$

The columns of the eigenvector matrix U_k span the *column* space of $\overline{Y}_k$, which is the same space spanned by the columns of the $M \times K$ outer product matrix $A_p c_p^T$, while the columns of V_k^T span the *row* space of $\overline{Y}_k$. An important condition is to have $k_p < M$.

To separate the sources, we can formulate a search algorithm that first sorts the sparse entries in each row of C in descending order of magnitude and then uses these to select the bases from Φ that best model the signal S_p. These bases are subsequently rank-ordered based on the degree to which they are able to preserve the signal subspace spanned by columns of A. For example, when $P = 1$, A is a single-column matrix. We therefore need to select a set of best bases, $\Im_p \subset \Phi$, where the dominant eigenvector set u_1^k, $k \in \{k_p\}$, computed using Eq. (2.31) is "the closest" to the single-column matrix A for each k. This amounts to finding the nearest manifold to the true one expressed by the matrix A. The associated dominant eigenvalue set, λ_1^k, $k \in \{k_p\}$, is sorted in decreasing order of magnitude to rank-order the basis set $\mathfrak{I}_p$. We refer to this approach as the multi-resolution analysis of signal subspace invariance technique, or MASSIT. It was proposed in brief form in Oweiss and Anderson (2001a,b), and detailed in Oweiss (2002c) and Oweiss and Anderson (2007).

We demonstrate the ability of the MASSIT algorithm to separate spike trains from multiple neurons through a number of examples. In each example, the dominant eigenvalue was calculated according to Eq. (2.31). The eigenvalues were assembled in a single K-dimensional vector, λ_1, and sorted in descending order of magnitude. A threshold was selected based on a statistical significance test calculated from the difference between adjacent entries in that vector. Basis indices corresponding to λ_1 entries above the threshold were selected as candidates for the best basis set $\mathfrak{I}_p \subset \Phi$. Each candidate basis was included in $\mathfrak{I}_p$ if the ℓ_2 norm $\left\| u_1^k - A_p \right\|_2$, $k = 1, \ldots, K$ was minimized. An initial estimate of A_p was obtained from $k = 0$ in Eq. (2.31) which is guaranteed to span the signal subspace if at least one spike event is present in the observation interval.

Figure 2.13 illustrates an example of this strategy for the case where $M = 4$, $P = 1$. Here $\overline{Y} = \hat{X}_1 = A_1 S_1^T$ and therefore it was sufficient to examine the principal eigenmode to estimate the signal vector.

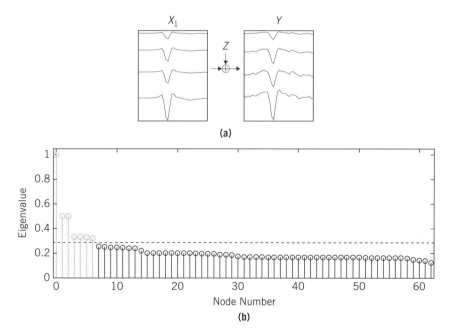

(a)

(b)

FIGURE 2.13

(*Continued*)

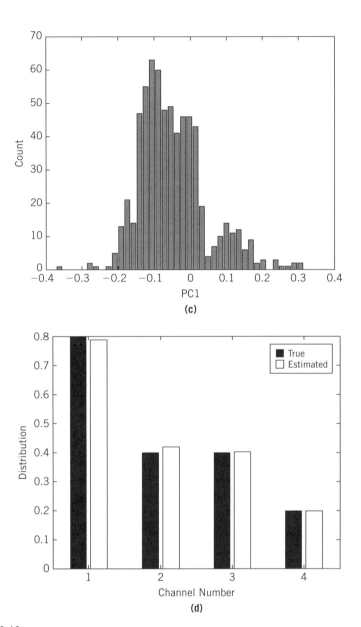

FIGURE 2.13

(*Continued*)

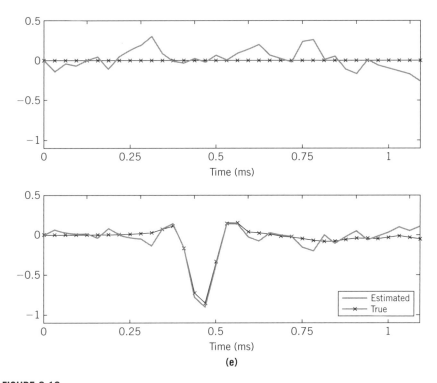

FIGURE 2.13

(a) Data snippet from a single source distributed across four channels ($M = 4, P = 1$).
(b) Dominant eigenvalue vector, λ_1, sorted in decreasing order of magnitude. Significant
entries are illustrated above the threshold (dashed line). (c) Distribution of the eigenvector
u_1^k across bases (projected onto its dominant principal component, $n = 100$ samples).
(d) True and estimated mixing matrix (A is single column). (e) (bottom): True and
reconstructed waveform using Eq. (2.26) (with the noise covariance inverse being the
identity matrix); (top): True and reconstructed waveform using the second eigenvector.
In this case, the second eigenvector spans the noise subspace.

Figure 2.14 illustrates the case where $M = 4$, $P = 2$. The distribution of
the dominant eigenvector across nodes shows two distinct clusters, indicative of the presence of two sources. The reconstructed waveforms are clearly
distinguishable from each other, even though one of them has a relatively
higher MSE.

In Figure 2.15, an example with $M = 4$, $P = 3$ is illustrated. The distribution of the dominant eigenvector across nodes demonstrates the existence of
three clusters, albeit one of them is not as highly probable as the other two.
This contributes to a larger MSE in its estimate as shown in the middle panel
of Figure 2.15(f). Additional information about this source is extracted by

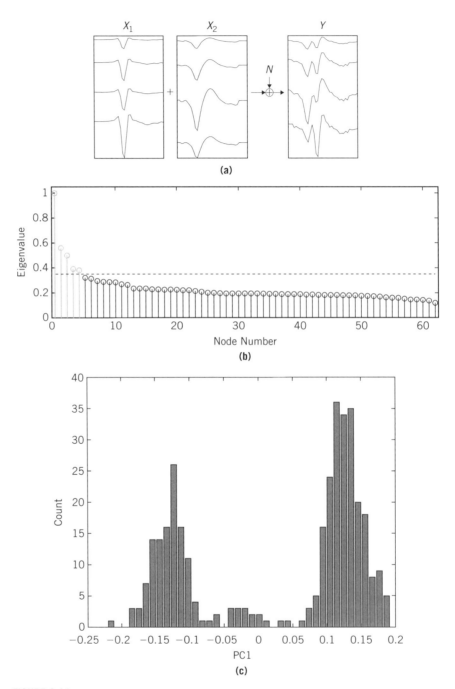

FIGURE 2.14

(Continued)

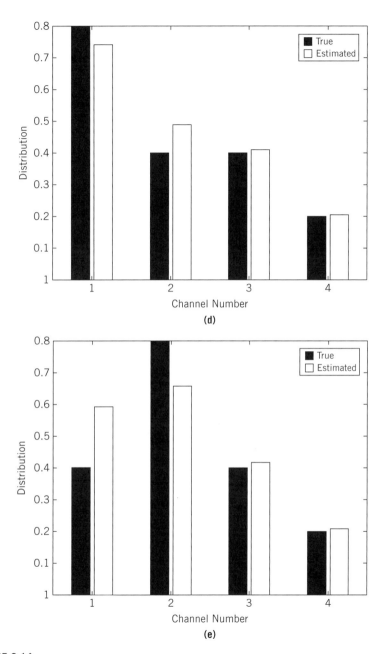

FIGURE 2.14

(*Continued*)

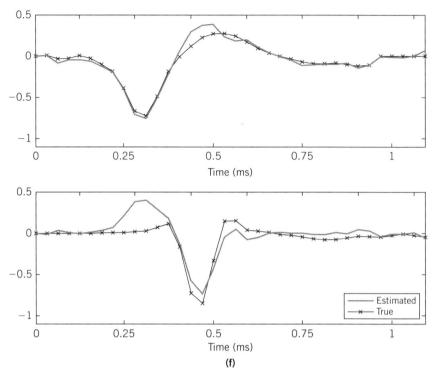

(f)

FIGURE 2.14

(a) Data snippet from two sources distributed across four channels ($M = 4$, $P = 2$). (b) Dominant eigenvalue vector λ_1 sorted in decreasing order of magnitude. Significant eigenvalues are illustrated above the threshold (dashed line). (c) Distribution of the dominant eigenvector u_1^k across bases (projected onto its dominant principal component, $n = 100$ samples). (d) and (e) True and estimated mixing matrix. (f) True and reconstructed spike waveforms using the two clusters of dominant eigenvectors.

repeating the analysis for the *second* dominant eigenvector. The estimate of that source can be obtained with a high degree of fidelity (Figure 2.15(i)).

Figure 2.16 demonstrates the more complex case of an undetermined system with $M = 2$ and $P = 3$. Here the principal eigenvectors are clustered into three disjoint clusters. Two sources were reasonably recovered (Figure 2.16(g)), while the first source was not. A possible explanation is the significant correlation that exists between spike waveform shapes and the fact that the basis used (*symlet4*) was not optimally selected to sparsify all of them simultaneously. Despite this diminished performance, the reconstructed waveforms can be reasonably distinguished from each other, which is our ultimate objective.

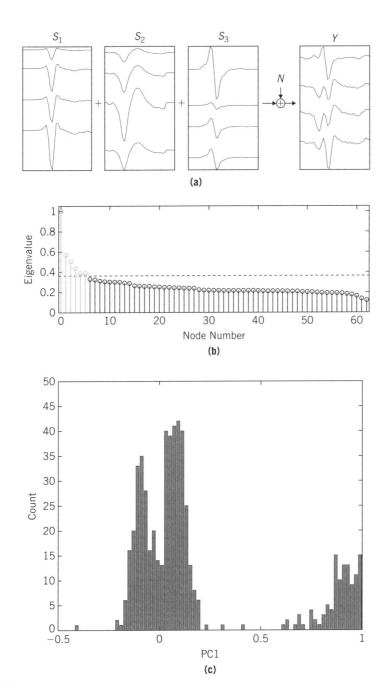

FIGURE 2.15

(*Continued*)

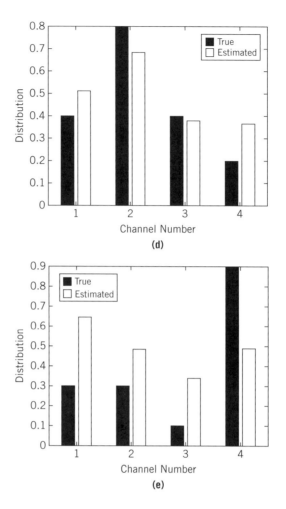

FIGURE 2.15

(Continued)

Compared to other BSS approaches, the MASSIT algorithm has a number of advantages: First, it can be generalized to any number of sources as long as the dictionary K is large enough. In theory, an upper bound on the number of sources that can be separated using this approach is $P \leq M \times K$. In practice, the quality of separation depends on the noise level and the degree of correlation between spike waveforms. The MASSIT algorithm can be viewed as a BSS algorithm that separates the subset of sources with the most sparse representation along the direction of a particular basis (fixed k). Alternatively, for a given source (fixed p), it can be viewed as a pattern recognition algorithm that

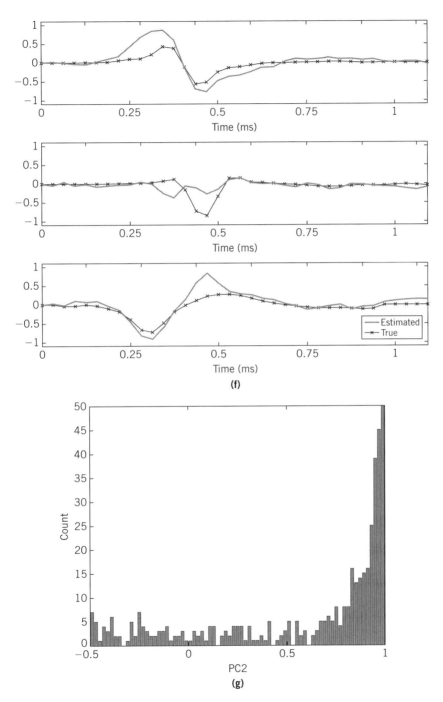

FIGURE 2.15

(*Continued*)

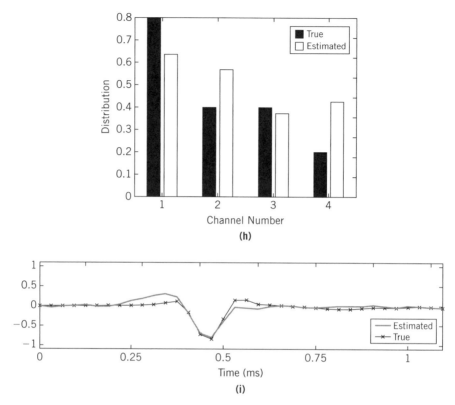

FIGURE 2.15

(a) Data snippet containing three events from three sources distributed across four channels ($M = 4$, $P = 3$). (b) Dominant eigenvalue λ_1 sorted across nodes in decreasing order of magnitude. Significant eigenvalues are illustrated above the threshold (dashed line). (c) Distribution of the dominant eigenvector u_1^k across bases (projected onto its dominant principal component, $n = 100$ samples). (d) and (e) True and estimated mixing matrix. (f) True and reconstructed waveforms using the two most probable clusters of principal eigenvectors in (c). (g) Distribution of the second dominant eigenvector u_2^k across bases (projected onto its dominant principal component, $n = 100$ samples). (h) True and estimated mixing matrix (i) True and reconstructed waveforms using the most probable cluster of second principal eigenvectors in (g). The first source was successfully recovered.

extracts source features along the direction of a specific set of bases, $\Im_p \subset \Phi$, for which the signal subspace spanned by the corresponding columns of A remains invariant. These features are the set of eigenvalues

$$\lambda_1^{\Im_p\{1\}} > \lambda_1^{\Im_p\{2\}} > \cdots > \lambda_1^{\Im_p\{k_p\}} \quad \text{such that } k = \underset{p}{\arg\min} \left\| u_p^k - A_p \right\| \quad (2.32)$$

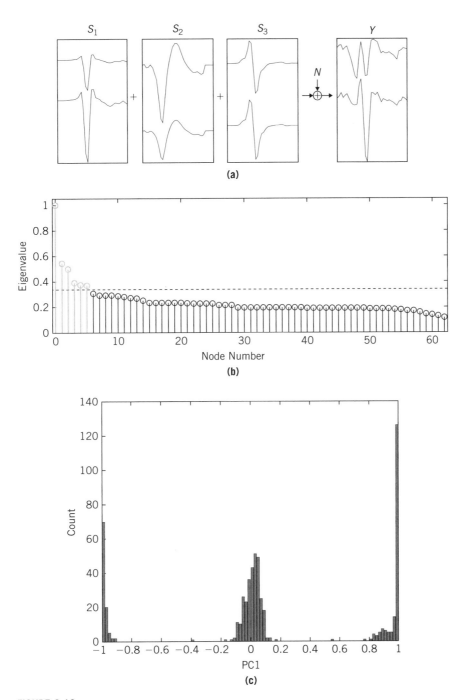

FIGURE 2.16

(*Continued*)

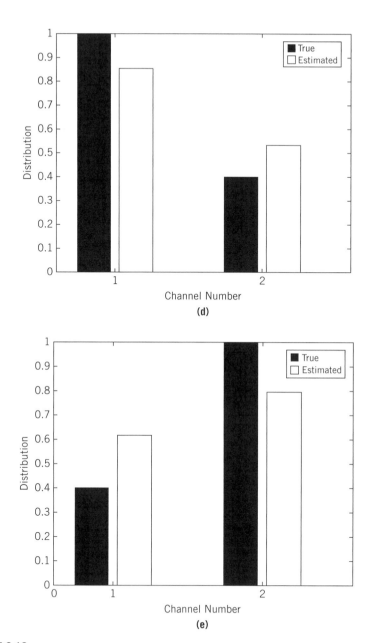

FIGURE 2.16

(Continued)

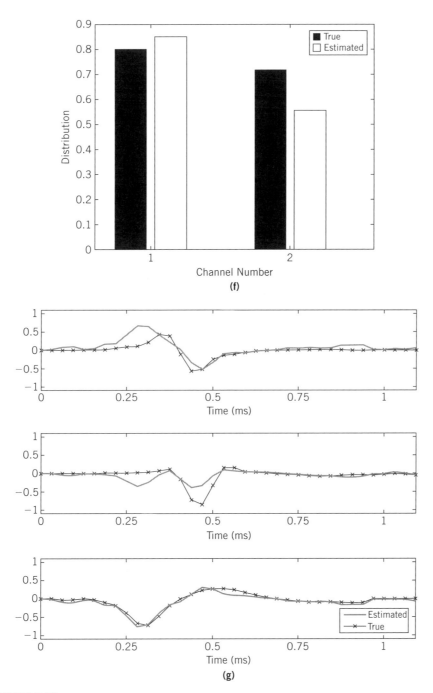

FIGURE 2.16

(*Continued*)

FIGURE 2.16

(a) Data snippet containing three events from three sources distributed across two channels ($M = 2$, $P = 3$). (b) Dominant eigenvalue λ_1 sorted across nodes in decreasing order of magnitude. Significant eigenvalues are illustrated above the threshold (dashed line). (c) Distribution of the dominant eigenvector u_1^k across bases (projected onto its dominant principal component). (d), (e), and (f) Estimate of the mixing matrix. (g) True and reconstructed waveforms using the most probable clusters of principal eigenvectors in (c). Despite that waveforms were recovered with variable degrees of accuracy, a considerable degree of separability between the waveforms is preserved, given the complex waveform shape in (a).

The sparsity introduced by the overcomplete basis representation in the N-dimensional space leads to a considerable dimensionality reduction of the observed mixture in the M-dimensional space. This compensates for the fact that $M \ll N$, which does not permit the use of 2D bases to sparsify both space and time as in image processing applications. It also permits enlarging the separation between the signal and noise subspaces by separating the signal coefficients from the noise coefficients. Because the representation of the spikes is not unique in an overcomplete basis, the ultimate goal is selecting the representation that *maximizes the separability* between spike classes by *learning* the data manifold in the selected basis (as opposed to selecting the ones that *best reconstruct* the spike).

2.4 PRACTICAL IMPLEMENTATION

Algorithms for spike detection and sorting have been in existence for over 20 years. Despite being already extensively studied, recent advances in large-scale neural recording using chronic implantable microelectrode arrays have renewed interest in exploring new algorithms. This is particularly the case for cortically controlled neuroprosthetic and BMI applications in which spike detection and sorting must be carried out in real time to permit instantaneous decoding of spike trains (Isaacs et al., 2000; Wessberg et al., 2000; Serruya et al., 2002; Hochberg et al., 2006; Velliste et al., 2008) and neuromodulation and neurostimulation (Kringelbach et al., 2007; Lehmkuhle et al., 2009; Nagel and Najm, 2009; Lega et al., 2010). In a pristine lab environment, computational resources with close to real-time performance are readily available to perform these tasks. However, many of these applications are intended for clinical use in which these tasks have to be implemented in fully implanted hardware platforms that process an ever-increasing number of data channels.

Here we are faced with yet another challenge to spike detection and sorting in clinically viable BMI systems. First, to be implantable these systems have to be very small, dissipate very low power in the surrounding brain tissue, and feature wireless communication to the outside world to minimize any risk of infection and discomfort to the patient. State-of-the-art systems, however, fall short of fulfilling these requirements because spike sorting *on chip* is computationally prohibitive while the alternative of transmitting the high-bandwidth spike data for *off-chip* sorting requires hardware platforms that do not conform to the strict constraints of implantable systems and that incur significant latency in the system as a whole.

In exploring the numerous techniques for extracting this information from the sparse spike representation, one important constraint is that the computational complexity of obtaining this representation has to scale linearly or sublinearly with the number of electrode channels and the length of the observation interval. Fortunately, the lifting approach for discrete wavelet transform computation fulfills this requirement because efficient architectures can be designed using size/power-optimized computational engines that can be sequentially reused to process a number of electrode channels in real time (Oweiss et al., 2007).

Sparse models of neural spikes, as in equation Eq. (2.15), can be extended to *sensible signal models* when the coefficients in the $K \times 1$ vector c are sorted in decreasing order of magnitude and are observed to follow a power law. If this is the case, one key design principle is to extract the *minimum amount of information* sufficient to classify the spikes by retaining only the coefficients that provide maximum separability between spike classes (Oweiss, 2006; Aghagolzadeh and Oweiss, 2009).

To explore this idea further, we define a set of thresholds, TH_p for $p = 1, \ldots, P$, for the *sensible signal models* that we want to discriminate. These thresholds (referred to herein as *sensing thresholds*) enable expressing the classification problem as a series of *P binary* hypothesis tests similar to Eq. (2.1). More precisely, neuron p is declared present in subband k if the alternative hypothesis H_1^k is in effect (i.e., the corresponding entry in c surpasses TH_p). For adequate selection of TH_p, coefficients representing spikes from other neurons have to fall under the null hypothesis H_0^k. It can be shown that the distance between the P hypotheses under the sparse representation is preserved based on the Johnson-Lindenstrauss lemma (Johnson and Lindenstrauss, 1984).

To quantitatively assess the performance of this strategy, we define *spike class separability* as a function of the number of coefficients retained per event:

$$\Gamma = \frac{\text{Between Cluster Separability}}{\text{Within Cluster Separability}} = \frac{\Gamma_B}{\Gamma_W} \tag{2.33}$$

for a set of clearly isolated clusters χ_k, where $\chi_k \leq P$. The *between-cluster* separability is given by

$$\Gamma_B = \sum_{k=1}^{K} \frac{\displaystyle\sum_{x \in \chi_k} \sum_{y \notin \chi_k} \| x - y \|}{|\chi_k| \displaystyle\sum_{k \neq j} |\chi_j|} \qquad (2.34)$$

where $|\chi_k|$ equals the number of elements in cluster χ_k, and x and y are elements from the data. The quantity in Eq. (2.34) provides a factor proportional to the *overall separation* between clusters. For improved separability, a large Γ_B is desired. On the other hand, the *within-cluster* separability is defined as

$$\Gamma_W = \sum_{k=1}^{K} \frac{\displaystyle\sum_{x \in \chi_k} \sum_{y \in \chi_k} \| x - y \|}{|\chi_k| (|\chi_k| - 1)} \qquad (2.35)$$

and is proportional to the overall spread within individual clusters. For improved separability, a small Γ_W is desired. Therefore, a large Γ indicates a greater overall separability.

An example of this approach is illustrated in Figure 2.17. Here we used a complete basis set for which $K = N$ (overcomplete basis, for example using the discrete wavelet packet transform DWPT (Coifman and Wickerhauser, 1992; Oweiss and Anderson, 2007), can also be used). Sensing thresholds were set to allow only $\|c\|_\infty = \max_k |c_k|$ to be retained. Based on this criterion, node 6 is an ideal candidate node for classifying events from neuron B. Its

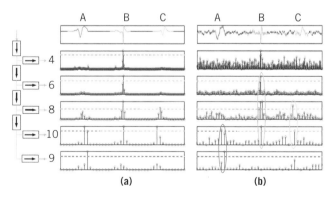

FIGURE 2.17

(Continued)

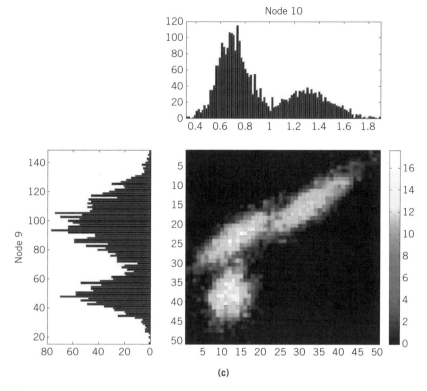

FIGURE 2.17

Strategy for using *sensible signal models* for sorting spikes from three units A, B, and C on a single channel. Sparse representation of the spikes in the (a) noiseless and (b) noisy case illustrated for the five wavelet decomposition levels (*symlet4* basis) indicated by the binary tree on the left. First-level high-pass coefficients (node 2) are omitted as they contain no information in the spectral band of the spikes. Sensing thresholds are set to allow only one coefficient/event to be retained in a given node. Any of nodes 4, 6, 8, and 10 can be used to mark events from unit B, yet node 6 provides the largest overall separability from other units' spikes. A similar argument can be made for node 9 and unit A. When single-feature separation is not feasible, joint features across nodes can be used, as illustrated by the 2D histogram in (c) for three units. When noise is present, the sensing threshold also serves as a denoising threshold. Please see this figure in color at the companion web site: www.elsevierdirect.com/companions/9780123750273
Source: Adapted from Aghagolzadeh and Oweiss (2009).

threshold is labeled TH_B because this particular node provides the *largest relative separation* between this neuron's events and events from the two other neurons. Another example can be seen in which node 9 can be used to mark events from unit A as it has the largest coefficient relative to those of other units within this particular node.

For a more strenuous evaluation, we applied this strategy to the experimental data of Figure 2.7. We compared it to two types of analyses: (1) manual, extensive *offline* sorting using hierarchical clustering of all features in the data—this mimics a pristine lab environment with relatively unlimited computational power; (2) automated, *online* pattern recognition using two PCA scores with EM cluster cutting. The latter mimics online spike detection and sorting with minimal user supervision. We computed a *separability ratio* (SR), defined as the ratio of separabilities computed using Eq. (2.33) for a binary hypothesis test, $\Gamma\{2\}$, along the direction of each basis. An SR ratio of 1 indicates equal degree of separability in both domains, while ratios larger than 1 indicate superior separability in the sparse representation domain. This later case implies that at least one unit can be separated in that node's feature space better than in the time domain's feature space. We compared these SRs to those of the PCA/EM features computed from the time domain.

The results of this analysis are illustrated in Figure 2.18. It can be seen that there are four putative units (units 1, 2, 3, and 5) occupying clusters that

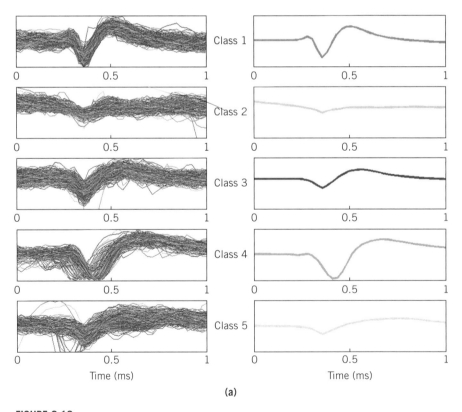

(a)

FIGURE 2.18

(Continued)

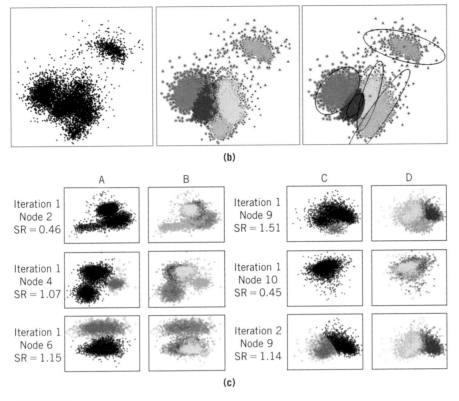

(b)

A   B   C   D

Iteration 1
Node 2
SR = 0.46

Iteration 1
Node 9
SR = 1.51

Iteration 1
Node 4
SR = 1.07

Iteration 1
Node 10
SR = 0.45

Iteration 1
Node 6
SR = 1.15

Iteration 2
Node 9
SR = 1.14

(c)

FIGURE 2.18

(a) (left): Events from five experimentally recorded units, aligned and superimposed on top of each other for comparison; (right) corresponding spike templates obtained by averaging all events from each unit in the left panel. (b) PCA 2D feature space: Dimensions represent the projection of spike events onto the two largest principal components: (left): unlabeled feature space; (middle): clustering result of manual, extensive offline sorting using hierarchical clustering with all features in the data; (right): clustering result using the two largest principal components and online sorting using EM cluster cutting with Gaussian mixture models (GMM). (c) Unit isolation quality of the data in (a). Each feature space in columns A and C shows the separation in the 2D feature space obtained using the binary hypothesis test in each node. The highest-magnitude coefficient that survives the sensing threshold for each event in a given node is marked with the dark gray cluster in each panel. The feature space of the sorted spikes using the manual, extensive offline spike sorting in (b) is redisplayed in columns B and D (illustrated with the same grayscale color code as in (b)) for comparison. For each node in the two class feature spaces (columns A and C) if the dark gray cluster matches *a single* cluster from the corresponding panel in columns B and D, respectively, this implies that the corresponding unit is well isolated in that node. Using this approach, three out of five units (units 1, 2, and 4) were isolated with one sweep in nodes 4, 6, and 9, respectively, leaving two units to be isolated with one additional sweep on node 9's second-largest coefficients. A sweep is obtained by cycling through all nodes once, eventually removing the largest coefficients corresponding to a single isolated cluster in a given node. In the first sweep, node 2 shows weak separation (SR = 0.45) between units. Unit 4 has larger separability in node 4 (SR = 1.07). Units 1 and 2 are isolated in nodes 6 and 9 with SR = 1.15 and 1.51, respectively. Units 3 and 5 are separated in node 9 with SR = 1.14. Please see this figure in color at the companion web site: www.elsevierdirect.com/companions/9780123750273

largely overlap. In addition, there are significant differences in the cluster shapes of these units between the manual, extensive offline sorting result and the automated online PCA/EM result. The sparse representation approach, on the other hand, permits each spike class to be identified in at least one node. For example, cluster 1 appears poorly isolated from cluster 5 in the time domain feature space, yet it is well separated from all the other classes in node 6.

Another rather surprising result is illustrated in Figure 2.19, where it can be seen that cluster separability monotonically *increases* as *fewer* signal coefficients are retained per spike. This demonstrates that basis selection based on optimizing classification outcome rather than optimal spike reconstruction is more effective. In theory, this strategy enables separating a maximum of $M \times K$ neurons if the L_∞ norm criterion is satisfied (i.e., there is a sufficiently large separation of coefficient magnitudes between spikes from different units). Uniform distribution of neurons across channels also helps attain this upper bound.

Figure 2.20 illustrates the performance for various SNRs, comparing sorting based on sparse representation features to sorting based on time domain features. Again, the performance demonstrates increased robustness of the sparse representation approach to spike sorting compared to the time domain approach.

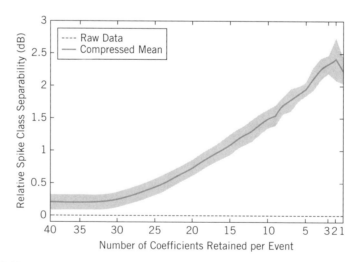

FIGURE 2.19

Quantitative analysis of spike class separability in the transform domain relative to that of the time domain versus the number of coefficients retained (i.e., degree of sparsity). The shaded area represents the standard deviation.

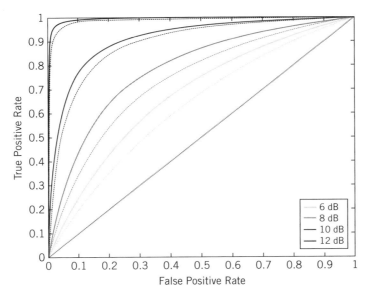

FIGURE 2.20

Comparison of the ROC curves obtained for spike sorting performance using time domain features (dotted lines) versus features obtained from the wavelet transform (solid lines) at various SNRs.

2.5 DISCUSSION AND FUTURE DIRECTIONS

In this chapter, a genuine attempt was made to provide an in-depth presentation of the problem of spike detection and sorting in extracellular neural recordings. Our main emphasis was to link the theoretical foundation that underlies detection and classification of signals in noisy observations to the practical considerations encountered in actual neurophysiological experiments.

For spike detection, we showed that the problem can be categorized based on the availability of prior information about the spike waveform to be detected and whether it is deterministic or stochastic. When the waveform is deterministic, the detection problem relies completely on the characterization of the probability distribution of the contaminating noise or, more specifically, on knowledge of its parameters (e.g., noise power). The matched filter detector was shown to be optimal under these conditions. When the waveform is stochastic, the detection problem becomes more complicated because knowledge of more parameters is needed. Specifically, we need to know the parameters that individually characterize the spike and noise distributions. Also needed is knowledge of the parameters governing how the spike and the noise *covary* together. We showed that, in the absence of prior knowledge, the coherent energy detector provides a near-optimal solution. Performance limits in each case were derived and compared to classical detection methods.

We then provided an in-depth discussion of the spike sorting problem, how it can be categorized, and the implications this categorization has on algorithm design. In particular, we demonstrated that, when viewed as a pattern recognition problem, an important condition for adequate separation of multiple units is the presence of clear disjoint clusters that can be identified in the feature space with minimal user supervision, which in turn depends on a number of factors such as unit stability, experimental SNR, and the effectiveness of the preceding spike detection step. Such conditions, however, are rarely satisfied in practice, particularly in long-term chronic recordings with implanted microelectrodes. Alternatively, spike sorting can be viewed as a blind source separation problem, particularly when an array of electrodes (a minimum of two are needed) is used to record the spikes. For the most part, this is useful when clusters in the feature space do not have clear separation, reminiscent of complex overlapping spikes that often result when two or more units fire nearly simultaneously. The BSS approach was shown to be capable of separating these complex events in a number of cases, provided that the individual and joint statistics of the signal and noise are known.

When addressing both problems, we focused explicitly on the sparse representation of spikes, as we believe this approach offers a good compromise between often incompatible algorithm design goals. In particular, sparse representation was shown to enhance the separation between the signal and noise subspaces, permitting effective denoising and subsequent detection to take place. It also allows reducing the dimensionality of the spike models in the N-dimensional space to a much smaller k-dimensional subspace, where $k \ll N$ determines the degree of sparsity. This strategy also permits formulating the classification problem of P neural sources as a *nested* series of P binary detection problems. This feature is particularly desirable when streamlining of the analysis is highly desired—for example, in BMI applications, where real-time neural decoding is needed. In such a case, the suitability of implementation in low-power, miniaturized electronics that can be easily implanted adds more appeal to this approach, particularly with an ever-increasing number of electrode channels.

Spike sorting is of fundamental importance to neural decoding algorithms that rely on simultaneously observed spike trains to estimate/predict an underlying behavior from the collective ensemble activity. Some recent work suggests that the inherent uncertainty and complexity of spike sorting may preclude its use in decoding, and that tuning of individual neurons to behavioral correlates can be an alternative approach—or an additional information channel—to neural decoding *with* or *without* waveform-based spike sorting (Ventura, 2008; Fraser et al., 2009). The idea is that tuning functions (or receptive fields) provide information about the tendency of a particular neuron to modulate its firing rate above or below its baseline level depending on the concurrent behavior. This approach is founded on a large body of experimental studies demonstrating neuronal specific tuning to a myriad of movement parameters such as direction (Georgopoulos et al., 1986), velocity (Moran and Schwartz, 1999), acceleration (Serruya et al., 2003), and endpoint (Carmena et al., 2003; Churchland et al., 2006). We note, however, that these tuning functions are typically obtained by *averaging* a large number of responses collected over multiple repeated trials. In these trials, animal subjects are *operant-conditioned* on specific tasks that require them to "optimize" the shape of these tuning curves over repeated trials to trigger reward actions. In contrast, single-trial responses in subjects performing volitional motor control during naturalistic behavior are typically very "noisy," with significant variability over a multitude of time scales (see for example Churchland et al. (2006) and Shadlen and Newsome (1998)).

While tuning functions may provide useful information about expected firing rates (Ventura, 2009), there is growing evidence that the noisiness of single-trial responses—also known as trial-to-trial variability—may reflect a *context-dependent* encoding mechanism that is *less predictable* (Cohen and Newsome, 2008, 2009; Kemere et al., 2008). This may be the result of, for example, the presence of unobserved inputs (to the observed neurons) that encode some other attention and cognitive aspects of the underlying behavior that are hard to measure (Shadlen and Newsome, 1994; Rolls et al., 1997; Chestek et al., 2007). On a global scale, these aspects may reflect not only movement execution but also the involvement of motor cortical areas in higher functions such as sensory-motor transformations, decision making, and motor planning related to action execution (Rizzolatti and Luppino, 2001). On a local scale, it may be the result as well of "recruiting" certain neurons with no specific tuning preferences but which are necessary to shape the *collective* response of the local population (Truccolo et al., 2010).

This provides an alternative account for the functional role of each neuron as being more *dynamic* than what is overtly implied by the neuronal tuning theory. In this account, the trial-to-trial variability rather reflects how the neuron orchestrates task encoding *in cooperation* with other elements of the ensemble (Eldawlatly et al., 2009; Aghagolzadeh et al., 2010), which is more consistent with the Hebbian cell assembly hypothesis (Hebb, 1949). This may explain the variability as being more dependent on the intrinsic and extrinsic contexts that govern neuronal excitability (experience with the task, level of fatigue, and so forth) (Cohen and Newsome, 2008; Sorger et al., 2009). An assessment of decoding results using a multivariate spiking model of motor population encoding with mixed tuning characteristics suggests that decoding *without* spike sorting maintains some information about movement parameters *on average*, but it is fraught with large decoding errors in the instantaneous case (Aghagolzadeh and Oweiss, 2009). It remains to be seen how robust this method is to plastic changes in neuronal tuning over prolonged periods of time.

With long-term, stable, chronic neural recording for which fully unsupervised spike detection and sorting are ultimate objectives, one key to a better understanding of the dynamics of trial-to-trial variability in a myriad of experiments—whether spike-dependent or tuning-dependent—is the ability to continuously record neural activity associated with a specific task across a broad range of conditions. Such capability does not currently exist, as most experiments provide only intermittent recording sessions because of the need to *tether* the subject to the recording equipment for finite periods. The practical implementation issues discussed in this chapter permit wireless continuous neural recording, and this should provide multiple avenues—in basic neuroscience experiments as well as in clinical applications—for adequate judgment of which spike sorting method is the best choice.

In summary, detection and classification of neuronal spikes is a very important, yet very challenging, problem in the analysis of microelectrode extracellular recordings. Its importance stems from the fact that all subsequent analysis of the neural data relies heavily on the reliability and accuracy of both steps. Despite being invasive, microelectrode extracellular recordings still remain the gold standard for examining cellular resolution neural activity, particularly in deep brain structures that cannot be assessed with noninvasive methods such as two-photon imaging of calcium dynamics (Stosiek et al., 2003). In contrast, other brain signal modalities such as surface electroencephalographic (sEEG) signals that are much easier to collect are more obscure in terms of the signal sources they originate from and are far more limited in terms of their spatial and temporal resolutions; moreover, their precise encoding of external covariates is not well understood. This makes them hard to model and consequently hard to detect or discriminate in a statistical sense. Nevertheless, the theory and techniques reviewed here for spike detection and sorting may be applicable to other neural signal modalities, perhaps with a different set of constraints. More robust and autonomous spike detection and sorting will probably be the overarching theme of next-generation algorithms and associated hardware platforms, particularly in clinical applications of brain–machine interface systems.

Acknowledgment

This work was supported by the National Institute of Neurological Disorders and Stroke. Grants number NS062031 and NS054148

References

Abeles, M., Gerstein, G., 1988. Detecting spatiotemporal firing patterns among simultaneously recorded single neurons. J. Neurophysiol. 60 (3), 909–924.

Aghagolzadeh, M., Eldawlatly, S., Oweiss, K., 2010. Synergistic coding by cortical neuronal ensembles. IEEE Trans. Inf. Theory 56 (2), 875–889.

Aghagolzadeh, M., Oweiss, K., 2009. Compressed and distributed sensing of neuronal activity for real time spike train decoding. IEEE Trans. Neural Syst. Rehabil. Eng. 17, 116–127.

Bankman, I.N., Johnson, K.O., Schneider, W., 1993. Optimal detection, classification, and superposition resolution in neural waveform recordings. IEEE Trans. Biomed. Eng. 40 (8), 836–841.

Bell, A.J., Sejnowski, T.J., 1995. An information-maximization approach to blind separation and blind deconvolution. Neural Comput. 7 (6), 1129–1159.

Blatt, M., Wiseman, S., Domany, E., 1996. Superparamagnetic clustering of data. Phys. Rev. Lett. 76 (18), 3251–3254.

Bokil, H.S., Pesaran, B., Andersen, R.A., Mitra, P.P., 2006. A method for detection and classification of events in neural activity. IEEE Trans. Biomed. Eng. 53 (8), 1678–1687.

Buzsaki, G., 2004. Large-scale recording of neuronal ensembles. Nat. Neurosci. 7 (5), 446–451.

Buzsáki, G., Geisler, C., Henze, D.A., Wang, X.J., 2004. Interneuron diversity series: circuit complexity and axon wiring economy of cortical interneurons. Trends Neurosci. 27, 186–193.

Caporale, N., Dan, Y., 2008. Spike timing dependent plasticity: a Hebbian learning rule. Annu. Rev. Neurosci. 31 (1), 25–46.

Cardoso, J.F., 1998. Blind signal separation: statistical principles. Proc. IEEE 86 (10), 2009–2025.

Carmena, J.M., Lebedev, M.A., Crist, R.E., O'Doherty, J.E., Santucci, D.M., Dimitrov, D.F., et al., 2003. Learning to control a brain-machine interface for reaching and grasping by primates. PLoS Biol. 1 (2), e2.

Chestek, C.A., Batista, A.P., Santhanam, G., Yu, B.M., Afshar, A., Cunningham, J.P., et al., 2007. Single-neuron stability during repeated reaching in macaque premotor cortex. J. Neurosci. 27 (40), 10742–10750.

Churchland, M.M., Afshar, A., Shenoy, K.V., 2006. A central source of movement variability. Neuron 52 (6), 1085–1096.

Cohen, M.R., Newsome, W.T., 2008. Context-dependent changes in functional circuitry in visual area MT. Neuron 60 (1), 162–173.

Cohen, M.R., Newsome, W.T., 2009. Estimates of the contribution of single neurons to perception depend on timescale and noise correlation. J. Neurosci. 29 (20), 6635–6648.

Coifman, R.R., Wickerhauser, M.V., 1992. Entropy-based algorithms for best basis selection. IEEE Trans. Inf. Theory 38 (2 Part 2), 713–718.

Comon, P., 1994. Independent component analysis, a new concept? Signal Processing 36 (3), 287–314.

Csicsvari, J., Henze, D.A., Jamieson, B., Harris, K.D., Sirota, A., Barthó, P., et al., 2003. Massively parallel recording of unit and local field potentials with silicon-based electrodes. Am. Physiol. Soc. 90, 1314–1323.

Dan, Y., Poo, M.M., 2004. Spike timing-dependent plasticity of neural circuits. Neuron 44, 23–30.

Del Marco, S., Weiss, J., et al., 1997. Improved transient signal detection using a wavepacket-based detector with an extended translation-invariant wavelet transform. IEEE Trans. Signal Process. 45 (4), 841–850.

Dempster, A.P., Laird, N., Rubin, D.B., 1977. Maximum likelihood from incomplete data via the EM algorithm. J. R. Stat. Soc. Series B Methodol. 39 (1), 1–38.

Dinning, G.J., Sanderson, A.C., 1981. Real-time classification of multiunit neural signals using reduced feature sets. IEEE Trans. Biomed. Eng. BME-28 (12), 804–812.

Donoho, D.L., 1995. De-noising by soft-thresholding. IEEE Trans. Inf. Theory 41 (3), 613–627.

Dzakpasu, R., Zochowski, M., 2005. Discriminating different types of synchrony in neural systems. Physica D. 208, 115–122.

Eggermont, J.J., 1993. Function aspects of synchrony and correlation in the auditory nervous system. Concepts Neurosci. 4 (2), 105–129.

Eldawlatly, S., Jin, R., Oweiss, K., 2009. Identifying functional connectivity in large-scale neural ensemble recordings: a multiscale data mining approach. Neural Comput. 21 (2), 450–477.

Fee, M.S., Mitra, P.P., Kleinfeld, D., 1996. Automatic sorting of multiple unit neuronal signals in the presence of anisotropic and non-Gaussian variability. J. Neurosci. Methods 69 (2), 175–188.

Feldman, D.E., 2009. Synaptic mechanisms for plasticity in neocortex. Annu. Rev. Neurosci. 32 (1), 33–55.

Fraser, G.W., Chase, S.M., Whitford, A., Schwartz, A.B., 2009. Control of a brain-computer interface without spike sorting. J. Neural Eng. 6, 055004.

Frisch, M., Messer, H., 1992. The use of the wavelet transform in the detection of an unknowntransient signal. IEEE Trans. Inf. Theory 38 (2 Part 2), 892–897.

Georgopoulos, A.P., Schwartz, A.B., Kettner, R.E., 1986. Neuronal population coding of movement direction. Science 233 (4771), 1416.

Gerstein, G.L., 2000. Cross correlation measures of unresolved multi-neuron recordings. J. Neurosci. Methods 100, 41–51.

Gerstein, G.L., Kirkland, K.L., 2001. Neural assemblies: technical issues, analysis, and modeling. Neural Netw. 14 (6–7), 589–598.

Gold, C., Henze, D.A., Koch, C., Buzsáki, G., 2006. On the origin of the extracellular action potential waveform: a modeling study. J. Neurophysiol. 95 (5), 3113–3128.

Goldberg, J.A., Boraud, T., Maraton, S., Haber, S.N., Vaadia, E., Bergman, H., 2002. Enhanced synchrony among primary motor cortex neurons in the 1-methyl-4-phenyl-1, 2, 3, 6-tetrahydropyridine primate model of Parkinson's disease. J. Neurosci. 22 (11), 4639.

Gozani, S.N., Miller, J.P., 1994. Optimal discrimination and classification of neuronal action potential waveforms from multiunit, multichannel recordings using software-based linear filters. IEEE Trans. Biomed. Eng. 41 (4), 358–372.

Gray, C.M., Maldonado, P.E., Wilson, M., McNaughton, B., 1995. Tetrodes markedly improve the reliability and yield of multiple single-unit isolation from multi-unit recordings in cat striate cortex. J. Neurosci. Methods 63 (1–2), 43–54.

Guillory, K.S., Normann, R.A., 1999. A 100-channel system for real time detection and storage of extracellular spike waveforms. J. Neurosci. Methods 91 (1–2), 21–29.

Harris, K.D., Henze, D.A., Csicsvari, J., Hirase, H., Buzsáki, G., 2000. Accuracy of tetrode spike separation as determined by simultaneous intracellular and extracellular measurements. J. Neurophysiol. 84 (1), 401–414.

Hebb, D.O., 1949. The Organization of Behavior; A Neuropsychological Theory. Wiley, New York.

Henderson, C.R., 1975. Best linear unbiased estimation and prediction under a selection model. Biometrics 31 (2), 423–447.

Hochberg, L.R., Serruya, M.D., Friehs, G.M., Mukand, J.A., Saleh, M., Caplan, A.H., et al., 2006. Neuronal ensemble control of prosthetic devices by a human with tetraplegia. Nature 442, 164–171.

Hulata, E., Segev, R., Ben-Jacob, E., 2002. A method for spike sorting and detection based on wavelet packets and Shannon's mutual information. J. Neurosci. Methods 117 (1), 1–12.

Hyvärinen, A., 1999. Fast and robust fixed-point algorithms for independent component analysis. IEEE Trans. Neural Netw. 10 (3), 626–634.

Hyvärinen, A., Oja, E., 2000. Independent component analysis: algorithms and applications. Neural Netw. 13 (4–5), 411–430.

Isaacs, R.E., Weber, D.J., Schwartz, A.B., 2000. Work toward real-time control of a cortical neural prothesis. IEEE Trans. Rehabil. Eng. [see also IEEE Trans. Neural Syst. Rehabil.] 8 (2), 196–198.

Jackson, A., Gee, V.J., Baker, S.N., Lemon, R.N., 2003. Synchrony between neurons with similar muscle fields in monkey motor cortex. Neuron 38 (1), 115–125.

Jain, A.K., Duin, R.P.W., Mao, J., 2000. Statistical pattern recognition: a review. IEEE Trans. Pattern. Anal. Mach. Intell. 22 (1), 4–37.

Johnson, W., Lindenstrauss, J., 1984. Extensions of Lipschitz mappings into a Hilbert space. Contemporary Math. 26 (189–206), 1–1.1.

Jung, H.K., Choi, J.H., Kim, T., 2006. Solving alignment problems in neural spike sorting using frequency domain PCA. Neurocomputing 69, 975–978.

Kandel, E.R., Schwartz, J.H., Jessell, T.M. (Eds.), 2000. Principles of Neural Science, fourth ed. McGraw-Hill, New York.

Kay, S.M., 1993. Fundamentals of statistical signal processing: estimation Theory. Prentice-Hall, New Jersey.

Kemere, C., Santhanam, G., Yu, B.M., Afshar, A., Ryu, S.I., Meng, T.H., et al., 2008. Detecting neural-state transitions using hidden Markov models for motor cortical prostheses. J. Neurophysiol. 100 (4), 2441–2452.

Kim, K.H., Kim, S.J., 2003. A wavelet-based method for action potential detection from extracellular neural signal recording with low signal-to-noise ratio. IEEE Trans. Biomed. Eng. 50 (8), 999–1011.

Kim, S., McNames, J., 2007. Automatic spike detection based on adaptive template matching for extracellular neural recordings. J. Neurosci. Methods 165 (2), 165–174.

Klema, V., Laub, A., 1980. The singular value decomposition: its computation and some applications. IEEE Trans. Autom. Control 25 (2), 164–176.

Kralik, J.D., Dimitrov, D.F., Krupa, D.J., Katz, D.B., Cohen, D., Nicolelis, M.A., 2001. Techniques for long-term multisite neuronal ensemble recordings in behaving animals. Methods 25 (2), 121–150.

Kringelbach, M.L., Jenkinson, N., Owen, S.L., Aziz, T.Z., 2007. Translational principles of deep brain stimulation. Nat. Rev. Neurosci. 8, 623–635.

Kulkarni, S., Bellary, S.V., 2000. Nonuniform M-band wavepackets for transient signal detection. IEEE Trans. Signal Process. 48 (6), 1803–1807.

Lega, B.C., Haipern, C.H., Jaggi, J.L., Baltuch, G.H., 2010. Deep brain stimulation in the treatment of refractory epilepsy: update on current data and future directions. Neurobiol Dis. 38 (3), 354–360.

Lehmkuhle, M.J., Bhangoo, S.S., Kipke, D.R., 2009. The electrocorticogram signal can be modulated with deep brain stimulation of the subthalamic nucleus in the hemiparkinsonian rat. J. Neurophysiol. 102 (3), 1811–1820.

Letelier, J.C., Weber, P.P., 2000. Spike sorting based on discrete wavelet transform coefficients. J. Neurosci. Methods 101 (2), 93–106.

Lewicki, M.S., 1998. A review of methods for spike sorting: the detection and classification of neural action potentials. Netw. Comput. Neural Syst. 9 (4), 53–78.

Liu, T.T., Fraser-Smith, A.C., 2000. Detection of transients in 1/f noise with the undecimated discretewavelet transform. IEEE Trans. Signal Process. 48 (5), 1458–1462.

McGill, K.C., Dorfman, L.J., 1984. High resolution alignment of sampled waveforms. IEEE Trans. BME-31, 462–470.

McNaughton, B.L., O'Keefe, J., Barnes, C.A., 1983. The stereotrode: a new technique for simultaneous isolation of several single units in the central nervous system from multiple unit records. J. Neurosci. Methods 8 (4), 391–397.

Moran, D.W., Schwartz, A.B., 1999. Motor cortical representation of speed and direction during reaching. J. Neurophysiol. 82 (5), 2676–2692.

Nagel, S.J., Najm, I.M., 2009. Deep brain stimulation for epilepsy. Neuromodulation 12 (4), 270–280.

Nenadic, Z., Burdick, J.W., 2005. Spike detection using the continuous wavelet transform. IEEE Trans. Biomed. Eng. 52 (1), 74–87.

Neyman, J., Pearson, E.S., 1933. On the problem of the most efficient tests of statistical hypothesis. Philos. Trans. R. Soc. Lond. A 231, 289–337.

Normann, R.A., Maynard, E.M., Rousche, P.J., Warren, D.J., 1999. A neural interface for a cortical vision prosthesis. Vision Res. 39 (15), 2577–2587.

Obeid, I., Nicolelis, M.A., Wolf, P.D., 2004. A multichannel telemetry system for single unit neural recordings. J. Neurosci. Methods 133 (1–2), 33–38.

Oweiss, K.G., 2002c. Multiresolution analysis of multichannel neural recordings in the context of signal detection, estimation, classification and noise suppression, PhD thesis. University of Michigan, Ann Arbor, MI.

Oweiss, K.G., 2006. A systems approach for data compression and latency reduction in cortically controlled brain machine interfaces. IEEE Trans. Biomed. Eng. 53 (7), 1364–1377.

Oweiss, K.G., Anderson, D.J., 2000. A new approach to array denoising. In: Proc. of the IEEE 34th Asilomar Conference on Signals, Systems and Computers, 1403–1407.

Oweiss, K.G., Anderson, D.J., 2001a. A new technique for blind source separation using subband subspace analysis in correlated multichannel signal environments. Proc. IEEE Int. Conf. Acoust. Speech Signal Process, 2813–2816.

Oweiss, K.G., Anderson, D.J., 2001b. MASSIT: Multiresolution analysis of signal subspace invariance technique: A novel algorithm for blind source separation. In: Proc. of the 35th IEEE Asilomar Conf. on Signals, Syst. and Computers, 819–823.

Oweiss, K.G., Anderson, D.J., 2002a. A multiresolution generalized maximum likelihood approach for the detection of unknown transient multichannel signals in colored noise with unknown covariance. Proc. IEEE Int. Conf. Acoust. Speech Signal Process. vol. 3, 2993–2996.

Oweiss, K.G., Anderson, D.J., 2002b. Independent component analysis of multichannel neural recordings at the micro-scale. Proc. of the 2nd Joint IEEE EMBS/BMES Conference, 1, 2012–2013.

Oweiss, K.G., Anderson, D.J., 2007. Tracking signal subspace invariance for blind separation and classification of nonorthogonal sources in correlated noise. EURASIP J. Adv. Signal Process. 2007.

Oweiss, K.G., Mason, A., Suhail. Y., Kamboh, A.M., Thomson, K.E., 2007. A scalable wavelet transform VLSI architecture for real-time signal processing in high-density intra-cortical implants. IEEE Trans. Circuits Syst. 54 (6), 1266–1278.

Pal, N.R., Bezdek, J.C., 1995. On cluster validity for the fuzzy c-means model. IEEE Trans. Fuzzy Syst. 3 (3), 370–379.

Papoulis, A., Pillai, S.U., 1965. Probability, random variables, and stochastic processes. McGraw-Hill, New York.

Perkel, D.H., Gerstein, G.L., Moore, G.P., 1967. Neuronal spike trains and stochastic point processes. II. Simultaneous spike trains. Biophys. J. 7 (4), 419–440.

Plonsey, R., 1974. The active fiber in a volume conductor. IEEE Trans. Biomed. Eng. BME-21, 371–381.

Poor, H.V., 1988. An introduction to signal detection and estimation. Springer Texts in Electrical Engineering, Springer, New York.

Quiroga, R., 2009. What is the real shape of extracellular spikes? J. Neurosci. Methods 177 (1), 194–198.

Quiroga, R.Q., Nadasdy, Z., Ben-Shaul, Y., 2004. Unsupervised spike detection and sorting with wavelets and superparamagnetic clustering. Neural Comput. 16 (8), 1661–1687.

Rizzolatti, G., Luppino, G., 2001. The Cortical Motor System. Neuron 31 (6), 889–901.

Rolls, E.T., Treves, A., Tovee, M.J., 1997. The representational capacity of the distributed encoding of information provided by populations of neurons in primate temporal visual cortex. Exp. Brain Res. 114 (1), 149–162.

Sakamoto, K., Mushiake, H., Saito, N., Aihara, K., Yano, M., Tanji, J., 2008. Discharge synchrony during the transition of behavioral goal representations encoded by discharge rates of prefrontal neurons. Cereb Cortex 18, 2036–2045.

Serruya, M., Hatsopoulos, N., Fellows, M., Paninski, L., Donoghue, J., 2003. Robustness of neuroprosthetic decoding algorithms. Biol. Cybern. 88 (3), 219–228.

Serruya, M.D., Hatsopoulos, N.G., Paninski, L., Fellows, M.R., Donoghue, J.P., 2002. Instant neural control of a movement signal. Nature 416, 141–142.

Shadlen, M.N., Newsome, W.T., 1994. Noise, neural codes and cortical organization. Curr. Opin. Neurobiol. 4 (4), 569–579.

Shadlen, M.N., Newsome, W.T., 1998. The variable discharge of cortical neurons: implications for connectivity, computation, and information coding. J. Neurosci. 18 (10), 3870–3896

Shoham, S., Fellows, M.R., Normann, R.A., 2003. Robust, automatic spike sorting using mixtures of multivariate t-distributions. J. Neurosci. Methods 127 (2), 111–122.

Sorger, B., Dahmen, B., Reithler, J., Gosseries, O., Maudoux, A., Laureys, S., et al., 2009. Another kind of 'BOLD Response': answering multiple-choice questions via online decoded single-trial brain signals. Prog. Brain Res. 177, 275–292.

Stosiek, C., Garaschuk, O., Holthoff, K., Konnerth, A., 2003. In vivo two-photon calcium imaging of neuronal networks. Proc. Natl. Acad. Sci. 100 (12), 7319.

Suhail, Y., Oweiss, K., 2005. Multiresolution bayesian detection of multiunit extracellular spike waveforms in multichannel neuronal recordings. Conf. Proc. IEEE Eng. Med. Biol. Soc. 1, 141–144.

Truccolo, W., Hochberg, L.R., Donoghue, J.P., 2010. Collective dynamics in human and monkey sensorimotor cortex: predicting single neuron spikes. Nat Neurosci 13 (1), 105–111.

Urkowitz, H., 1967. Energy detection of unknown deterministic signals. Proc. IEEE 55 (4), 523–531.

Van Trees, H.L., 2002. Detection, estimation and modulation theory: optimum array processing, vol. 4, Wiley, New York.

Velliste, M., Perel, S., Spalding, M.C., Whitford, A.S., Schwartz, A.B., 2008. Cortical control of a prosthetic arm for self-feeding. Nature 453 (7198), 1098–1101.

Ventura, V., 2009. Traditional waveform based spike sorting yields biased rate code estimates. Proc. Natl. Acad. Sci. 106 (17), 6921.

Ventura, V., 2008. Spike train decoding without spike sorting. Neural Comput. 20 (4), 923–963.

Wessberg, J., Stambaugh, C.R., Kralik, J.D., Beck, P.D., Laubach, M., Chapin, J.K., et al., 2000. Real-time prediction of hand trajectory by ensembles of cortical neurons in primates. Nature 408 (6810), 361–365.

Wheeler, B.C., Heetderks, W.J., 1982. A comparison of techniques for classification of multiple neural signals. IEEE Trans. Biomed. Eng. 29, 752–759.

Wheeler, B.C., Smith, S.R., 1988. High-resolution alignment of action potential waveforms using cubic spline interpolation. J. Biomed. Eng. 10 (1), 47.

Wise, K.D., Anderson, D.J., Hetke, J.F., Kipke, D.R., Najafi, K., 2004. Wireless implantable microsystems: high-density electronic interfaces to the nervous system. Proc. IEEE 92, 76–97.

Wolf, M.T., Burdick, J.W., 2009. A Bayesian clustering method for tracking neural signals over successive intervals. IEEE Trans. Biomed. Eng. 56 (11), 2649–2659.

Wood, F., Black, M.J., Vargas-Irwin, C., Fellows, M., Donoghue, J.P., 2004. On the variability of manual spike sorting. IEEE Trans. Biomed. Eng. 51 (6), 912–918.

Yu, S., Huang, D., Singer, W., Nikolic, D., 2008. A small world of neuronal synchrony. Cereb Cortex 18, 2891–2901.

Information-Theoretic Analysis of Neural Data

3

Don H. Johnson

Department of Electrical and Computer Engineering Rice University, Houston, Texas

3.1 INTRODUCTION

Shannon's classic work (Shannon, 1948) founded the field we know today as information theory. His foundational work and enhancements made over the years determine the ultimate fidelity limits that communication and signal processing systems can achieve. It is the generality of his results that hallmarks information theory: No specific cases need be assumed or special constraints imposed for information theory to apply, as the theory establishes clear benchmarks for optimal performance behavior. Shannon's work concentrated on digital communication, determining how well a signal can be represented by a sequence of bits and how well bit streams can be communicated through a noisy channel. This chapter summarizes this well-known theory. Most important for neuroscience, however, is Shannon's more general viewpoint that encompasses all communication and processing systems regardless of whether the signals involved are bit streams, analog signals, or spike trains.

Despite the unquestioned importance of Shannon's ideas and results in establishing performance limits, extending them to the analysis of *real* systems, which are probably operating suboptimally, is fraught with difficulties. How does one examine measurements to determine how close to the limits some particular system is operating? For man-made systems, such empirical questions have limited utility: Instead of measuring, design the system correctly from the outset. However, for natural systems like the brain, how well a system works is very relevant and for systems as overwhelmingly complex and adaptive as the brain, there are no design guidelines other than evolution's selective pressures. The somewhat vague question of how information is encoded, can, in an information theory context, be refined into "What is the most effective way of representing acoustic or visual stimulus?" To analyze an existing system, Shannon's work suggests the quantities that can be used to assess effectiveness but not how to interpret them or how they can be measured. The latter issue is not a question of finding good measurement techniques; a much deeper problem surfaces. The key information-theoretic quantities—capacity and the rate-distortion function—are asymptotic solutions to mathematical optimization problems where the givens are signal and system models. How can we solve these problems when we don't have sufficiently detailed models? Furthermore, information theory is silent on how to judge a given system's performance on a general basis relative to these milestones. Despite this quandary, information-theoretic tools have been developed that can help interpret neural coding and processing.

Figure 3.1 shows the fundamental model underlying classic information theory and how it can be translated to the neuroscience context. An information source produces an information-bearing signal S that, from a neuroscience perspective, is the stimulus. Note that we can take S to express some aspect of the actual stimulus that results from previous neural processing. Consequent of encoding, transmission and decoding are represented by $\widehat{S}$, the stimulus estimate viewed here as the estimated information or percept. The signal X represents—encodes—the stimulus to preserve the *pertinent* information contained in the stimulus. Mathematically, we assume the encoder's transformation is invertible; otherwise, the information could never be accurately inferred. In neuroscience, the encoder represents neural coding: how the stimulus is represented in the firing pattern of one or more neurons. Note that the encoder can involve preprocessing that removes some aspects of the stimulus, thereby allowing the encoder to focus its resources.

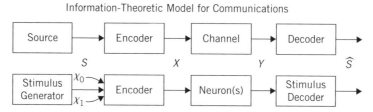

FIGURE 3.1

The classic information theory model ubiquitously applied to communications problems is shown as well as its translation to a neuroscience context. The extraneous signals χ_0 and χ_1 represent a difference in the two models. In neuroscience, more than the signal of interest is frequently also encoded and must be removed by subsequent processing.

In the auditory system, for example, the cochlea serves as a bandpass filter, which means that relatively narrowband acoustic signals are represented by individual auditory-nerve fibers. In the visual system, receptive fields gather visual signals from a specific region of visual space spatially filtered by the receptive field. We can also have the situation wherein signals extraneous to estimating the stimulus are encoded as well. The encoder's output X we take to be the instantaneous rate $\lambda(t)$ of the neural discharges produced by a single neuron or a population.

The encoded signal passes through the channel, which disturbs the signal in such a way that the channel's input X cannot be precisely determined from its output Y. In neuroscience, the channel represents the conversion of an instantaneous discharge rate into spikes, wherein "channel disturbances" arise from the stochastic behavior of neural responses. A distinct difference between communication systems and neural systems is that the channel may involve processing of its own, suppressing irrelevant information (such as the signals χ_0 and χ_1 in Figure 3.1) that helps subsequent processing systems focus on important stimulus features. Current thinking is that attention mechanisms are at work here, meaning that the focus attention is often adaptive. Here, Y represents spike train(s), sometimes represented in neuroscience papers as the response R (Borst and Theunissen, 1999). The decoder, which represents the final processing stage, produces $\widehat{S}$, an estimate of the stimulus. Unfortunately, whether $\widehat{S}$ can be measured or not is unclear in the neuroscience context. In neural recordings, how does one show that the recordings represent a percept and not some intermediary? Can $\widehat{S}$ be gleaned only from psychophysical measurements? Despite not having $\widehat{S}$ available, information theory can make statements about the ultimate achievable quality of the signal estimate.

This chapter defines the fundamental quantities of classic information theory and discusses some of the empirical and philosophical problems that occur. The dominant theme is that, while Shannon's entropy, mutual information, and capacity define fundamental communication and processing limits, these quantities have limited utility in analyzing neural data. We then turn to a somewhat different way of looking at neural systems that embraces new information-theoretic quantities, not the classic ones.

3.2 THE ENCODER

The first stage in the basic model (Figure 3.1) is the encoder that, in a well-designed system, presumably represents an effective (and efficient) way of converting the stimulus into a signal that can be transmitted through the channel. A fundamental assumption of information theory is that all stimulus signals are stochastic and described by their probability distribution $p_S(s)$. Furthermore, this signal *is* the information, which means that information is a stochastic quantity. At first look, this assumption would seem to be acceptable, but does it impact experimental design? Does the experimenter need to "randomize" the sequence to apply information-theoretic results? More on this issue once information theory's viewpoint becomes clearer. What is already clear is that an encoder, when faced with representing its input as it comes, cannot depend on whether the stimulus is randomized or not. Encoders do depend on stimulus complexity: More complex signals require more encoder resources to represent them accurately. In classic information theory, complexity is directly related to the *entropy* of the stimulus probability distribution.

Let S be a stimulus that can be one of M values, $S \in \{s_1, \ldots, s_M\}$, and let $P(s_m)$ denote the probability with which each stimulus occurs. The entropy of the stimulus, $H(S)$, is defined according to

$$H(S) \overset{\Delta}{=} -\sum_m P(s_m) \log_2 P(s_m) \tag{3.1}$$

By using base-two logarithms, we express entropy in units of bits. Entropy must be non-negative: It equals zero when one stimulus has unit probability and achieves its upper bound—$\log_2 M$—when all stimuli are equally likely. Also, when Shannon defined entropy as a measure of stimulus complexity in his classic work, he sought a formula that had the so-called *additive property*: If a source consists of two statistically independent random variables, the complexity should equal the sum of the individual complexities. He showed that the entropy formula (Eq. (3.1)) uniquely satisfies this and other properties. Entropy generally characterizes stimulus uncertainty: The greater the entropy, the greater the uncertainty (the more uniform the stimulus probabilities) in a logarithmic fashion. But because discrete-valued stimulus sets usually consist of a predetermined selection of continuous-valued signals (such as a series of sinusoidal gratings at several distinct spatial frequencies, orientations, and contrasts), entropy does *not* in fact express stimulus complexity. Stimulus set entropy depends *only* on stimulus probabilities, not on the nature of the component stimuli. This discrepancy illustrates the first (but not the last) mismatch in translating information theory into neuroscience's domain.

Shannon showed that entropy provides an operational bound on the minimum length (number of bits) of any binary code needed to accurately represent the stimulus *index*.

Source Coding Theorem. *The smallest number of bits* B *required to represent a discrete-valued stimulus* S *without incurring decoding errors satisfies*

$$H(S) \leq B < H(S) + 1 \qquad (3.2)$$

The lower bound in the source coding theorem defines a fundamental limit: If fewer bits than stimulus entropy are used, at least one stimulus having a nonzero probability cannot be correctly decoded, even if communicated via a noiseless channel. The upper bound means that no more than one bit beyond entropy need be used to have a decodable source coding algorithm. In this way, entropy has taken the role of assessing the demands a stimulus set makes on an encoder, which in turn must be related to stimulus uncertainty. For the reasons just outlined, this interpretation has little application in neuroscience. First of all, this result assumes that the decoder knows a priori what the stimulus set is, allowing it to decode the index and relate that index to the correct stimulus component. Second, sensory system encoders, such as the cochlea in the auditory system and the eye in the visual system, do not change their characteristics depending on the experimental situation. Instead, their representational abilities are honed to the complexity of naturally occurring signals, not to the uncertainty of particular stimulus sets. Consequently, the way entropy is used in information theory does not measure stimulus complexity from a neuroscience viewpoint.

Unfortunately, extending the definition of entropy to the case of continuous-valued stimuli (a piece of music, for example) proves problematic, so much so that the revised definition has a different name and symbol to distinguish it from the definition for discrete stimulus sets. The so-called *differential entropy* is defined for continuous-valued random variables analogously to the definition of entropy given above for discrete-valued ones.

$$h(S) \stackrel{\Delta}{=} - \int p_S(s) \log_2 p_S(s) \, ds \qquad (3.3)$$

While the entropy of a discrete-valued random variable is always nonnegative, differential entropy can assume any real value, positive or negative. For example, the differential entropy of a Gaussian random variable (having mean m and variance σ^2) equals $\frac{1}{2} \log_2 2\pi e \sigma^2$. Depending on whether $2\pi e \sigma^2$ is greater than or less than one, the differential entropy can be positive or negative, even zero if the variance precisely equals $1/2\pi e$. Moreover, when the stimulus is continuous-valued, the source coding theorem *cannot* apply: The number of bits required to encode a continuous-valued quantity without error must be infinite, regardless of the underlying probability distribution.

Consequently, the differential entropy has decidedly less importance in information theory and, as we shall show, introduces ambiguity in using entropy in neuroscience problems.

Several potential roles for entropy have been sought in neuroscience. For example, can we use entropy to determine when a neural signal is representing information and, by extension, can we determine the complexity of the represented stimulus by measuring the entropy of the response? For efficient information-theoretic encoders—those that encode a discrete-valued stimulus and obey the source coding theorem—this idea works: The entropy of an encoder's output comes within a bit of the source entropy. However, spike trains are not bit sequences, which brings to the fore an even more basic dilemma: How do we calculate the entropy of a spike train?

Neural discharges recorded from sensory systems can be well described as stochastic point processes (Johnson, 1996), which are complicated mathematical objects, described by the joint distribution of event times (continuous variables) and the number of events that occur (a discrete random variable). The entropy of a point process can be defined (McFadden, 1965), but issues arise because of the mixed continuous-discrete random variable character of joint probability distribution. First of all, since spike times are continuous variables, the source coding theorem certainly does not apply to encoders that vary spike times to represent stimuli. But even applying entropy to spike trains encoded by rate codes, wherein the number of spikes occurring in some time interval is all that matters, is problematic. Consider a Poisson process having a constant rate λ over the interval $[0, T)$. The probability distribution of this process can be written two ways, both of which express the fact that, in a rate code, event times do not matter.

- The probability distribution of the number of events that occur, $N_{\{0 \leq t < T\}}$, is given by

$$p_{N_{\{0 \leq t < T\}}}(n) = \frac{(\lambda T)^n e^{-\lambda T}}{n!} \tag{3.4}$$

The event count is a discrete-valued random variable and its entropy is non-negative. Using Eq. (3.1) to calculate the entropy yields

$$H(N_{\{0 \leq t < T\}}) = \frac{\lambda T (1 - \ln \lambda T) - \mathsf{E}[\ln N!]}{\ln 2}$$

The expected value $\mathsf{E}[\ln N!]$ has no closed-form expression but is easily computed numerically. According to this result, the greater the event rate, the greater the entropy regardless of what the spike train represented.

- The joint probability distribution of the number of events and the event times equals

$$p_{\mathbf{w}, N_{\{0 \le t < T\}}}(\mathbf{w}, n) = \lambda^n e^{-\lambda T} \tag{3.5}$$

Here, $\mathbf{w} = [w_1, \ldots, w_{N_{\{0 \le t < T\}}}]$, $0 \le w_1 \le w_2 \le \cdots \le w_{N_{\{0 \le t < T\}}} \le T$, is the vector of event times. The fact that the joint distribution does not depend on event times indicates that they do not matter. Because this joint distribution contains both discrete- and continuous-valued random variables, we must use both entropy (Eq. (3.1)) and differential entropy (Eq. (3.3)) formulas. Accordingly,

$$h(\mathbf{w}, N_{\{0 \le t < T\}}) = \frac{\lambda T (1 - \ln \lambda)}{\ln 2}$$

which is an expression significantly different from the first.

Because these two models describe the same stochastic process but cannot be described by the same entropy function, *entropy cannot be defined unambiguously for Poisson processes* and, for that matter, *any* point process. Thus, entropy stands out from other common manipulations of probability distributions: Regardless of which model is used, the same results obtain for the expected number of events, the variance, the maximum-likelihood estimate of λ, and the accompanying Cramér-Rao bound.

Measuring entropy from spike train recordings encounters a similar problem. The standard approach to spike train analysis is to segment the time axis into a sequence of abutting bins of duration Δt and count how many events occur in each bin (Johnson, 1996). This approach amounts to an analog-to-digital converter for spike trains. Assuming a small bin duration, either one or zero spikes occur in each bin, which means that the spike train has been converted into a Bernoulli (binary-valued), discrete-time random process. Under a Poisson model for the spike train, the probability of a spike in a bin equals $\lambda \Delta t$. Since a Bernoulli random variable is discrete-valued, its entropy can be computed according to (Eq. (3.1)). Moreover, since the event occurrence in each bin is statistically independent of the others under the Poisson model, the entropy of the spike train can be found by simply adding the entropies of the individual bins. Accordingly, for small values of $\lambda \Delta t$, we obtain

$$H(N_{\{0 \le t < T\}}) \approx \frac{\lambda T (1 - \ln \lambda) - \lambda T \ln \Delta t}{\ln 2}. \tag{3.6}$$

This result most closely resembles the differential entropy consequent of the second model, but it is always positive since it is the entropy of a discrete-valued variable. However, the dangling $\ln \Delta t$ term means that the result depends on bin width and diverges in the small bin width limit, becoming infinite asymptotically. Note that this bin width term cannot be removed by normalization. Similar results emerge when non-Poisson processes are considered.

With three different answers for the entropy of a simple point process, we must question the utility of using entropy alone to characterize the "information" conveyed by neural responses. More fundamentally, when analyzing information flow, no rule governs how the entropy of a signal changes as it passes through a system: Examples show that entropy can increase, decrease, or stay the same.[1] Thus, when entropy changes do occur, they cannot be interpreted to mean that information has been extracted or inserted.

One viable application of entropy to spike train analysis is testing for statistical independence of simultaneous recordings. If $\mathbf{Y} = [Y_1, \ldots, Y_N]$ represents recordings from N neurons, the joint entropy can be no larger than the sum of the component entropies, the result that obtains when the recordings are statistically independent.

$$H(\mathbf{Y}) \leq \sum_n H(Y_n)$$

This holds for differential entropy as well and, regardless of how the spike trains are digitized, this property summarizes statistical dependence. Another way of stating this property is that statistical dependence reduces entropy. The result is quite powerful: Joint entropy equals the sum of individual entropies if and only if the underlying quantities are statistically independent. However, one should not conclude that statistically independent responses best represent stimuli. As shown subsequently, more penetrating analyses using measures other than entropy show that such simple interpretations are not correct.

The most commonly used entropy estimation method is the "direct method," in which the number of spikes that occur in each bin are taken to be the values of discrete-valued random variables. The method then forms "words"—a sequence of bits—from these values both over time and across jointly recorded neurons. The probability of a word occurring is estimated with

---

[1] Consider a Gaussian process passing through an amplifier. Passing a signal through a noiseless amplifier should not affect its information content. However, since the amplifier's gain affects the variance, the expression for a Gaussian random variable's entropy given earlier shows that it can increase or decrease.

a simple histogram, and the quantity of interest is calculated by "plugging" the estimated distributions into the desired formulas (Strong et al., 1998). Many of these estimators have significant bias and variance. We have seen that the choice of bin width directly affects the measured entropy value, which means that comparing one response's entropy with another must be done carefully. Several lines of work have attempted to reduce these effects. Paninski (2003) provides a nice discussion of estimator quality. A number of methods devised by the neuroscience community to estimate entropy are generally customized to deal with different experimental scenarios (repeated trials versus many single trials, for example) and types of signals (binned spike times versus continuous spike times or nonspike waveforms). Victor (2006) provides a thorough review of the many different estimation methods.

3.3 THE CHANNEL

Much more interesting to neuroscience is the way information theory characterizes the channel, especially its ability to convey information despite the disturbances it introduces. The channel's input-output relation is defined by the conditional probability distribution $p_{Y|X}(y|x)$. Shannon's characterization of a channel begins by defining the *mutual information* between two jointly defined random quantities, X and Y. It has similar forms for both discrete- and continuous-valued versions, exemplified here in integral form.

$$I(X;Y) = \iint p_{X,Y}(x,y) \log_2 \frac{p_{X,Y}(x,y)}{p_X(x)p_Y(y)} \, dx \, dy \qquad (3.7)$$

The discrete version can be obtained by replacing the integration by summation over the values taken by the channel's input and output.[2] Again, mutual information has units of bits because of the base-two logarithm. In both discrete and continuous versions, mutual information is non-negative and its properties are much less sensitive to the nature of the random variables. This expression can be simplified by writing it as the difference of two entropies, which also has the effect of making explicit the channel's

---

[2] In this and subsequent expressions, we are deliberately vague as to what kind of entities x and y represent. These expressions and the properties of the information-theoretic quantities thus defined do not depend on such details: They could represent scalars, vectors, or stochastic processes without affecting the results.

input-output relationship.

$$I(X;Y) = \iint p_{Y|X}(y|x)p_X(x)\log_2 \frac{p_{Y|X}(y|x)}{p_Y(y)}\,dx\,dy = h(Y) - h(Y|X) \quad (3.8)$$

Here, the so-called *conditional entropy* $h(Y|X)$ is defined to be

$$h(Y|X) \overset{\Delta}{=} -\int p_X(x)\left(\int p_{Y|X}(y|x)\log_2 p_{Y|X}(y|x)\,dy\right)dx$$

Writing mutual information as the difference of entropies (Eq. (3.8)) shows that it depends on *both* the channel's input-output relationship *and* the probability distribution of the input. The dependence on the input is disguised: The probability distribution of the output $p_Y(y)$ equals $\int p_{Y|X}(y|x)p_X(x)\,dx$, showing that $h(Y)$ depends on both the channel's characteristics and the input's probability distribution. The mutual information can be interpreted as the reduction of entropy (uncertainty) in the output because of its relationship to the input via the channel. Loosely speaking, mutual information indicates how much information is shared between the input and the output, an interpretation that gives rise to its name.

Because of its dependence on the input, mutual information does not summarize the behavior of the channel. The mutual information between the stimulus and a response depends on neural processing, the stimuli, and the stimulus probabilities. Especially when the stimulus set is chosen according to probabilities determined by the experimenter, mutual information does not constitute an objective measure of information processing relevant to the organism's ecological behavior. Returning to the mathematics, if the output does not depend on the input, which makes the output statistically independent of the input, $h(Y|X) = h(Y)$, this case becomes one that minimizes mutual information. If the output equals the input (the input and output are maximally dependent), maximum mutual information results. Aside from these information-theoretic interpretations, mutual information is a powerful measure of statistical dependence between two random variables. $I(X;Y) = 0$ only when X and Y are statistically independent, and it achieves its maximal value when $X = Y$: $I(X;X) = H(X)$ in the discrete-valued case and $I(X;X) = +\infty$ in the continuous-valued case. Thus, mutual information completely characterizes the degree of similarity between the statistical properties of two random quantities.

The difficulties in reconciling entropy between discrete- and continuous-valued cases do not carry over to mutual information. To illustrate, perhaps the simplest model for a rate code has a constant rate λ serving as the input to a Poisson process generator. To calculate the mutual information, we assume that the event rate is a random variable having the probability distribution $p_\lambda(\lambda)$. We can interpret both mathematical descriptions for the neuron's input-output relationship (Eqs. (3.4) and (3.5)) as being $p_{N_{\{0\leq t<T\}}|\lambda}(n|\lambda)$ and $p_{\mathbf{w},\,N_{\{0\leq t<T\}}|\lambda}(\mathbf{w},n|\lambda)$, respectively, with λ now a random variable. In contrast to entropy, the same, but difficult to evaluate, result emerges from using either description.

$$I(\lambda;N_{\{0\leq t<T\}}) = \sum_{n=0}^{\infty} \int \frac{(\lambda T)^n e^{\lambda T}}{n!} p_\lambda(\lambda) \log_2 \frac{\lambda^n e^{-\lambda T}}{\int \alpha^n e^{-\alpha T} p_\lambda(\alpha)\,d\alpha}\,d\lambda$$

This result can also be derived from the Bernoulli approximation used to derive Eq. (3.6) as the inherent bin width–related error that plagues entropy estimates does not carry over to mutual information: The offending bin width–related term in Eq. (3.6) cancels because mutual information equals the difference of entropies (Eq. (3.8)). When the rate is allowed to vary with time as a stochastic process, the expression for mutual information is much more complicated (Brémaud, 1981).

Shannon sought a fundamental quantity that characterized the channel regardless of the input. One way to eliminate the dependence of mutual information on the input's probability distribution is to optimize mutual information over all input probability distributions. Maximizing in this way yields the *channel capacity*, a fundamental quantity that characterizes *any* channel.

$$C = \lim_{T\to\infty} \max_{p_X(\cdot)\in\mathscr{C}} \frac{1}{T} I(X;Y) \tag{3.9}$$

As mutual information usually increases proportionally with the length of the observation interval, we divide by T, which results in capacity having units of bits/second. The maximization is usually restricted to probability distributions for the channel input that has characteristics defined by the constraint class $\mathscr{C}$. For example, in neuroscience we might want to constrain the spike rate to lie within some range.

In his landmark 1948 paper, Shannon showed that the channel capacity captures everything required to determine when a digital communications channel conveying an encoded source sequence at a rate of R bits/second can *reliably* communicate information.

Noisy Channel Coding Theorem. *If the data rate produced by a discrete-valued source and its encoder is less than capacity* (R < C), *there exists a channel coding scheme so that **all** errors incurred in the transmission of information can be corrected. Furthermore, the converse is true: If* R > C, *no scheme can prevent errors from occurring.*

The converse makes capacity a fundamental quantity: Capacity uniquely defines a sharp boundary between reliable (error-free) and unreliable digital communication. In conjunction with the source coding theorem, Shannon's result completely determines (theoretically) when a discrete source can be communicated over a digital channel without error. Results like this typify the fundamental nature of information theory: that it defines the achievable limits for communication and processing in general scenarios. Shannon's proof of the noisy channel coding theorem and the other proofs that followed are not constructive: The channel coding scheme that results in reliable communication over an unreliable channel is not known to this day. The noisy channel coding theorem is restricted to digital communication, which limits its generality. Later, we will show that capacity proves to be the important quantity in a much broader range of circumstances.

As we shall see, to capture how well any neural code can express a stimulus set, we need to know the capacities of a single neuron and of a population. To calculate capacity for situations relevant to neuroscience, we must model discharge patterns. The most common mathematical model is the point process (Johnson, 1996), wherein the occurrence of a spike depends on the stimulus and on when previous events from the same or other neurons occurred. Considering first a single neuron, the point process's intensity has the general form $\mu(t; \mathcal{H}_t)$, where $\mathcal{H}_t$ denotes the neuron's discharge pattern history (what occurred before time t). The intensity summarizes how discharge rate at any moment depends on previous events and on the stimulus. The simplest point process is the Poisson process, wherein the intensity does not depend on previous events, allowing us to express it as an instantaneous rate function: $\mu(t; \mathcal{H}_t) = \lambda(t)$.

Despite the fact that most ecological stimuli are modeled by a continuous probability distribution, laboratory experiments often (out of necessity) use a finite set of test stimuli. Finding the capacity for a simple case illustrates the problems induced by using such a random discrete-stimulus set. Again consider the constant-rate Poisson channel model expressed by Eq. (3.4), where the rate λ assumes one of M values. We can maximize mutual information using a constraint set $\mathcal{C}$ that specifies the number of stimuli and the rate range $[\lambda_{min}, \lambda_{max}]$. To maximize mutual information to find capacity according to Eq. (3.9), we need to find the stimulus probabilities that achieve capacity. Stein (1967) solved this problem, finding that equally likely stimuli are not optimal and that the rates should lie on an approximately quadratic curve. However, his result depends on the number of stimuli (rate levels) in a very nonlinear manner. Once the number of stimuli exceeds a threshold number that depends on the rate range, the "extra" stimuli should be assigned zero probability, which means that presenting them anyway lowers mutual information from its maximal value (Johnson, 2002). Furthermore, the capacity does not depend on which stimuli are selected. Thus, just as with entropy considerations, imposing the discrete-stimulus model on an information-theoretic analysis does not result in an approximation to the more interesting and realistic situation with a continuous stimulus distribution.

Fortunately, Kabanov (1978) derived the capacity of the single point process channel, and Johnson and Goodman (2008) extended his result to the multiple Poisson process case. Neither of these derivations restricted the probability distribution of the channel input to have any particular form (as was done in Stein's calculation). Kabanov imposed minimal and maximal constraints on the instantaneous rates: $\mathscr{C} = \{\mu(t; \mathscr{H}_t) : \lambda_{\min} \leq \mu(t; \mathscr{H}_t) \leq \lambda_{\max}\}$. If the maximal rate were not constrained, the capacity would be infinite. Note that this maximal rate equals the peak rate a given neuron can produce, even though it may be only capable of doing so transiently. Otherwise, the probability distribution of the input is not constrained: It could be discrete- or continuous-valued and, more important, the input could be a stochastic process. Kabanov found that $C^{(1)}$, the capacity of the single-neuron channel, is attained by a Poisson process and is related to the constraints according to

$$C^{(1)} = \frac{\lambda_{\min}}{\ln 2}\left[\frac{1}{e}\left(\frac{\lambda_{\max}}{\lambda_{\min}}\right)^{\frac{\lambda_{\max}}{\lambda_{\max}-\lambda_{\min}}} - \ln\left(\frac{\lambda_{\max}}{\lambda_{\min}}\right)^{\frac{\lambda_{\max}}{\lambda_{\max}-\lambda_{\min}}}\right]$$

Typically, a neuron's rate of discharge can dip to zero, making $\lambda_{\min} = 0$. In this case, the expression for capacity simplifies greatly, and we frequently use this result in subsequent expressions.

$$C^{(1)} = \frac{\lambda_{\max}}{e \ln 2} \tag{3.10}$$

Frey (1991) showed that when the maximal rate varies with time, as it does to describe a neuron's response to a suddenly applied stimulus, Kabanov's capacity result easily generalizes.

$$C^{(1)} = \frac{1}{e \ln 2} \lim_{T \to \infty} \frac{1}{T} \int_0^T \lambda_{\max}(t) \, dt$$

The input that achieves capacity is known as a random telegraph wave—that is, a square wave that randomly switches between the minimal and maximal rates in such a way that the probability of the input equaling $\lambda_{\max}$ at any random time is $1/e$. This signal has infinite bandwidth, which may not be realistic for neuroscience. Restricting the constraint class to band-limited rate variations has only yielded capacity bounds without specifying the capacity-achieving rate probability distribution (Shamai and Lapidoth, 1993).

Kabanov further showed that no non-Poisson process satisfying the constraints can have a larger capacity. Our recent work elaborates this result (Johnson, 2010), showing that more realistic point process models that embrace refractory effects and other dependencies on previous events have a strictly smaller capacity neatly given by the expression

$$C^{(1)} = \max_{\mu(t; \mathcal{H}_t) \in \mathcal{C}} \frac{\mathsf{E}[\mu(t; \mathcal{H}_t)]}{e \ln 2}$$

where $\mathsf{E}[\mu(t; \mathcal{H}_t)]$ equals the expected value of intensity with respect to all possible histories. For a Poisson process that has no history effects, this quantity equals $\lambda_{\max}$. For more realistic models of single-neuron discharge characteristics, the expected intensity is strictly less than $\lambda_{\max}$, thereby giving a smaller capacity. For example, if absolute refractory effects occur (refractory interval Δ), the capacity under a maximal rate constraint of $\lambda_{\max}$ is given by

$$C^{(1)} = \frac{1}{1 + \lambda_{\max} \Delta} \cdot \frac{\lambda_{\max}}{e \ln 2} \tag{3.11}$$

Capacity results are quite limited for multi-channel point processes that describe the joint discharge characteristics of neural populations. In contrast with Kabanov's general single point process result, multi-neuron results are confined to jointly Poisson process descriptions of a restricted type (Johnson and Goodman, 2008). Two issues arise in an information-theoretic analysis of a population's capacity. The first is input innervation: Do the neurons comprising the population have individual innervation, do they share a single common input, or do they have a combination of the two? Secondly, are there lateral connections among the population's members, inducing connection-dependent statistical dependencies unrelated to the stimulus? Interestingly, when no lateral connections exist, input innervation (at least for the two extreme cases) does *not* affect the capacity result. For a population of N neurons, each of which is modeled as a Poisson process having a maximal firing rate of λ_{max}, the population capacity $C^{(N)}$ simply equals $NC^{(1)}$. Thus, the capacity is simply proportional to the population size, which, we will learn subsequently, makes a population far more capable than a single neuron of encoding a stimulus to a high degree of fidelity. When lateral connections among population members do exist, capacity depends on the degree of connection-induced correlation and on the input innervation. If a common input serves the population, increased coupling between neurons results in a capacity smaller than $NC^{(1)}$. However, if each neuron has its own input, a different result emerges: Capacity increases with increased coupling, attaining values that can double it from the uncoupled baseline.

This last result puts on a sound theoretical basis the empirical finding that correlated populations can express information better than a population of independently functioning neurons (Brenner et al., 2000; Reich et al., 2001; Schneidman et al., 2003; Latham and Nirenberg, 2005): From an information-theoretic perspective, the whole can be greater than the sum of its parts. In neuroscience, "synergy" means that $C^{(N)} > NC^{(1)}$; "redundancy" means that $C^{(N)} < NC^{(1)}$. For the Poisson population model, synergy and redundancy depend on both innervation and lateral connections. Only these extremes of innervation were considered by Johnson and Goodman (2008); whether synergy can be sustained with partial sharing of inputs was not determined. Another issue is the jointly Poisson assumption made in this work. It can be shown that jointly Poisson processes cannot be negatively correlated, a situation seen in many neural systems. Consequently, extrapolating the Poisson result to more realistic descriptions of population discharges may not be justified.

But can capacity apply more directly to neuroscience problems? The capacity-achieving input, which is equivalent to the neuron's discharge rate, in both the single Poisson process Poisson population cases, is the random telegraph wave described earlier. Theory suggests that effective communications systems should have channel-input probability distributions mimicking the capacity-achieving distribution (Shamai and Verdú, 1997). We know of no measured firing rate resembling a stochastic square wave, which means that neural systems viewed from an information-theoretic paradigm are operating inefficiently to some degree. When the input probability distribution does not correspond to its capacity-achieving form, information theory is silent about the resulting fidelity losses. Ignoring this issue, capacity seemingly can be estimated because it depends only on $p_{Y|X}(y|x)$, the stimulus–response characteristic of the neural system in question. However, to perform the required maximization, this quantity is needed for all conceivable stimuli, a daunting empirical task. If the stimulus–response characteristics were available, the Blahut-Arimoto algorithm (Cover and Thomas, 2006) could be used to compute capacity. The primary difficulty arises because the quantities of interest depend critically on the joint distribution $p_{X,Y}(x,y)$ over very-high-dimensional stimuli and responses spaces with limited observations of $p_{Y|X}(y|x)$ and highly restricted choices of $p_X(x)$.

One of the first mutual information estimation methods used in analyzing neural data was the "reconstruction method." In this approach, the best linear predictor of the stimulus given the spike response is found and the mutual information between the stimulus and the reconstruction is calculated (Bialek et al., 1991). While this technique gives only a lower bound on capacity, it is sometimes possible to derive upper bounds independently based on physiological constraints, thereby reducing the uncertainty in estimating the true mutual information. Several researchers have taken a different tack by decomposing mutual information into quantities that express how much rate, timing, and interneuronal correlations contribute to the overall value (Panzeri et al., 1999; Panzeri and Schultz, 2001). This approach provides information-theoretic insight into the contributions made by various components of a population response to mutual information.

But, more fundamentally, neurons do not encode binary code words and represent a discrete set of stimuli. The relevance of the noisy channel coding theorem becomes problematic when the digital communication model does not apply. Shannon had similar issues with analog communication schemes prevalent in his day, such as AM and FM radio. Furthermore, the digital communication framework discussed above addresses only communicating information from point to point and does not encompass the information processing that underlies the fundamental task of neural systems. We must ask the question: "Is channel capacity relevant to neuroscience?" In response to his own questions about generalizing the communication paradigm, Shannon developed his all-encompassing result: rate-distortion theory.

3.4 RATE-DISTORTION THEORY

Together, the source coding and noisy channel coding theorems provide fundamental limits on reliable communication for discrete-valued sources. Rate-distortion theory embraces all situations—discrete- and continuous-valued sources and channels—and provides a unifying theory of entire communication systems. This theory begins by introducing a *distortion measure* $d(S,\widehat{S})$ that expresses how to assess the fidelity of the communication (or signal processing) system. As in Figure 3.1, S represents the stimulus and $\widehat{S}$ its "estimate." Presumably, the distortion (error) increases as the stimulus and its reconstruction become more disparate. Interestingly, Shannon's framework allows *any* reasonable distortion measure ($d(S,\widehat{S}) \geq 0$, equaling zero when $\widehat{S} = S$). It can be chosen according to whatever criteria are used to evaluate the meaning of effective communication or processing in any situation. For example, the common distortion measure used in communications and signal processing is the squared error: $d(S,\widehat{S}) = (\widehat{S} - S)^2$. More relevant to sensory neuroscience would be a perceptual error measure, such as one that reflects Weber's law. The distortion measure could also incorporate some desired stimulus processing, such as feature extraction, in which $\widehat{S}$ is an estimate of some feature of S.

Next, Shannon defined the average distortion $\overline{D}$ as the expected value of the distortion measure with respect to the joint distribution of the stimulus and its estimate.

$$\overline{D} \triangleq \mathsf{E}\left[d(S,\widehat{S})\right] = \iint d(s,\hat{s}) p_{\widehat{S}|S}(\hat{s}|s) p_S(s) \, ds \, d\hat{s}$$

The conditional distribution $p_{\widehat{S}|S}(\hat{s}|s)$ depends on virtually everything in a neural coding scenario: how the stimulus is encoded, the neuron's spiking characteristics, and how the decoder works. Recall that we rarely know (or can measure) $\widehat{S}$, the estimated stimulus. Despite this difficulty, we will show that Shannon's rate-distortion theory is very relevant. Shannon proceeded by defining the *rate-distortion function* $\mathscr{R}(D)$ to be the minimal mutual information between the stimulus and its estimate over *all* possible channels, encoders, and decoders that yield an average distortion smaller than D.

$$\mathscr{R}(D) \triangleq \lim_{T \to \infty} \frac{1}{T} \min_{p_{\widehat{S}|S}(\cdot|\cdot):\overline{D} \leq D} I(S;\widehat{S}) \tag{3.12}$$

Note that the minimization is calculated over *all* possible relationships between a stimulus and its estimate, not just the one under study. In this way, we do not need to specify the estimator. The rate-distortion function has units (assuming base-two logarithms) of bits/second. It is *not* the bit rate of some equivalent digital scheme; rather, it becomes the intermediate variable, a universal unit of information, of evaluating how well communication and processing systems can perform according to the specified distortion criterion.

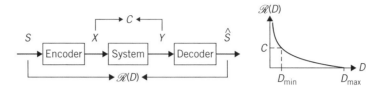

FIGURE 3.2

The key quantities in Shannon's classic information theory—capacity and the rate-distortion function—are defined in the context of the standard model shown in Figure 3.1. Capacity (C) summarizes the channel that presumably introduces disturbances into the communication process. The rate-distortion function $\mathscr{R}(D)$ depends solely on stimulus characteristics. Shannon's rate-distortion theorem relates these two quantities, showing that the smallest possible distortion $D_{\min}$ is determined by $C = \mathscr{R}(D_{\min})$.

Rate-distortion functions are notoriously difficult to calculate, with only a few results known. If the stimulus source is a band-limited Gaussian random process having maximal power $\mathscr{P}$ and bandwidth W, and the distortion measure is squared error, the rate-distortion function equals (Cover and Thomas, 2006)

$$\mathscr{R}(D) = \begin{cases} W \log_2 \frac{\mathscr{P}}{D}, & D \leq \mathscr{P} \\ 0, & D > \mathscr{P} \end{cases} \tag{3.13}$$

This result, shown in Figure 3.2, illustrates the properties that *all* rate-distortion functions satisfy.

- $\mathscr{R}(D)$ is a strictly decreasing and convex function.

- $\mathscr{R}(D)$ equals zero at some distortion, equaling zero for any larger values. This critical value is known as the maximal distortion $D_{\max}$, which corresponds to the decoder's best guess of what the stimulus might be with no data. For example, in the Gaussian case, mean square distortion is minimized by guessing $\widehat{S} = \mathsf{E}[S]$, which makes $D_{\max} = \mathscr{P}$. In the binary case, the minimum-probability-of-error decoder having no data to process simply flips a coin biased to the stimulus probability p. Consequently, no combination of encoder, channel and decoder should yield a distortion larger than $D_{\max}$.

Shannon's crowning result, the *rate-distortion theorem*, unifies all of his findings.

Rate-Distortion Theorem. *The distortion at which the rate-distortion function equals the channel capacity defines the smallest possible distortion* D_{min}

any *encoder and decoder can obtain for a given source and channel (see Figure 3.2):* $\mathcal{R}(D_{min}) = C$.

The rate-distortion function depends only on stimulus characteristics and on the desired level of distortion. Capacity summarizes the properties of the channel and determines via the rate-distortion function the smallest possible distortion that any encoder and decoder can achieve for a given source. The value of D_{min} defined by the source and the channel determines how well a given system can perform, thereby serving as a benchmark.

The more one considers Shannon's rate-distortion theorem and its implications, the greater one's appreciation for how important and general it is. Rate-distortion theory applies when *no* digital scheme is involved. Quite surprisingly, "bits/second" is the fundamental unit of exchange in *any* communication or signal processing system, whether it communicates discrete- or continuous-valued signals. Information-theoretic rate (not to be confused with spike rate) serves as the intermediary. First, the result applies to *any* distortion measure. Even though a neuroscientist does not know what Nature's measure of distortion might be, *all* rate-distortion functions look like the one portrayed in Figure 3.2. Second, an important, general concept emerges that highlights the importance of capacity: The larger the capacity, the smaller the achievable distortion in *all* cases. Therefore, if capacity can be increased, by increasing the population size for example, then the estimate can have smaller error *if* the appropriate encoder and decoder can be found.

In our Gaussian stimulus example (Eq. (3.13)), the smallest possible distortion decreases exponentially with capacity.

$$D_{min} = D_{max} 2^{-C/W} \tag{3.14}$$

Note that this result applies no matter what channel intervenes between encoder and decoder. It could be a radio channel, cable television, or a group of neurons. We can determine the effectiveness of *any* single neuron encoding/decoding scheme for Gaussian stimuli simply by plugging the capacity formula from Eq. (3.10) into Eq. (3.14) to discover that $D_{min} = D_{max} \exp\{-\lambda_{max}/eW\}$. Thus, the maximal rate needs to be several times the bandwidth to obtain significant distortion reductions from the performance D_{max} provided by the intelligent but data-blind decoder. Even for signals with a modest bandwidth (visual signals having a bandwidth of about 30 Hz, for example), maximal firing rates would need to be well over 100 Hz for a single neuron to represent temporal stimulus changes accurately. In the auditory system, the situation is much worse. Auditory-nerve fibers having a center frequency of 1 kHz have a bandwidth of about 500 Hz. Thus, a single neuron would need to be capable of a maximal discharge rate of several thousand spikes/second to achieve a high-fidelity representation. At such high rates, refractory effects would come into play and, as Eq. (3.11) shows, higher rates may not suffice: Refractory effects limit capacity to no more than $1/(\Delta t \cdot e \ln 2)$. We arrive at our first result directly applicable to neuroscience: *Only with population coding can measured perceptual distortions be achieved.* Using $C^{(N)} = N\lambda_{\max}/(e \ln 2)$ for a population's capacity, the best possible distortion decreases *exponentially* in population size: $D_{\min} = D_{\max} \exp\{-N\lambda_{\max}/eW\}$. Now, almost without regard to its input innervation pattern or to stimulus bandwidth, a sufficiently large population has the *capability* to accurately render a stimulus. When synergy occurs, capacity increases, which means that smaller distortions can occur with the same size population.

As fundamental and general as these results are, they do not address all of the information processing questions found in neuroscience. First of all, they only define the limits to which a stimulus can be estimated. Real systems, like the brain, probably operate less efficiently than the theoretical maximum, yielding a distortion larger than $D_{\min}$. Furthermore, it is unclear (from an information-theoretic viewpoint) how to assess information processing efficiency unless the distortion can be measured. As we have repeatedly noted, $\widehat{S}$ is rarely available, making distortion measurements impossible. More telling is the point that the result of neural processing is not just a reconstruction (estimate) of the stimulus. A perceived stimulus could be a gestalt of features, not a unified whole. Also, actions consequent of perception placed in the context of environment and experience can be the result of information processing. Information theory is silent on these more complicated, but pertinent, scenarios. A broader, more applicable framework is needed to examine how the stimulus results in meaningful actions, not just how it is communicated and processed.

3.5 POST–SHANNON INFORMATION THEORY

Despite the unquestioned importance of Shannon's classic theory, how does it concern "information"? Shannon equated information with the statistical character of the signal produced by the source. He explicitly divorced meaning from the definition of information. Consequently, classical information theory concentrates on the conditions for reliably communicating *signals*, not just the information that they might convey. Many cases, including neural processing, go well beyond simple signal communication. For example, what matters in transcribing a speech signal, for example, is the fidelity of the text in representing the meaning of what was said. The traditional approach would be to consider the transcription as a channel: The input signal is the speech and the channel is the transcription process. To find the transcription system's capacity, one needs the mutual information between the speech and its transcription, which in turn requires their joint probability distribution, an impossible quantity to capture in a formula. Furthermore, the primary quantities—capacity and the rate-distortion function—cannot be measured; they are solution-of-optimization problems wherein the results are examined over an exhaustive set of inputs and transcription systems. Therefore, the transcription system cannot be characterized using classical information theory. Even if one did measure the mutual information, it is not clear what the result would mean: It would be a lower bound on capacity without any notion of how the deficit would impact performance.

One post-Shannon result is very important.

Data Processing Theorem. *Consider a cascade of two systems symbolically described by* $X \rightarrow Y \rightarrow Z$ *that indicates the relationships the systems impose on the signals* X, Y *and* Z. *Then* $I(X;Y) \geq I(X;Z)$ *regardless of the kind of signals involved.*

Recalling that smaller mutual information means less statistical similarity, this result indicates that any processing system, say the one that has Y as its input and Z as its output, can maintain, but in most cases will lessen how well signal X is expressed by Y. For example, the binning process reduces the knowledge we have about spike timing and the transcription process cannot enhance the expressivity of the speech signal's content.

Inspired by the way experiments are conducted, we developed a new approach to quantify how well any information processing system works that uses the data processing theorem as its cornerstone. Our theory rests on several fundamental assumptions (Johnson et al., 2001; Sinanović and Johnson, 2007). We use a standard communication-like model as a backdrop (Figure 3.1).

- *Information can have any form and need not be stochastic.*
  In general, we make no assumption about what form information takes. We represent it by the symbol α that we simply take to be a member of a set (either countable or uncountable). Information is not assumed to result from a probabilistic generative process.

- *Signals encode information.*
  Information does not exist in tangible form; rather, it is always *encoded* into a signal, such as text in the transcription example. Signals representing the same information can take on very different forms.

- *What may or may not be information is determined by the information sink.*

  The recipient of the information—the information sink—determines what is informative about the signal it receives. Additional, often extraneous, information χ_0, χ_1 is represented by signals as well; it is the information sink that determines whether these are informative or not. An immediate consequence of this assumption is that *no single objective measure can quantify the information contained in a signal.* "Objective" here means analysis of a signal out of context and without regard to the recipient.

- *When systems act on their input signal(s) to produce output signal(s), they indirectly perform information processing.*

  Systems map signals from their input space to their output space. Because information does not exist as an entity, systems affect the information encoded in their inputs by their input-output map. We assess the effect a system has on information bearing signals by comparing how well the input and output signals express information.

The immediate impact of these assumptions is that measures like entropy and mutual information, which just measure the statistical property(s) of signals, are not as important. How do we determine whether a system, such as the transcription system described previously, has modified the relevant information? The key idea is to mimic the way neurophysiological experiments are performed: We measure whether the system's output signal changes when some aspect of the encoded information changes. More concretely, if the stimulus (information) changes from α_0 to α_1, how well does a signal (or collection of signals), no matter where it occurs in the system—the encoder, the decoder, or the channel—represent that change? We do assume that the signals representing information are stochastic, which means that they are completely described by their probability distributions. Consequently, to measure how much two signals differ in a general way, we need a measure of how two probability distributions differ.

Much work in information theory has demonstrated the importance of the Kullback-Leibler distance.

$$\mathscr{D}_X(\alpha_1 \| \alpha_0) = \int p_X(x; \alpha_1) \log_2 \frac{p_X(x; \alpha_1)}{p_X(x; \alpha_0)} \, dx. \qquad (3.15)$$

The semicolons in the probability functions denote that α_0 and α_1 are parameters of the distribution, not random variables. While mutual information defines how similar two random variables are, the Kullback-Leibler distance characterizes disparity: how much one random variable's probability distribution changes when its parameters change from α_0 to α_1. Only when the probability distributions are unchanged does this quantity equal zero; in all other cases, the Kullback-Leibler distance is positive. This distance can be infinite, expressing that the two distributions are maximally different. Note that calling this quantity a "distance" is to misname it: It is not a symmetric function of the two probability distributions, which is a technical requirement for any quantity to be called a distance. Consequently, we don't use "distance" in the strict sense here. Note that mutual information Eq. (3.7) can be expressed as the Kullback-Leibler distance between a joint distribution and the product of the marginals.

$$I(X;Y) = \mathscr{D}\left(p_{X,Y}(x,y) \| p_X(x)p_Y(y)\right)$$

The Kullback-Leibler distance has emerged as *the* information-theoretic quantity for assessing how two probability distributions differ. Furthermore, it can be related to how well the parameters can be estimated and how well the parameter change can be discerned with an optimal detector (Sinanović and Johnson, 2007). The bigger the Kullback-Leibler distance for some parameter change, the smaller the error in estimating the parameter and the easier that change is to detect. Thus, this single quantity characterizes how well information *could* be extracted from a signal by an optimal system.

The importance of the Kullback-Leibler distance becomes evident when we recast the data processing theorem. Doing so reveals how stimulus fidelity—the ability to extract information—changes when a system acts on an information-bearing signal.

Data Processing Theorem Redux. *The Kullback-Leibler distance between two inputs $X(\alpha_0)$ and $X(\alpha_1)$ to any system must exceed that between the corresponding outputs $Y(\alpha_0)$ and $Y(\alpha_1)$, regardless of what forms the input and output signals may have.*

$$\mathscr{D}_X(\alpha_1 \| \alpha_0) \geq \mathscr{D}_Y(\alpha_1 \| \alpha_0) \quad \text{or} \quad \gamma_{X,Y}(\alpha_0,\alpha_1) \triangleq \frac{\mathscr{D}_Y(\alpha_1 \| \alpha_0)}{\mathscr{D}_X(\alpha_1 \| \alpha_0)} \leq 1$$

Successive stages of processing only maintain or, more likely, decrease the fidelity to which information can be extracted. The closer the ratio $\gamma_{X,Y}(\alpha_0,\alpha_1)$ is to one, the more the system has preserved in its output the fidelity of the parameter encoding inherent in the input. Note that, as opposed to classical information theory, which seeks the ultimate limits of fidelity, the Data Processing Theorem can be used to assess whether existing systems are performing optimally with respect to any stimulus change.

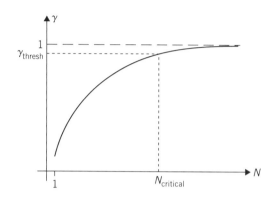

FIGURE 3.3

A population's Kullback-Leibler distance ratio increases monotonically with the population size N according to either of the formulas given in Eq. (3.16).

A system theory based on Kullback-Leibler distance analysis complements classic information-theoretic considerations (Sinanović and Johnson, 2007). In particular, a sequence of processing stages lessens the fidelity to which the ultimate output reflects input information.

We have also shown that, for a population innervated with a common input, the ratio $\gamma_{X,Y}(\alpha_0, \alpha_1)$ increases monotonically with population size N, regardless of what makes up the population (Johnson, 2004) (Figure 3.3).

$$\gamma_{X,Y}(\alpha_0, \alpha_1) = \begin{cases} 1 - \frac{k}{N} & X \text{ continuous-valued} \\ 1 - k_1 \exp\{-k_2 N\} & X \text{ discrete-valued} \end{cases} \tag{3.16}$$

This result parallels the population capacity results described earlier—$C^{(N)} = N C^{(1)}$—and the decreasing nature of all rate-distortion functions. The Kullback-Leibler distance approach is more direct, indicating that once the population size exceeds a critical value, say N_{critical}, as in Figure 3.3, further increases do not provide significant performance increases but do afford redundancy. One application of this result occurs in the auditory periphery of mammals, wherein about twenty auditory-nerve fibers innervate each inner hair cell. Theoretically, this innervation pattern indicates both that the population capacity is twenty times that of one fiber and that, even so, the fidelity of the estimated acoustic stimulus does not increase proportionally.

The Kullback-Leibler distance and the input-output ratio of distances was developed expressly for empirical studies. Many of the statistical techniques developed for measuring entropy and mutual information apply to the Kullback-Leibler distance as well. In particular, an algorithm very similar to the direct method described above has been successfully applied to estimating Kullback-Leibler distances when paired with statistical resampling techniques to reduce bias (Johnson et al., 2001; Rozell et al., 2004). In Johnson et al. (2001), biophysical simulations of particular central-auditory-system neurons and subsequent Kullback-Leibler analysis indicated that these neurons encoded both stimulus amplitude and the direction from which the sound came. The fidelity of this coding was determined from the distance measured from the neuron's output spike train. That analysis indicated that location was encoded with constant fidelity—the Kullback-Leibler distance was a constant—when measured during different temporal segments of the response. The fidelity of amplitude was high during the initial response but decreased rapidly afterwards. These results fostered the conclusion that these neurons encoded the direction of the acoustic source and secondarily encoded sound amplitude. In the second case, signal processing methods allowed the analog input to sustaining fibers found in the crayfish visual system and their spiking output to be teased apart from intracellular recordings (Rozell et al., 2004). Across several different stimulus conditions, the largest measured value of $\gamma_{X,Y}(\alpha_0, \alpha_1)$ was very small: on the order of 10^{-4} to 10^{-3}. Theoretical analysis indicated that, because of the high input signal-to-noise ratio and the low output spike rate, this was the largest ratio of Kullback-Leibler distances that could be achieved. Consequently, the sustaining fiber system was performing as well as possible under its physiological constraints. Stimulus fidelity could be increased if additional, independently functioning sustaining fibers were to be incorporated into a population.

3.6 DISCUSSION

Shannon's information theory is one of the greatest technical achievements of the twentieth century. In fact, once the power of Shannon's results became evident, the title of his work changed from "*A* mathematical theory of communication" to "*The* mathematical theory…" (A good, modern reference for information theory is Cover and Thomas (2006).) Shannon's theory defines the ultimate fidelity limits that communication and information processing systems can attain under a wide variety of situations. Be that as it may, how relevant such an approach is to understanding how the brain processes information can certainly be questioned.

Information theory provides few answers when it comes to analyzing a preexisting system whose components and constraints are only vaguely known. In neuroscience applications, the simple act of applying a stochastic stimulus does little to respond to the issues. More basic and interesting would be how to assess how well a stimulus or its features are represented by a neural response and how well a neural group extracts features from its input. The Kullback-Leibler distance is the information-theoretic suggestion, but estimating it for a population response can require much data: Producing accurate estimates even for small populations could demand more data than could be reasonably measured. The reason for the "data-hungry" nature of estimating this and other information-theoretic quantities is also the same reason for employing them: their generality. They can cope with any communication and processing scenario, but this capability comes at an experimental price of demanding data to fill in the details.

Do note that we have followed in the footsteps of many other authors in ignoring the existence of feedback and its corollary adaptivity. Feedback is a notoriously difficult concept to handle with the traditional tools of information theory, but it would be foolish to discount the role of feedback connections when they are so prominent anatomically even in the most peripheral sensory pathways. Some recent work has sparked interest in feedback in the information theory community (Massey, 1990; Venkataramanan and Pradhan, 2005); however, none of it has yet been translated to a neural setting. Furthermore, how can systems that adapt to their previous inputs (learn from them) fit into the rather static formulation of classic information theory? If this could be accomplished, it would be interesting to determine the performance limits of adaptive systems. Information theorists are actively examining this topic. Results on this front would help define the backdrop for appreciating the brain's performance capabilities, but may not shed light on how well it actually works.

References

Bialek, W., Rieke, F., de Ruyter van Steveninck, R.R., Warland, D., 1991. Reading a neural code. Science 252, 1852–1856.

Borst, A., Theunissen, F.E., 1999. Information theory and neural coding. Nat. Neurosci. 2, 947–957.

Brémaud, P., 1981. Point Processes and Queues. Springer-Verlag, New York.

Brenner, N., Strong, S.P., Koberle, R., Bialek, W., de Ruyter van Steveninck, R.R., 2000. Synergy in a neural code. Neural Comput. 12, 1531–1552.

Cover, T.M., Thomas, J.A., 2006. Elements of Information Theory, second ed. Wiley, New York.

Frey, M.R., 1991. Information capacity of the Poisson channel. IEEE Trans. Inf. Theory 37, 244–256.

Johnson, D.H., 1996. Point process models of single-neuron discharges. J. Comput. Neurosci. 3, 275–299.

Johnson, D.H., 2002. Four top reasons mutual information does not quantify neural information processing. In: Computational Neuroscience '02. Chicago, IL.

Johnson, D.H., 2004. Neural population structures and consequences for neural coding. J. Comput. Neurosci. 16, 69–80.

Johnson, D.H., 2010. Information theory and neural information processing. IEEE Trans. Inf. Theory 56, 653–666.

Johnson, D.H., Goodman, I.N., 2008. Inferring the capacity of the vector Poisson channel with a Bernoulli model. Netw.: Comput. Neural Syst. 19, 13–33.

Johnson, D.H., Gruner, C.M., Baggerly, K., Seshagiri, C., 2001. Information-theoretic analysis of neural coding. J. Comput. Neurosci. 10, 47–69.

Kabanov, Y.M., 1978. The capacity of a channel of the Poisson type. Theory Probab. Appl. 23, 143–147.

Latham, P.E., Nirenberg, S., 2005. Synergy, redundancy, and independence in population codes, revisited. J. Neurosci. 25, 5195–5206.

Massey, J.L., 1990. Causality, feedback and directed information. In: Proceedings of 1990 International Symposium on Information Theory and its Applications, Waikiki, Hawaii, 303–305.

McFadden, J.A., 1965. The entropy of a point process. SIAM J. Appl. Math. 13, 988–994.

Paninski, L., 2003. Estimation of entropy and mutual information. Neural Comput. 15, 1191–1253.

Panzeri, S., Schultz, S.R., Treves, A., Rolls, E.T., 1999. Correlations and the encoding of information in the nervous system. Proc. R. Soc. Lond. 266, 1001–1012.

Panzeri, S., Schultz, S.R., 2001. A unified approach to the study of temporal, correlational, and rate coding. Neural Comput. 13, 1311–1349.

Reich, D.S., Mechler, F., Victor, J.D., 2001. Independent and redundant information in nearby cortical neurons. Science 294, 2566–2568.

Rozell, C.J., Johnson, D.H., Glantz, R.M., 2004. Measuring information transfer in crayfish sustaining fiber spike generators: methods and analysis. Biol. Cybern. 90, 89–97.

Schneidman, E., Bialek, W., Berry II M.J., 2003. Synergy, redundancy, and independence in population codes. J. Neurosci., 23, 11539–11553.

Shamai, S., Lapidoth, A., 1993. Bounds on the capacity of a spectrally constrained Poisson channel. IEEE Trans. Inf. Theory 39, 19–29.

Shamai, S., Verdú, S., 1997. Empirical distribution for good codes. IEEE Trans. Inf. Theory 43, 836–846.

Shannon, C.E., 1948. A mathematical theory of communication. Bell Syst. Tech. J. 27, 379–423, 623–656.

Sinanović, S., Johnson, D.H., 2007. Toward a theory of information processing. Signal Processing 87, 1326–1344.

Stein, R.B., 1967. The information capacity of nerve cells using a frequency code. Biophys. J. 7, 67–82.

Strong, S.P., Koberle, R., de Ruyter van Steveninck, R.R., Bialek, W., 1998. Entropy and information in neural spike trains. Phys. Rev. Lett. 80, 197–200.

Venkataramanan, R., Pradhan, S.S., 2005. Directed information for communication problems with common side information and delayed feedback/feedforward. In: Proceedings of Allerton Conference on Communication, Control and Computing.

Victor, J.D., 2006. Approaches to information-theoretic analysis of neural activity. Biol. Theory 1, 302–316.

Identification of Nonlinear Dynamics in Neural Population Activity

4

Dong Song[*†], **Theodore W. Berger**[*†**]

*Department of Biomedical Engineering, †Center for Neural Engineering,
**Program in Neuroscience, University of Southern California, Los Angeles, California*

4.1 INTRODUCTION

Brain regions underlying cognitive function, like the hippocampus and the neocortex, are massively parallel, multiple-input, multiple-output (MIMO), nonlinear dynamic systems. Communication between any two brain regions is based on millions of axonal projections and synaptic connections. In a given brain region, information is represented in the ensemble firing of populations of neurons (Georgopoulos et al., 1986; Salinas and Abbott, 1994; Deadwyler and Hampson, 1995; Pouget et al., 2003)—that is, spatio-temporal patterns of electrophysiological activity. Brain regions process the information by transforming the incoming spatiotemporal patterns into outgoing spatio-temporal patterns, and this transformation depends on highly nonlinear dynamic mechanisms, such as ligand-dependent synaptic transmission and voltage-dependent conductance (Berger et al., 1988; Sclabassi et al., 1988). Thus, the information processing underlying cognition involves transformations of neural representations that are dynamic, nonlinear, and often nonstationary (time-varying). While recent advances in multi-electrode technology have made it possible to record the simultaneous activities of populations of neurons in behaving animals, modeling such complex system behavior still remains one of the most challenging tasks in computational neuroscience (Brown et al., 2004).

There are two main approaches to modeling a biological system: parametric and nonparametric. A parametric model is "internal" or "bottom up." It utilizes specific model structures defined by known or postulated physiological mechanisms and associated parameters that can be related to certain biological processes. Such a model seeks to uncover the underlying

Statistical Signal Processing for Neuroscience and Neurotechnology. DOI: 10.1016/B978-0-12-375027-3.00004-1

103

physiological mechanisms and their influences on system behavior. A good example is the Hodgkin-Huxley ionic-channel-based, morphologically realistic compartmental model of neurons (Carnevale and Hines, 2006). By contrast, a nonparametric model is "external" or "top down." It is obtained directly from system input-output data collected under broadband experimental conditions or natural behaviors, with a general model structure that does not rely on a priori knowledge or assumptions about the system. The main purpose of a nonparametric model is to capture the linear/nonlinear dynamical properties of the system and predict system output based on system input. The process of developing a nonparametric model is termed *system identification*. Common nonparametric models include linear regression models, Volterra-Wiener series, and artificial neural networks (Marmarelis, 2004).

Since parametric and nonparametric models emphasize different aspects of the modeled system, they have different relative advantages and disadvantages (Berger et al., 1994; Song et al., 2009a,b). The main advantage of a parametric model is its biological interpretability, in the sense that the model parameters can be related to biological mechanisms/processes and that the modeling results may further lead to testable biological hypotheseis. Its main disadvantage is its fixed model structure based on partial knowledge that is subject to potential biases. When applied to complex systems such as a brain region, parametric modeling also often involves a large number of open parameters that cannot be sufficiently constrained by input-output data and thus results in semi-quantitative models determined with multilevel data from different experimental preparations and conditions. On the other hand, a nonparametric model has a general but scalable model structure that can potentially capture arbitrary system input-output properties. Its model parameters/coefficients are selected and estimated directly from input-output data, and provide a straightforward representation of a system's input-output properties. The main disadvantage of a nonparametric model is its lack of direct mechanistic interpretation due to its descriptive nature. The particular modeling approach used for any given case should be chosen based on modeling aims and experimental limitations.

4.2 PROBLEM STATEMENT

The purpose of our modeling study of the hippocampus is twofold. First, we aim to build quantitative models of neural population activity for the development of hippocampal prostheses. In cortical regions, including the hippocampus, information is represented in the ensemble spiking activity of neuron populations (Georgopoulos et al., 1986; Schwartz et al., 1988;

Deadwyler and Hampson, 1995; Pouget et al., 2003; Puchalla et al., 2005). The hippocampus has long been known to be responsible for the formation of declarative memories (Squire, 1992; Deadwyler and Hampson, 1995; Deadwyler et al., 1996; Eichenbaum, 1999; Burgess et al., 2002). Damage to it disrupts the propagation of spatio-temporal patterns of activity through hippocampal internal circuitry, resulting in severe anterograde amnesia. Developing a neural prosthesis for the damaged hippocampus requires restoring this MIMO transformation of spiking activity. In other words, the model needs to predict the output spike trains based on the ongoing input spike trains of the modeled brain region.

Our second purpose is to gain a better understanding of the functional organization of the hippocampal neuronal network. This requires the model to provide interpretable representations of the neurons' nonlinear dynamical properties and the functional connectivity between neurons. Model variables should be related to at least the principal biological processes of the system.

Because of the duality of our goal, we combined both parametric and nonparametric modeling methods—spiking neuron modeling and Volterra series—to build a nonlinear dynamical model of neural population activity (Song et al., 2007, 2009c). The model not only captures the nonlinear dynamics underlying the input-output spike train transformation but also provides physiologically plausible variables that can be used for further in-depth interpretations of the observed system dynamics. This chapter will focus on this modeling strategy and describe issues about model configuration, parameter estimation, model validation, and model prediction.

4.3 NONLINEAR MODEL OF NEURAL POPULATION DYNAMICS

The model has both parametric and nonparametric components. First, the overall model structure is parameterized to be "neuron-like." It captures the stereotypical features of spiking neurons and explicitly includes variables that can be interpreted as principal cellular processes, such as post-synaptic potential, spike-triggered after-potential, pre-threshold noise, and spike-generating threshold. This configuration partitions the system dynamics in a physiologically plausible manner and thus facilitates comparison with intracellular recording results. The more versatile features of spiking neurons (i.e., the transformation from input spikes to post-synaptic potentials and the transformation from the output spike to the after-potential), on the other hand, are modeled nonparametrically with the Volterra series, taking advantage of its flexibility in capturing nonlinear dynamics. All model parameters/coefficients are estimated from the timings of input-output spikes.

4.3.1 Model Configuration

A MIMO model (Figure 4.1(a)) consists of a set of MISO models (Figure 4.1(b)). Within each MISO model, the output spike train is predicted based on the input spike trains considered. The MISO model structure is inspired by the electrophysiological properties of single spiking neurons and consists of five components (Figure 4.1(c)):

1. A feedforward block K transforming the input spike trains x to a continuous hidden variable u that can be interpreted as the synaptic potential.
2. A feedback block H transforming the *preceding* output spikes to a continuous hidden variable a that can be interpreted as after-potential.
3. A noise term ε that captures the system uncertainty caused by both the intrinsic neuronal noise and the unobserved inputs.
4. An adder generating a continuous hidden variable w that can be interpreted as pre-threshold potential.
5. A threshold function generating an output spike when the value of w crosses θ (Song et al., 2007; Song et al., 2009c).

The model can be expressed by the following equations:

$$w = u(k, x) + a(h, y) + \varepsilon(\sigma) \tag{4.1}$$

$$y = \begin{cases} 0 & \text{when } w < \theta \\ 1 & \text{when } w \geq \theta \end{cases} \tag{4.2}$$

K and H can take any mathematical form as long as it sufficiently captures the nonlinear dynamic transformations between the variables: x to u and y to a. In our approach, K takes the form of a Volterra model in which u is expressed, in terms of the inputs x by means of the Volterra series expansion, as

$$u(t) = k_0 + \sum_{n=1}^{N} \sum_{\tau=0}^{M_k} k_1^{(n)}(\tau) x_n(t-\tau) + \sum_{n=1}^{N} \sum_{\tau_1=0}^{M_k} \sum_{\tau_2=0}^{M_k} k_{2s}^{(n)}(\tau_1, \tau_2) x_n(t-\tau_1) x_n(t-\tau_2)$$

$$+ \sum_{n_1=1}^{N} \sum_{n_2=1}^{n_1-1} \sum_{\tau_1=0}^{M_k} \sum_{\tau_2=0}^{M_k} k_{2x}^{(n_1, n_2)}(\tau_1, \tau_2) x_{n_1}(t-\tau_1) x_{n_2}(t-\tau_2) + \cdots \tag{4.3}$$

The zeroth-order kernel, k_0, is the value of u when the input is absent. The first-order kernels $k_1^{(n)}$ describe the linear relation between the n*th* input x_n and u as functions of the time intervals (τ) between the present time and the past time. The second-order self-kernels $k_{2s}^{(n)}$ describe the second-order nonlinear relation between the n*th* input x_n and u. Second-order cross-kernels $k_{2x}^{(n_1, n_2)}$

(a)

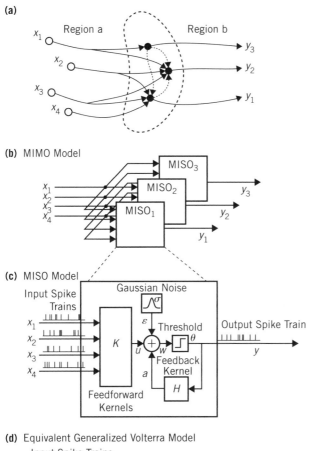

(b) MIMO Model

(c) MISO Model

(d) Equivalent Generalized Volterra Model

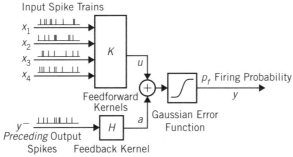

FIGURE 4.1

Multiple-input, multiple-output (MIMO) model for neural population dynamics.
(a) Schematic of spike train propagation between two brain regions. (b) MIMO model as a series of multiple-input, single-output (MISO) models. (c) Structure of a MISO model.
(d) MISO model is equivalent to a generalized Volterra model (GVM) with a *probit* link function.

describe the second-order nonlinear interactions between each unique pair of inputs (x_{n_1} and x_{n_2}) as they affect u. N is the number of inputs. Mk denotes the memory length of the feedforward process. Higher-order kernels (e.g., third- and fourth-order) are not shown in this equation.

Similarly, H takes the form of a first-order Volterra model as in

$$a(t) = \sum_{\tau=1}^{M_h} h(\tau)y(t-\tau) \tag{4.4}$$

where h is the linear feedback kernel, and M_h is the memory length of the feedback process (note that τ starts from 1 instead of 0 to avoid predicting the current output with itself).

One of the major challenges in Volterra modeling is the large number of open parameters (coefficients) to be estimated. The total number increases exponentially with input dimension and model order. The MISO model in this study involves 2D (time and index of the input neurons) input and second-order nonlinearity. The number of parameters easily becomes unwieldy even in a moderately large model (e.g., the 24-input MISO model shown in Table 4.1). To solve this problem, we employed a Laguerre expansion of the Volterra kernel (LEV) and statistical model selection techniques.

Using LEV, both k and h are expanded with orthonormal Laguerre basis functions b (Marmarelis, 1993; Marmarelis, 2004; Zanos et al., 2008; Song et al., 2009a,b). With input and output spike trains x and y convolved with b,

$$v_j^{(n)}(t) = \sum_{\tau=0}^{M_k} b_j(\tau)x_n(t-\tau), \quad v_j^{(h)}(t) = \sum_{\tau=1}^{M_h} b_j(\tau)y(t-\tau) \tag{4.5}$$

Table 4.1 Number of Coefficients to be Estimated in MISO Models

Model	k_0	k_1	K_{2s}	k_{2x}	h	Total
V	1	NM_k	$NM_k(M_k+1)/2$	$N(N-1)M_kM_k/2$	M_h	72,018,301
		12,000	3,006,000	69,000,000	300	
LEV	1	NL	$NL(L+1)/2$	$N(N-1)L2/2$	L	2704
		72	144	2484	3	
rLEV	1	18	36	9	3	67

V: Volterra kernel model. LEV: Laguerre expansion of the Volterra kernel model. rLEV: reduced LEV model with only significant inputs and cross-terms. The numbers are calculated with $N = 24$, $L = 3$, $M_k = 500$, and $M_h = 300$. Numbers of rLV models are from an example with six significant inputs and one significant cross-term.

Equations (4.3) and (4.4) can be rewritten as

$$u(t) = c_0 + \sum_{n=1}^{N}\sum_{j=1}^{L} c_1^{(n)}(j)v_j^{(n)}(t) + \sum_{n=1}^{N}\sum_{j_1=1}^{L}\sum_{j_2=1}^{j_1} c_{2s}^{(n)}(j_1,j_2)v_{j_1}^{(n)}(t)v_{j_2}^{(n)}(t)$$

$$+ \sum_{n_1=1}^{N}\sum_{n_2=1}^{n_1-1}\sum_{j_1=0}^{L}\sum_{j_2=0}^{L} c_{2x}^{(n_1,n_2)}(j_1,j_2)v_{j_1}^{(n)}(t)v_{j_2}^{(n)}(t) + \cdots \tag{4.6}$$

$$a(t) = \sum_{j=1}^{L} c_h(j)v_j^{(h)}(t) \tag{4.7}$$

Here $c_1^{(n)}$, $c_{2s}^{(n)}$, $c_{2x}^{(n_1,n_2)}$, and c_h are the sought Laguerre expansion coefficients of $k_1^{(n)}$, $k_{2s}^{(n)}$, $k_{2x}^{(n_1,n_2)}$, and h, respectively (C_0 is simply equal to k_0). Since the number of basis functions (L) can be made much smaller than the memory length (M_k and M_h), the number of open parameters is greatly reduced by the expansion.

The noise term ε is modeled as Gaussian white noise with standard deviation σ.

4.3.2 Model Estimation

With recorded input and output spike trains x and x, model parameters can be estimated using a maximum likelihood method. The negative log-likelihood function L can be expressed as

$$L(y|x,k,h,\sigma,\theta) = - \sum_{t=0}^{T} \ln P(y|x,k,h,\sigma,\theta) \tag{4.8}$$

where T is the data length, and P is the probability of generating the recorded output x:

$$P(y|x,k,h,\sigma,\theta) = \begin{cases} \text{Prob}(w \geq \theta|x,k,h,\sigma,\theta) & \text{when } y=1 \\ \text{Prob}(w < \theta|x,k,h,\sigma,\theta) & \text{when } y=0 \end{cases} \tag{4.9}$$

Since ε is assumed to be Gaussian, the conditional firing probability intensity function P_f (the conditional probability of generating a spike, that is, $\text{Prob}(w \geq \theta|x,k,h,\sigma,\theta)$ in Eq. (4.9), at time t can be calculated with the

Gaussian error function (the integral of the Gaussian function) erf:

$$P_f(t) = 0.5 - 0.5 erf \left(\frac{\theta - u(t) - a(t)}{\sqrt{2}\sigma} \right) \tag{4.10}$$

where

$$erf(s) = \frac{2}{\sqrt{\pi}} \int_0^s e^{-t^2} dt \tag{4.11}$$

P at time t can then be calculated as

$$P(t) = \begin{cases} p_f(t) & \text{when } y = 1 \\ 1 - p_f(t) & \text{when } y = 0 \end{cases} \tag{4.12}$$

or

$$P(t) = 0.5 - \left[y(t) - 0.5 \right] erf \left(\frac{\theta - u(t) - a(t)}{\sqrt{2}\sigma} \right) \tag{4.13}$$

Model coefficients c can now be estimated by minimizing the negative log likelihood function L:

$$\tilde{c} = \arg\min(L(c)) \tag{4.14}$$

It is instructive to point out that the model just described is mathematically equivalent to a generalized linear model (GLM) with x dependent variables, the convolutions of Laguerre basis functions with inputs x (v in Eqs. (4.6) and (4.7)) as well as the products of these convolutions (vv in Eq. (4.6)) as independent variables, and c as unknown parameters (Figure 4.1(d)). The GLM link function is the probit function (inverse cumulative distribution function of the normal distribution) because the latter is defined as

$$probit(y) = \sqrt{2} erf^{-1}(2y - 1) \tag{4.15}$$

Given this important equivalence, model coefficients c and their covariance matrices can be estimated using the iteratively reweighted least squares method, the standard algorithm for fitting GLMs (McCullagh and Nelder, 1989; Truccolo et al., 2005). For the same reason, this model can be referred to as the generalized Volterra model (GVM) (Song et al., 2008). Since u, a, and n are dimensionless variables, without loss of generality, both θ and σ can be set to unity value; only c are estimated. σ is later restored, and θ remains unity value.

4.3.3 Model Selection

Theoretically, the model can be used to estimate arbitrary MISO models. However, in practice model complexity often must be reduced by selecting an optimal subset of model parameters (coefficients). This procedure, termed model selection, is particularly necessary and desirable in modeling the Neural Population dynamics for the following reasons.

First, neurons are often sparsely connected. In a brain region, an output neuron is seldom affected by all input neurons. The full Volterra kernel model as described in Eq. (4.6) is not the most efficient or interpretable way to represent a system like this. More important, the number of coefficients to be estimated in a full Volterra kernel model grows rapidly with the number of inputs and the model order. Estimation of such a model, especially higher-order ones, can easily become unwieldy (Table 4.1). Furthermore, a model with too many open parameters (coefficients) tends to fit the noise instead of the signal in the training data. An overfitting model results in poor generalization of the training data and bad predictions of novel data. Consequently, interpretation becomes problematic. To solve this problem, the following statistical model selection method is applied to the configuration and estimation of GVMs.

Before model selection, the input-output data set is partitioned into two subsets. One subset (the training set) is used for model estimation; the other (the testing set) is retained for validation of the results from the training set. Results from the two subsets are referred to as in–sample and out-of–sample, respectively.

The model starts from the zeroth order. A zeroth-order model contains only c_0, which is equivalent to the standard deviation of the pre-threshold Gaussian noise (see Section 4.3.4). It essentially models system output as a homogeneous Poisson process (constant firing probability intensity). The minimal negative likelihood (L) of a zeroth-order model provides a starting point for model selection. In the second step, feedback terms (as described by Eq. (4.7)) are added. The output spike train is predicted by the preceding output spikes without considering any input. If L decreases in both the in-sample and out-of-sample results, the feedback term is added to the model. In the third step, inputs are selected using a forward stepwise selection procedure (Kutner et al., 2004). Self-terms involving first-order and second-order kernels are constructed for all inputs. With the zeroth-order term and feedback term (if selected in the previous step) in the model, the values of L with and without each input are calculated. The input that decreases L the most is then added into the model. With the newly selected input added, selection is performed on the remaining inputs. Repeating this procedure, inputs are sequentially added. The selection is stopped when out-of-sample L starts to increase (in-sample L always decreases as more terms are added), indicating the occurrence of overfitting. In the fourth step, cross-terms involving cross-kernels are selected. Cross-terms are first constructed for every unique pair of selected inputs and then chosen following the just described forward stepwise and cross-validation procedures.

4.3.4 Kernel Reconstruction and Interpretation

The final coefficients $\hat{c}$ and $\hat{\sigma}$ can be obtained from estimated Laguerre expansion coefficients, $\tilde{c}$, with a simple normalization/conversion procedure:

$$\hat{c}_0 = 0 \tag{4.16}$$

$$\hat{c}_1^{(n)} = \frac{\tilde{c}_1^{(n)}}{1 - \tilde{c}_0} \tag{4.17}$$

$$\hat{c}_{2s}^{(n)} = \frac{\tilde{c}_{2s}^{(n)}}{1 - \tilde{c}_0} \tag{4.18}$$

$$\hat{c}_{2x}^{(n_1, n_2)} = \frac{\tilde{c}_{2x}^{(n_1, n_2)}}{1 - \tilde{c}_0} \tag{4.19}$$

$$\hat{c}_h = \frac{\tilde{c}_h}{1 - \tilde{c}_0} \tag{4.20}$$

$$\hat{\sigma} = \frac{1}{1 - \tilde{c}_0} \tag{4.21}$$

Feedforward and feedback kernels then can be reconstructed as

$$\hat{k}_0 = 0 \tag{4.22}$$

$$\hat{k}_1^{(n)}(\tau) = \sum_{j=1}^{L} \hat{c}_1^{(n)}(j) b_j(\tau) \tag{4.23}$$

$$\hat{k}_{2s}^{(n)}(\tau_1,\tau_2) = \sum_{j_1=1}^{L}\sum_{j_2=1}^{j_1} \frac{\hat{c}_{2s}^{(n)}(j_1,j_2)}{2}\left[b_{j_1}(\tau_1)b_{j_2}(\tau_2)+b_{j_2}(\tau_1)b_{j_1}(\tau_2)\right] \quad (4.24)$$

$$\hat{k}_{2x}^{(n_1,n_2)}(\tau_1,\tau_2) = \sum_{j_1=1}^{L}\sum_{j_2=1}^{j_1} \hat{c}_{2x}^{(n_1,n-2)}(j_1,j_2)b_{j_1}(\tau_1)b_{j_2}(\tau_2) \quad (4.25)$$

$$\hat{h}(\tau) = \sum_{j=1}^{L}\hat{c}_h(j)b_j(\tau) \quad (4.26)$$

Threshold θ is equal to 1 in this normalized representation.

The normalized kernels provide an intuitive representation of the system input-output nonlinear dynamics. Each input's single-pulse and paired-pulse response functions (r_1 and r_2) can be derived using (Song et al., 2009a)

$$r_1^{(n)}(\tau) = \hat{k}_1^{(n)}(\tau) + \hat{k}_{2s}^{(n)}(\tau,\tau) \quad (4.27)$$

and

$$r_2^{(n)}(\tau_1,\tau_2) = 2\hat{k}_{2s}^{(n)}(\tau_1,\tau_2) \quad (4.28)$$

$r_1^{(n)}$ is simply the response in u elicited by a single spike from the nth input neuron; $r_2^{(n)}$ describes the joint nonlinear effect of pairs of spikes from the n*th* input neuron in addition to the summation of their first-order responses—that is, $r_1^{(n)}(\tau_1) + r_1^{(n)}(\tau_2) \cdot \hat{k}_{2x}^{(n_1,n_2)}(\tau_1,\tau_2)$ represents the joint nonlinear effect of pairs of spikes with one spike from neuron n_1 and one spike from neuron n_2. h represents the output spike-triggered after-potential on u (Figure 4.2)

4.3.5 Model Validation and Prediction

Two methods are used to evaluate the goodness-of-fit of the estimated GVM models. The first directly evaluates the continuous firing probability intensity predicted by the model with the recorded output spike train. According to the Time-Rescaling Theorem, an accurate model should generate a conditional firing intensity function P_f that can rescale the recorded output spike train into a Poisson process with unit rate (Brown et al., 2002; Song et al., 2007). By further variable conversion, inter-spike intervals should be rescaled into independent uniform random variables on the interval (0, 1). The model goodness-of-fit can then be assessed with a Kolmogorov-Smirnov (KS) test, in which the rescaled intervals are ordered from the smallest to the largest

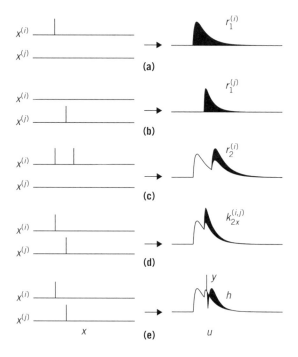

FIGURE 4.2

Interpretations of the feedforward and feedback kernels. $r_1^{(i)}$ is the response in u elicited by a single spike from the ith input neuron; $r_2^{(i)}$ describes the joint nonlinear effect of pairs of spikes from the ith input neuron in addition to the linear summation of their first-order responses. $k_{2x}^{(i,j)}$ represents the joint nonlinear effect of pairs of spikes from neurons i and j. h represents the output spike-triggered after-potential on u. The black areas represent the effect of each kernel on u.

and then plotted against the cumulative distribution function of the uniform density. If the model is correct, all points should lie on the 45-degree line of the KS plot within the 95% confidence bounds.

The second method quantifies the similarity between the recorded output spike train y and the predicted output spike train $\hat{y}$ after a smoothing process. First, $\hat{y}$ is realized through simulation: u is calculated with inputs x and the estimated feedforward kernels. This forms the deterministic part of pre-threshold potential w. A Gaussian random sequence with standard deviation $\hat{\sigma}$ is then generated and added to u, rendering w stochastic. At each time t, if w crosses threshold ($\theta = 1$), a spike is generated and added to $\hat{y}$; also, a feedback process a is triggered and added to the future values of w. The calculation then moves on to time $t + 1$ with updated w until it reaches the end of the record. In the second step, the point process signals $\hat{y}$ and y

are smoothed to continuous signals $\hat{y}_{\sigma_g}$ and y_{σ_g} through convolution with a Gaussian kernel having standard deviation σ_g. Correlation coefficients r are then calculated as

$$r(\sigma_g) = \left(\sum_{t=0}^{T}\hat{y}_{\sigma_g}(t)y_{\sigma_g}(t)\right)\bigg/\sqrt{\left(\sum_{t=0}^{T}y_{\sigma_g}(t)y_{\sigma_g}(t)\right)\left(\sum_{t=0}^{T}\hat{y}_{\sigma_g}(t)\hat{y}_{\sigma_g}(t)\right)}$$

(4.29)

Because $\hat{y}_{\sigma_g}$ and y_{σ_g} are both positive vectors, r is a quantity between 0 and 1 that measures the similarity between $\hat{y}$ and y as a function of the "smoothness parameter" $\sigma_g \cdot \sigma_g$ essentially determines the temporal resolution used in comparing the predicted spike train with the actual spike train. A large value means low temporal resolution; a small value means high temporal resolution. This parameter does not influence model estimation because the estimation is carried out by maximizing the likelihood function defined with a fixed 2-ms bin size. In this study, σ_g varies from 2 to 100 ms. The mean and standard deviation of r is estimated with 32 trials of simulation.

4.4 RESULTS: APPLICATION TO HIPPOCAMPAL CA3-CA1 POPULATION ACTIVITY

The modeling methodology has been successfully applied to hippocampal CA3-CA1 population dynamics. The experimental data were collected and provided by Deadwyler and Hampson's laboratory at Wake Forest University. More details on the behavioral and neural procedures can be found in our joint publications (Song et al., 2007, 2009c).

4.4.1 Behavioral Task

Male Long-Evans rats are trained to perform a two-lever, spatial delayed-nonmatch-to-sample (DNMS) task with randomly occurring variable-delay intervals (Deadwyler et al., 1996; Hampson et al., 1999). Animals perform the task by pressing a single lever presented in one of the two positions in the sample phase (left or right); this behavioral event is called the "sample response." The lever is then retracted and the delay phase initiated; for the duration of the delay phase, the animal is required to nose-poke into a lighted space in the opposite wall. Following termination of the delay, the nose-poke light is extinguished, both levers are extended, and the animal is required to press the lever opposite to the sample lever; this act is called the "nonmatch response." If the correct lever is pressed, the animal is rewarded and the trial is completed (Figure 4.3(a)).

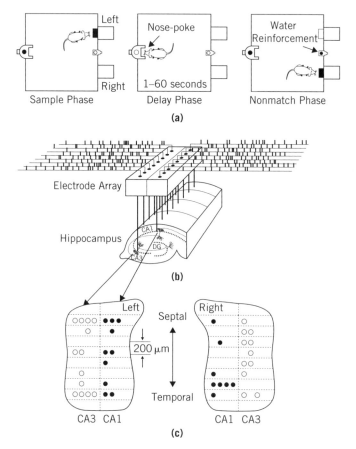

FIGURE 4.3

Input and output spike trains recorded from hippocampal CA3 and CA1 regions during a delayed nonmatch-to-sample (DNMS) behavior task. (a) Schematic of the DNMS task. (b) CA3 and CA1 spike trains are recorded using a multi-electrode array during the task (not real data). (c) Anatomical locations of input (CA3) and output (CA1) neurons indicated on a foldout map of the hippocampus (Swanson et al., 1978). For this sample MIMO data set, there are 24 inputs (white symbols) and 18 outputs (black symbols).

4.4.2 Data Preprocessing

Spike trains are obtained with multisite recordings from different septotemporal regions of the hippocampus of rats performing the DNMS task (Figure 4.3(b)). For each hemisphere of the brain, eight pairs of recording electrodes are positioned along the septotemporal axis of the hippocampus at 200-μm intervals over a total expanse of 1.6 mm. One electrode of each pair is targeted for CA3; the other is targeted for CA1 (Hampson et al., 1999). Each electrode has the capacity to record from as many as four discriminable units (Figure 4.3(c)). Spikes are sorted and timestamped with a 25-μs resolution.

Data sets from 25 rats were analyzed (Figure 4.3). One to four recording sessions were selected from each rat. A session included approximately 100 successful DNMS tasks that each consisted of two of the four behavioral events: right sample (RS) and left nonmatch (LN), or left sample (LS) and right nonmatch (RN).

Spike trains are prescreened based on mean firing rate and peri-event histogram. Neurons with mean firing rates in the range of 0.5 to 15 Hz and an identifiable peri-event histogram are included in further analyses. Low (<0.5 Hz) and high (>15 Hz) mean rate recordings are rejected because they can represent artifacts or mixtures of action potentials from two or more neurons (Berger et al., 1983; Christian and Deadwyler, 1986). Both presumed principal (pyramidal) neurons and interneurons are included in the data sets analyzed. Peri-event (−2-s to +2-s) spike trains of the four behavioral events are extracted from each session and then concatenated to form the data sets.

First, the timestamped data are discretized with a 2-ms bin size. With such a narrow width, each bin can contain a maximum of only one spike event. To facilitate model estimation, the data length is further reduced with an inter-spike binning (ISB) method (Figure 4.4). In ISB, bins that contain a spike are retained; those within inter-spike intervals are merged with a larger bin size

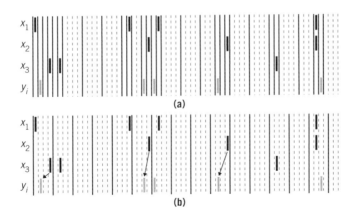

FIGURE 4.4

Comparison of inter-spike binning and conventional binning for spike trains. Black bars represent input spikes; gray bars represent output spikes. (a) with inter-spike binning (solid lines), the causal relations between input spikes and output spikes are retained. (b) with conventional binning (solid lines), future inputs may influence present outputs in analysis and cause spurious causal relations between inputs and outputs (arrows). Dashed lines in parts a and b represent 2-ms bin.

(Song et al., 2009c). This method preserves the accurate timing of every spike but reduces the data length. To preserve the causal relation between input and output spikes, bins with an output spike are checked to see whether they also contain one or more input spikes. If there is no other spike, no operation is performed; if a spike is found, the 2-ms bin is further divided between the input spike and the output spike (the latter case is extremely rare because of the low firing rate of hippocampal neurons). For most of the data sets, inter-spike bin size can be as large as 40 ms without introducing a noticeable change in the coefficient estimates.

4.4.3 **CA3-CA1 MIMO Model**

MIMO data sets consist of 6 to 32 CA3 spike trains and 5 to 24 CA1 spike trains, depending on the number of neurons recorded from each animal and the number of recorded neurons meeting the selection criteria. Each MIMO data set is divided into MISO data sets for MISO model estimation. Thirty-two MIMO datasets from 25 rats are analyzed. Figure 4.3(c) shows the approximate anatomical locations of neurons in one representative MIMO data set (circles). Each "compartment" represents one recording site along the septotemporal axis of the hippocampus. The position of each neuron in the cell indicates only the indices (from 1 to 4) of this neuron within recording and so does not carry anatomical information.

With the preprocessed data set, MISO models are obtained using the previously described estimation and selection method. Figure 4.5 illustrates the selection of inputs and cross-terms of one representative MISO model. As described in Section 4.3.3, the model starts from the zeroth order. The feedback term is then included in the model because it decreases L (negative log-likelihood) in both training and testing data sets (Figure 4.5(a)). With the zeroth-order and feedback terms, inputs are added to the model in a forward stepwise fashion (Figure 4.5(b)). The in-sample L decreases monotonically as more terms are added. However, the out-of-sample L starts to increase from the seventh input. Six inputs are selected based on this cross-validation result (Figure 4.5(a)). With them, 15 cross-terms are then constructed and added to the model (Figure 4.5(c)). Again, in-sample L decreases monotonically; out-of-sample L starts to increase from the second cross-term, so only one cross-term is selected by cross-validation (Figure 4.5(a)). Table 4.1 summarizes the number of coefficients in a Volterra model, a Laguerre expansion of a Volterra (LEV) model, and the rLEV after model selection. It is evident that the model selection procedure greatly reduces model complexity and thus allows reliable model estimation.

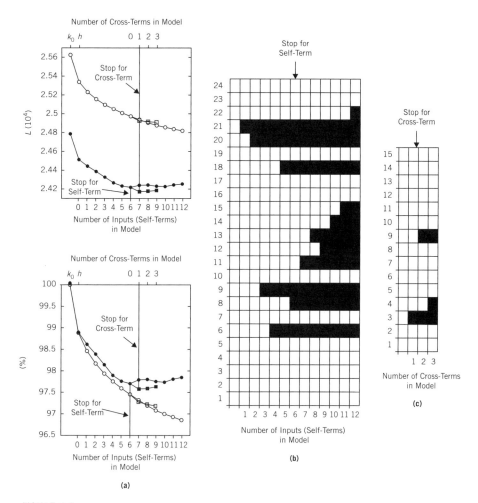

FIGURE 4.5

Selecting significant inputs and cross-terms of a MISO model. (a) The model starts from the zeroth order (first data points). Feedback kernels are then selected and included in the model (second data points). Six inputs (first- and second-order self-terms) and one cross-term are selected based on cross-validation. Open circles represent in-sample results; closed circles represent out-of-sample results; top: absolute negative log-likelihood (L); bottom: normalized L. (b) Selection path of the inputs. (c) Selection path of the cross-terms. The 15 cross-terms are constructed with the selected 6 inputs only.

The estimated and normalized kernels provide intuitive representations of the system input-output properties. Figure 4.6 illustrates the first- and second-order response functions for the six selected inputs, one second-order cross-kernel for inputs 6 and 9, and the feedback kernel. These functions

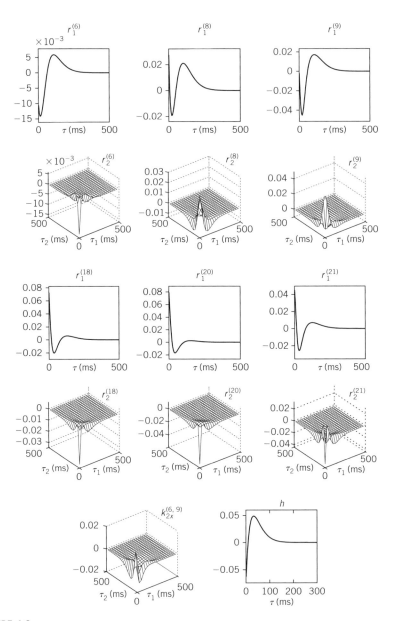

FIGURE 4.6

Feedforward and feedback kernels of a MISO model. r_1 represents the single-pulse response functions. r_2 represents the paired-pulse response functions for the same input neuron. k_{2x} represents cross-kernels for pairs of neurons. h is the feedback kernel.

quantitatively describe how synaptic potential is influenced by a single spike and by pairs of spikes from a single input neuron or pairs of input neurons. The noise standard deviation is estimated to be 0.418, which reflects a 4.19-Hz spontaneous (baseline) firing rate of this CA1 neuron (see Eq. (4.10) for calculation) while the mean output firing rate is 7.44 Hz.

The estimated model is validated using the out-of-sample KS test. The summation of synaptic potential u and after-potential a is calculated using the input-output spike trains of the testing set and the kernels estimated from the training set (Figure 4.7(a), first row). Firing probability intensity P_f is then calculated using the Gaussian error function (Figure 4.7(a), second row). The KS plot shows that all data points are within the 95% confidence bound (Figure 4.7(c)).

The output spike train is also predicted using the recursive simulation method described in Section 4.3.4. Input spike trains of the testing set and kernels of the training set are used for prediction, and the predicted model output is then compared with the output spike train of the testing set. Figure 4.7(b) shows one realization of the output spike train with a Gaussian random noise path. The similarities between the predicted and actual output spike trains are quantified with correlation coefficients r. Figure 4.7(d) illustrates the mean and standard deviation of r as functions of the standard deviation of the smoothing Gaussian kernel (σ_g). It shows that the model can generate output highly correlated with the actual output for a large range of σ_g.

Finally, all individually estimated and validated MISO models are concatenated to form the MIMO model for hippocampal CA3-CA1 neural population dynamics. Figure 4.8 compares the actual CA1 spatio-temporal pattern with that predicted by one representative MIMO model. Despite the difference in fine details (mostly due to the stochastic nature of the system), the MIMO model replicates the salient features of the actual CA1 spatio-temporal pattern.

4.5 DISCUSSION

We have formulated a Volterra kernel-based MIMO modeling method for evaluating neural population dynamics and have applied it successfully to the identification of hippocampal CA3-CA1 spike train transformations during learned behavior. Our approach accomplishes this task in a neurobiological context that includes multisite dendritic integration and filtering, somatic summation, thresholding functions, spike generation, and spike-dependent feedback. It inherits from the nonparametric Volterra modeling method the capability of modeling high-order nonlinear dynamic systems, at the same time incorporating several critical parametric modifications/improvements

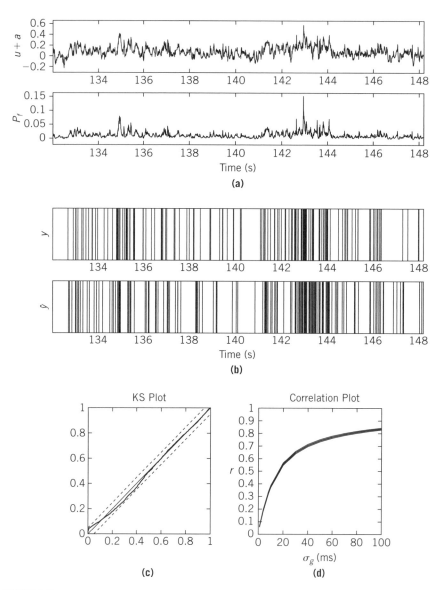

FIGURE 4.7

Model validation and prediction. (a) Firing probability intensity P_f is calculated from the summation of synaptic potential u and after-potential a. (b) Actual output spike train y and predicted output spike train $\hat{y}$. (c) Validating P_f with the actual output spike train y using a KS test. Dashed lines represent 95% confidence bounds. (d) Correlation coefficient r is calculated as a function of the standard deviation of the Gaussian smoothing kernel. Thin lines show the standard deviation of r.

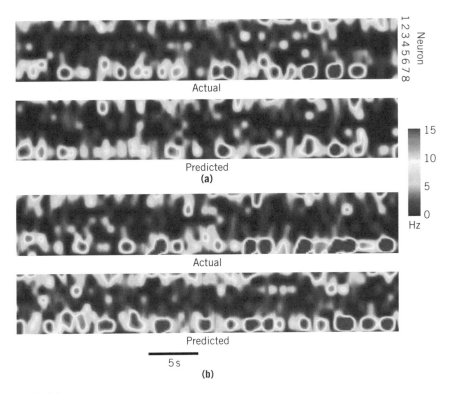

FIGURE 4.8

Predicting the CA1 output spatiotemporal pattern with a MIMO model. Eight CA1 neurons are included. Examples a and b are both out-of-sample results. Spike trains are spatially (piecewise cubic Hermite interpolating polynomial) and temporally (nonoverlapping 200-ms bin size) interpolated for better visualization. Please see this figure in color at the companion web site: www.elsevierdirect.com/companions/9780123750273

specifically for the modeling of neural population activity. With feedforward Volterra kernels, the model can capture the single-pulse and paired-pulse (non-linear) dynamic effects of input spikes to the output neuron and it can explicitly represent them in the form of kernel functions.

Unlike an ordinary Volterra model (Marmarelis and Berger, 2005; Song et al., 2009a,b), ours includes hidden variables representing the internal states of the system. Point process outputs (spikes) are considered as realizations of a firing probability intensity function determined by those hidden variables. This configuration allows simultaneous estimation of all model parameters (i.e., kernel coefficients and noise variance) directly from the input and output spike trains. The noise term and the threshold function may be considered "soft" because they map to and generate output spikes based on the estimated continuous probability intensity function and thus avoid dichotomizing such continuous variables with a cutpoint that depends highly on the relative frequencies of spikes (1s) and nonspikes (0s) in the outputs (Kutner et al., 2004). The resulting model structure is equivalent to a generalized linear model (GLM) with a *probit* link function and a binomial distribution function.

Taking advantage of the concave likelihood function (L) and the well-established estimation methods of GLM, model parameters can be reliably (avoiding local minima) and efficiently estimated through an iterative reweighted least squares method (Paninski et al., 2004; Truccolo et al., 2005). Furthermore, a feedback kernel (h) is added to the model. This auto-regressive component captures the effect of preceding output activity on current output. Much evidence has shown that the output spike-triggered after-potential can profoundly influence the spiking activity of a neuron (Alger and Nicoll, 1982; Berger et al., 1994; Keat et al., 2001; Gerstner and Kistler, 2002; Song et al., 2002; Jolivet et al., 2004; Paninski et al., 2004; Storm, 1987). As shown in our results (Figure 4.5), the feedback component can account for a significant portion of the output variance. In our model selection scheme, the feedback term is not only included but also selected prior to the selection of inputs and cross-terms. This is akin to the Granger causality analysis, in which the moving averages of inputs are statistically tested with the auto-regressive component built in (Nedungadi et al., 2009). Such a scheme differentiates the output-dependent dynamics from the input-dependent dynamics and thus provides more certain assessments of the causal relations between inputs and output.

The feedforward Volterra model is chosen to be second order in this study. However, with the expense of more open parameters and the computational cost, the model can also include higher-order terms—for example, third-order self-kernels and cross-kernels. Our previous analysis showed that adding third-order self-kernels only marginally improves model performance (Song et al., 2007).

The model described in this paper was developed within the context of hippocampal prostheses and CA3-CA1 population dynamics. For example, to replace a CA1 cell field selectively damaged in association with stroke—a common consequence of even brief periods of anoxia—the prosthetic device has to reinstate the output signal (e.g., CA1 spikes) based purely on the signals recorded in an upstream region (e.g., CA3) since the former are unavailable in such a scenario. For this reason, our model does not include other CA1 neurons as inputs in predicting the activity of any one CA1 neuron and thus can be considered an *inter-region* population model (i.e., no intra-CA1 region dynamics are included). The inter-region configuration used here caneasily be extended by taking all recorded neuronal activities (from both CA3 and CA1 neurons) as inputs and then performing model selection/estimation to more thoroughly analyze a given neural population and reveal both the intra- and inter-region functional connectivity in it (Okatan et al., 2005; Eldawlatly et al., 2009; Stevenson et al., 2009).

4.6 FUTURE DIRECTIONS

The underlying assumption of this modeling work is that the MIMO nonlinear dynamics of the hippocampus are based on stationary processes. This is probably a valid assumption for the learned behavior of well-trained animals (e.g., during the DNMS task) when the spiking representations of extrinsic cues in both input and output regions are formed and stabilized. However, an important characteristic of the cortical regions, especially the hippocampus, is experience-dependent plasticity—for example, long-term potentiation (LTP) and long-term depression (LTD). Thus, the MIMO nonlinear dynamics of the hippocampus are likely to evolve during learning and memory formation. The logical next step is to extend the nonlinear dynamic modeling of the hippocampus to such *nonstationary* cases. Input-output (e.g., CA3-CA1) spike trains should be recorded from animals *learning* the behavior task. Volterra kernels (k and h), which represent the MIMO nonlinear dynamics of the hippocampus, should be estimated recursively from these recordings. The emergence of and change in the nonlinear dynamics would then be tracked by the temporal evolution of the kernels. Preliminary simulation results show that this can be achieved by combining the nonlinear dynamical model with the point process Kalman filter technique (Eden et al., 2004; Chan et al., 2008).

The change in the MIMO nonlinear dynamics of a given hippocampal region results from the experiences of the animal. These experiences are internally represented as the flow of the input-output spatiotemporal patterns of spike trains. A further, fundamental question is whether it is possible to reconstruct the emergence of and change in the MIMO nonlinear dynamics of a hippocampal region, which are characterized by the nonstationary modeling, based on the region's input-output activities and a LTP/LTD-like *learning rule* defining how to modify the MIMO nonlinear dynamics. Such a learning rule is critical for (1) understanding the *in vivo* synaptic modification of the hippocampus during behavior and (2) developing novel algorithms for the next-generation, adaptive cortical neural prostheses essential for the patient to adjust to what can be expected to be continual changes in his or her environment.

Acknowledgments

This work was supported by the National Science Foundation (NSF) Engineering Research Center for Biomimetic MicroElectronic System Center, by the DARPA Human-Assisted Neural Devices (HAND) Program, and by the USC Biomedical Simulations Resource, supported by the National Institute of Biomedical Imaging and BioEngineering (NIBIB).

D.S. was partially supported by the James H. Zumberge Faculty Research and Innovation Fund at the University of Southern California.

References

Alger, B.E., Nicoll, R.A., 1982. Pharmacological evidence for two kinds of GABA receptor on rat hippocampal pyramidal cells studied in vitro. J. Physiol. 328, 125–141.

Berger, T.W., Ahuja, A., Courellis, S.H., Deadwyler, S.A., Erinjippurath, G., Gerhardt, G.A., et al., 2005. Restoring lost cognitive function. IEEE Eng. Med. Biol. Mag. 24 (5), 30–44.

Berger, T.W., Baudry, M., Brinton, R.D., Liaw, J.S., Marmarelis, V.Z., Park, A.Y., et al., 2001. Brain-implantable biomimetic electronics as the next era in neural prosthetics. Proc. IEEE 89 (7), 993–1012.

Berger, T.W., Chauvet, G., Sclabassi, R.J., 1994. A biological based model of functional properties of the hippocampus. Neural Netw. 7, 1031–1064.

Berger, T.W., Eriksson, J.L., Ciarolla, D.A., Sclabassi, R.J., 1988. Nonlinear systems analysis of the hippocampal perforant path-dentate projection. II. Effects of random impulse train stimulation. J. Neurophysiol. 60, 1076–1094.

Berger, T.W., Rinaldi, P.C., Weisz, D.J., Thompson, R.F., 1983. Single-unit analysis of different hippocampal cell types during classical conditioning of rabbit nictitating membrane response. J. Neurophysiol. 50 (5), 1197–1219.

Brown, E.N., Barbieri, R., Ventura, V., Kass, R.E., Frank, L.M., 2002. The time-rescaling theorem and its application to neural spike train data analysis. Neural Comput. 14 (2), 325–346.

Brown, E.N., Kass, R.E., Mitra, P.P., 2004. Multiple neural spike train data analysis: state-of-the-art and future challenges. Nat. Neurosci. 7, 456–461.

Burgess, N., Maguire, E.A., O'Keefe, J., 2002. The human hippocampus and spatial and episodic memory. Neuron 35 (4), 625–641.

Carnevale, N.T., Hines, M.L., 2006. The NEURON Book. Cambridge University Press, Cambridge, UK.

Chan, R.H.M., Song, D., Berger, T.W., 2008. Tracking temporal evolution of nonlinear dynamics in hippocampus using time-varying volterra kernels. Conf. Proc. IEEE Eng. Med. Biol. Soc. 2008, pp. 4996–4999.

Christian, E.P., Deadwyler, S.A., 1986. Behavioral functions and hippocampal cell types: evidence for two nonoverlapping populations in the rat. J. Neurophysiol. 55 (2), 331–348.

Deadwyler, S.A., Bunn, T., Hampson, R.E., 1996. Hippocampal ensemble activity during spatial delayed-nonmatch-to-sample performance in rats. J. Neurosci. 16 (1), 354–372.

Deadwyler, S.A., Hampson, R.E., 1995. Ensemble activity and behavior – What's the code? Science 270 (5240), 1316–1318.

Eden, U.T., Frank, L.M., Barbieri, R., Solo, V., Brown, E.N., 2004. Dynamic analysis of neural encoding by point process adaptive filtering. Neural Comput. 16, 971–998.

Eichenbaum, H., 1999. The hippocampus and mechanisms of declarative memory. Behav. Brain Res. 103 (2), 123–133.

Eldawlatly, S., Jin, R., Oweiss, K.G., 2009. Identifying functional connectivity in large-scale neural ensemble recordings: a multiscale data mining approach. Neural Comput. 21 (2), 450–477.

Georgopoulos, A.P., Schwartz, A.B., Kettner, R.E., 1986. Neuronal population coding of movement direction. Science 233 (4771), 1416–1419.

Gerstner, W., Kistler, W.M., 2002. Spiking Neuron Models: Single Neurons, Populations, Plasticity. Cambridge University Press, Cambridge, UK.

Hampson, R.E., Simeral, J.D., Deadwyler, S.A., 1999. Distribution of spatial and nonspatial information in dorsal hippocampus. Nature 402 (6762), 610–614.

Jolivet, R., Lewis, T.J., Gerstner, W., 2004. Generalized integrate-and-fire models of neuronal activity approximate spike trains of a detailed model to a high degree of accuracy. J. Neurophysiol. 92 (2), 959–976.

Keat, J., Reinagel, P., Reid, R.C., Meister, M., 2001. Predicting every spike: a model for the responses of visual neurons. Neuron 30 (3), 803–817.

Kutner, M.H., Nachtsheim, C.J., Neter, J., Li, W., 2004. Applied Linear Statistical Models, fifth ed. McGraw-Hill/Irwin, Boston.

Marmarelis, V.Z., 1993. Identification of nonlinear biological systems using Laguerre expansions of kernels. Ann. Biomed. Eng. 21 (6), 573–589.

Marmarelis, V.Z., 2004. Nonlinear Dynamic Modeling of Physiological Systems. Wiley-IEEE Press, Hoboken.

Marmarelis, V.Z., Berger, T.W., 2005. General methodology for nonlinear modeling of neural systems with Poisson point-process inputs. Math. Biosci. 196 (1), 1–13.

McCullagh, P., Nelder, J.A., 1989. Generalized Linear Models, second ed. Chapman & Hall/CRC, Boca Raton, FL.

Nedungadi, A.G., Rangarajan, G., Jain, N., Ding, M., 2009. Analyzing multiple spike trains with nonparametric granger causality. J. Comput. Neurosci. 27, 55–64.

Okatan, M., Wilson, M.A., Brown, E.N., 2005. Analyzing functional connectivity using a network likelihood model of ensemble neural spiking activity. Neural Comput. 17 (9), 1927–1961.

Paninski, L., Pillow, J.W., Simoncelli, E.P., 2004. Maximum likelihood estimation of a stochastic integrate-and-fire neural encoding model. Neural Comput. 16 (12), 2533–2561.

Pouget, A., Dayan, P., Zemel, R.S., 2003. Inference and computation with population codes. Annu. Rev. Neurosci. 26, 381–410.

Puchalla, J.L., Schneidman, E., Harris, R.A., Berry, M.J., 2005. Redundancy in the population code of the retina. Neuron 46 (3), 493–504.

Salinas, E., Abbott, L.F., 1994. Vector reconstruction from firing rates. J. Comput. Neurosci. 1, 89–107.

Schwartz, A.B., Kettner, R.E., Georgopoulos, A.P., 1988. Primate motor cortex and free arm movements to visual targets in 3-dimensional space. 1. Relations between single cell discharge and direction of movement. J. Neuroscie. 8 (8), 2913–2927.

Sclabassi, R.J., Eriksson, J.L., Port, R.L., Robinson, G.B., Berger, T.W., 1988. Nonlinear systems analysis of the hippocampal perforant path-dentate projection. I. Theoretical and interpretational considerations. J. Neurophysiol. 60, 1066–1076.

Song, D., Chan, R.H., Marmarelis, V.Z., Hampson, R.E., Deadwyler, S.A., Berger, T.W., 2007. Nonlinear dynamic modeling of spike train transformations for hippocampal-cortical prostheses. IEEE Trans. Biomed. Eng. 54, 1053–1066.

Song, D., Chan, R.H., Marmarelis, V.Z., Hampson, R.E., Deadwyler, S.A., Berger, T.W., 2009c. Nonlinear modeling of neural population dynamics for hippocampal prostheses. Neural Netw. 22, 1340–1351.

Song, D., Hendrickson, P., Marmarelis, V.Z., Aguayo, J., He, J., Loeb, G.E., et al., 2008. Predicting EMG with generalized Volterra kernel model. Conf. Proc. IEEE Eng. Med. Biol. Soc. 2008, 201–204.

Song, D., Marmarelis, V.Z., Berger, T.W., 2009a. Parametric and non-parametric modeling of short-term synaptic plasticity. Part I: Computational study. J. Comput. Neurosci. 26 (1), 1–19.

Song, D., Wang, Z., Berger, T.W., 2002. Contribution of T-type VDCC to TEA-induced long-term synaptic modification in hippocampal CA1 and dentate gyrus. Hippocampus 12 (5), 689–697.

Song, D., Wang, Z., Marmarelis, V.Z., Berger, T.W., 2009b. Parametric and non-parametric modeling of short-term synaptic plasticity. Part II: Experimental study. J. Comput. Neurosci. 26 (1), 21–37.

Squire, L.R., 1992. Memory and the hippocampus – a synthesis from findings with rats, monkeys, and humans. Psychol. Rev. 99 (2), 195–231.

Stevenson, I.H., Rebesco, J.M., Hatsopoulos, N.G., Haga, Z., Member, L.E.M., Kording, K.P., 2009. Bayesian inference of functional connectivity and network structure from spikes. IEEE Trans. Neural Syst. Rehabil. Eng. 17 (3), 203–213.

Storm, J.F., 1987. Action potential repolarization and a fast after-hyperpolarization in rat hippocampal pyramidal cells. J. Physiol. 385, 733–759.

Swanson, L.W., Wyss, J.M., Cowan, W.M., 1978. An autoradiographic study of the organization of intrahippocampal association pathways in the rat. J. Comp. Neurol. 181, 681–715.

Truccolo, W., Eden, U.T., Fellows, M.R., Donoghue, J.P., Brown, E.N., 2005. A point process framework for relating neural spiking activity to spiking history, neural ensemble, and extrinsic covariate effects. J. Neurophysiol. 93 (2), 1074–1089.

Zanos, T.P., Courellis, S.H., Berger, T.W., Hampson, R.E., Deadwyler, S.A., Marmarelis, V.Z., 2008. Nonlinear modeling of causal interrelationships in neuronal ensembles. IEEE Trans. Neural Syst. Rehabil. Eng. 16 (4), 336–352.

Graphical Models of Functional and Effective Neuronal Connectivity

5

Seif Eldawlatly*, Karim Oweiss*†

*Electrical and Computer Engineering Department,
†Neuroscience Program, Michigan State University, East Lansing, Michigan

5.1 INTRODUCTION

Cortical neurons are known to mediate many complex brain functions through their coordinated interaction along numerous parallel and sequential pathways with highly intricate network structures. The massive size of these networks suggests the vastly complex information processing mechanism that underlies their operation (Bressler, 1995; Varela et al., 2001). Many nervous system diseases and disorders are also known to result from abnormal changes in the function and/or structure of these networks (Catani and Fftyche, 2005; Liu et al., 2008; Supekar et al., 2008). Deciphering the neural circuitry underlying the observed behavior is therefore of utmost importance in systems and in clinical neuroscience.

System identification techniques aimed at characterizing widely distributed functional cortical networks require simultaneous monitoring of neural constituents while subjects carry out certain functions. Many functional neuroimaging studies have demonstrated selective coupling between distinct cortical areas during behavioral tasks (Greicius et al., 2003; Logothetis, 2003; Caclin and Fonlupt, 2006; Tsao et al., 2008; Verhagen et al., 2008). While these studies have yielded much insight into the brain's functional connectivity, it is widely believed that the temporal and spatial resolutions of current neuroimaging techniques remain a significant obstacle to the study of functional connectivity at the single-cell and population levels (Logothetis, 2003, 2008).

Techniques for simultaneous recording of multiple single-unit activities seem to hold the promise of increased resolution, although they are still limited to local areas and shallow cortical depths. Two-photon calcium imaging techniques (Stosiek et al., 2003) or chronically implanted high-density microelectrodes (Nicolelis, 1999; Normann et al., 1999; Csicsvari et al., 2003; Wise et al., 2004) have been used in the last few years to simultaneously monitor ensembles of spiking neurons with the hope of deciphering their population-coding properties. This has provided more insight into how functional neuronal networks form in response to dynamic complex stimuli (Averbeck et al., 2006; Quiroga et al., 2007) or to signify movement intention and execution (Chapin et al., 1999; Serruya et al., 2002; Hochberg et al., 2006). These techniques have also been used to identify markers of structural plasticity (Wagenaar et al., 2006; Wallach et al., 2008) or to monitor dendritic spine formation during motor learning and adaptation (Baeg et al., 2007; Xu et al., 2009).

There is increasing evidence that fine-scale connectivity between neurons exists in local circuits and may span multiple adjacent cortical layers (Yoshimura and Callaway, 2005) consistent with segregated subcolumnar cortical structures. There is also increasing evidence that selective activation (or inactivation) of local areas via microstimulation triggers sparse patterns of activity (Histed et al., 2009), indicating highly localized, nonrandom connectivity patterns between adjacent neurons (Song et al., 2005). The functional role of this organization is poorly understood, and only in the last ten years have methods of large-scale neural ensemble recordings in awake behaving subjects started to emerge, despite the inherent difficulty in implanting microelectrode arrays in brain areas of interest for sustained periods of time.

Methods for inferring connectivity among simultaneously recorded neurons date back to the early work of Perkel, Gerstein, and Moore (1967), when cross-correlation methods were introduced, while more recent methods, briefly surveyed by Brown et al. (2004), include spike pattern classification, likelihood, and frequency domain. Almost all of these methods were developed for analyzing single-electrode recordings and are therefore constrained by the number of neurons that a single-depth electrode can pick, mostly pairs or triplets (Gerstein, 2000; Buzsaki, 2004). When large-scale recording of multiple single units became more common, usage of these methods to infer a complex dynamic network structure showed itself to be nontrivial. In this chapter, we focus on reviewing current techniques for inferring neuronal connectivity from simultaneously recorded spike trains. We provide an in-depth presentation of graphical methods, as they seem to enjoy a number of advantages, particularly for inferring large-scale networks. Although our discussion is specific to neuronal spike trains, the methods reviewed here can be applied to other signal modalities. We demonstrate their performance and compare them to classical methods when applied to simulated and experimental spike train data.

5.2 BACKGROUND AND OVERVIEW

The problem of inferring connectivity between neurons can be stated as follows: We observe the spike trains of a population of N neurons, denoted $\mathbf{S} = [S_i]_{N \times T}$, during an interval of length T. We want to identify the set of neurons, π_i, in the set of observed neurons, N, that affect the firing of neuron i, S_i, in a statistical sense. We assume that $\mathbf{S}$ is expressed in the form of *digital* spike trains with a very small bin width Δ (usually 1–3 ms). This form assumes that at time bin m, $S_i(m\Delta) = 1$ indicates a spike while $S_i(m\Delta) = 0$ indicates no spike.

We first define two types of connectivity that are frequently used in the literature: *functional* and *effective*. *Functional connectivity* refers to the statistical dependence observed between the spike trains (rows of $\mathbf{S}$). This can result from the presence of a synaptic link between the neurons, or it can be observed when two unlinked neurons respond to a common driving input (e.g., an unobserved neuron). *Effective connectivity*, on the other hand, refers to the causal relationships governing this dependence and involves identifying the pre- and post-synaptic neurons in the observed mixture. There has been extensive research in the past few decades on discovering both types of connectivity (e.g., Gerstein and Perkel, 1969; Aertsen et al., 1989; Friston, 1994). We briefly survey these methods here.

5.2.1 The Crosscorrelogram

The crosscorrelogram is one of the oldest and most widely used methods for inferring neuronal connectivity (Perkel, Gerstein, and Moore, 1967; Gerstein, 2000). It is computed as a histogram of spike counts for a given neuron within a bin of predefined width ω, relative to the spike count of another neuron. For a given pair of neurons (i,j) and a given lag $k\omega$, the crosscorrelogram CC_{ij} is expressed as

$$CC_{ij}(k\omega) = \sum_{m=1}^{T} S_i(m\Delta) \sum_{l=1+(k-1)\omega/\Delta}^{k\omega/\Delta} S_j((m+l)\Delta) \qquad (5.1)$$

The *peak* of this histogram at a certain lag indicates excitation, while a *trough* indicates inhibition. An extension to the crosscorrelogram method is the joint peri-stimulus time histogram (JPSTH), which measures the lagged synchrony between pairs of neurons relative to the onset of an external stimulus (Gerstein and Perkel, 1969). JPSTH results in a 2-dimensional histogram in which the value at a given element (t_1, t_2) represents the number of times the first neuron fired at time t_1 *and* the second neuron fired at time t_2, as shown in Figure 5.1. Both t_1 and t_2 are measured relative to the stimulus onset. The further the distance from the diagonal, the more lagged the synchrony.

Both methods have a number of limitations. First, the analysis bin size is fixed, which means that neuronal firing is assumed stationary during this interval. Frequency domain methods such as partial directed coherence (PDC) attempt to circumvent bin width dependence at the expense of lost temporal resolution (Jarvis and Mitra, 2001). There is ample evidence, however, suggesting that the time scale of neuronal excitability is largely time-varying,

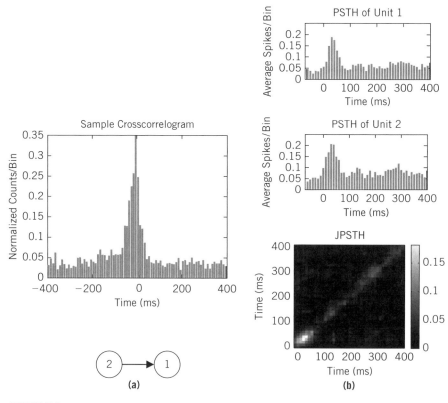

(a) **(b)**

FIGURE 5.1

(a) Crosscorrelogram. (b) PSTH and JPSTH. All three were computed from the spike trains of two neurons in the rat somatosensory (barrel) cortex in response to whisker stimulation. Bin size was set to 10 ms in parts a and b. The crosscorrelogram indicates an excitatory interaction between the two neurons as indicated by the significant narrow peak around 10 ms. The JPSTH shows that the significant synchrony between the two neurons occurs about 30 ms post-stimulus presentation and therefore may indicate that the stimulus represents the *cause* of the synchrony.

possibly to maintain a principled excitation–inhibition homeostatic balance (Maass and Natschlager, 2000; Frerking et al., 2005). Most important, recent studies on spike timing–dependent plasticity (STDP) have demonstrated that neurochemical receptors can generate excitatory post-synaptic potentials (EPSPs) at resting potential that can last a few hundred milliseconds, which elongates the time window required for synaptic integration and eventually alters the temporal length for neuronal excitability (Markram et al., 1997; Tsodyks and Markram, 1997; Bi and Poo, 1998; Roberts and Bell, 2002; Destexhe and Marder, 2004; Dzakpasu and Zochowski, 2005; Frerking et al., 2005). Second, both the CC and JPSTH methods consider only correlation between pairs of neurons. Consideration of the effect of the entire population becomes important when quantifying information transmission in large networks that is not accounted for by pairwise association measures such as cross-correlation. Third, cross-correlation can be interpreted as a measure of statistical dependence only when the pairwise responses are jointly Gaussian, which may not always be the case. Lastly, the analysis is typically limited to very short coincidence intervals above which the effect of covariation of firing rates in response to a common driving input cannot be ruled out (Smith and Fetz, 2009).

5.2.2 Information-Theoretic Methods

Information-theoretic (IT) methods attempt to circumvent the limitations of crosscorrelograms by considering measures such as mutual information (Cover and Thomas, 2006) that rely on higher-order statistics but may differ in the neuronal response property used. The advantage here is that the analysis can be carried out independently from the detailed system properties because it relies on building empirical histogram estimates of joint response distributions. In some cases, precise spike timing is used to estimate these distributions (Yamada et al., 1996; Bettencourt et al., 2007), thereby revealing the tendency of neurons to synchronize their firing. In others, a binned version of the spike train is used, thereby measuring stimulus-dependent *covariation* of firing rates. One example is the transfer entropy (TE) measure defined as (Gourevitch and Eggermont, 2007):

$$TE_{j \to i} = I\left(S_i^f; S_j^h \middle| S_i^h\right) = \sum_{s_i^f} \sum_{s_j^h} \sum_{s_i^h} p\left(s_i^f, s_j^h, s_i^h\right) \log \frac{p\left(s_i^f \middle| s_j^h, s_i^h\right)}{p\left(s_i^f \middle| s_i^h\right)}, \quad (5.2)$$

where s_i^f, s_j^h, and s_i^h are binary variables, $I\left(S_i^f; S_j^h \middle| S_i^h\right)$ is the mutual information between the future firing of neuron i, S_i^f and the past firing of neuron

j, S_j^h that cannot be accounted for by the past firing of neuron i, S_i^h. This measure can also be computed using binned versions of the spike trains.

IT methods can also be viewed as pattern detection methods that search for frequent patterns of spikes (symbols) appearing more than at a chance level (Abeles and Gerstein, 1988; Grün et al., 2002). The size of the pattern (word length) is an important feature that has to be identified to determine if the frequency of occurrence of a certain pattern above chance is indicative of a functional relationship. In a few studies (Averbeck et al., 2006; Schneidman, 2006), the choice of time scale was found to yield strongly correlated network states, even though pairwise correlations computed over a fixed bin width were statistically insignificant. One potential caveat regarding IT methods is the need for large amounts of data for which the connectivity can be assumed stationary to provide unbiased estimates of the joint distributions of the patterns. Neural plasticity may affect these estimates if it occurs over short time scales (Gourevitch and Eggermont, 2007).

The most popular IT method is the maximum entropy model (MaxEnt), rooted in statistical physics, in which the joint density of the population is expressed as an exponential function of the weighted individual and pairwise activities:

$$p(S_1, S_2, \ldots, S_N) = \frac{1}{Z} \exp \left(\sum_{i=1}^{N} \delta_i S_i + \frac{1}{2} \sum_{i=1}^{N} \sum_{\substack{j=1 \\ j \neq i}}^{N} \gamma_{ij} S_i S_j \right) \tag{5.3}$$

where Z is a normalization factor, and δ_i and γ_{ij} are the unknown parameters of the model. These parameters are estimated using an iterative scaling algorithm so that the expected values of the individual and pairwise responses match the measurable information in the experiment (Tang et al., 2008)—the mean spike count within a fixed window and pairwise correlations.

Pairwise MaxEnt models have been shown to fit experimental data *in vitro* for small systems such as retinal ganglion cell cultures (Pillow et al., 2008). Their performance for larger systems ($> \sim 20$ neurons) remains questionable (Roudi et al., 2009). Recently, a minimum entropy distance (MinED) algorithm was proposed to alleviate this *curse of dimensionality* when characterizing larger systems (Aghagolzadeh et al., 2010). It attempts to find the network structure that minimizes the distance between the entropy of single trials and the true entropy of the data, estimated from a large number of repeated trials. MinED has been tested on data from biologically plausible models of spiking neurons with natural adaptation mediated by an STDP mechanism (Dan and Poo, 2004). It demonstrated good generalization capability given the limited data available.

5.2.3 **Granger Causality**

Granger causality is a popular method for studying casual links between random variables (Granger, 1969). Specifically, suppose that the spike train of neuron i at time bin m can be predicted given the neuron's own firing history and that of another neuron j using the bivariate auto-regressive model:

$$S_i(m\Delta) = \sum_{k=1}^{K} A_{ii}(k)S_i((m-k)\Delta) + \sum_{k=1}^{K} A_{ij}(k)S_j((m-k)\Delta) + \varepsilon_{i|j}(m\Delta)$$

(5.4)

where K is the maximum number of lags (model order) and A represents the linear regression coefficients obtained by minimizing the squared prediction error $\varepsilon_{i|j}$ when S_j is used to predict S_i. Neuron j is said to *Granger-cause* neuron i if the inclusion of S_j in Eq. (5.4) reduces the variance of the prediction error. This bivariate model can be extended to a multivariate (MVAR) model by summing the effect of other neurons in the population in the second and third terms of Eq. (5.4) (Kamiski et al., 2001). Extending the time domain model in that equation to the frequency domain yields the partial directed coherence (PDC) and directed transfer function (DTF) approaches (Franaszczuk et al., 1994; Sameshima and Baccala, 1999). The main advantage is the abolition of the time domain models' dependence on a specific bin size, under the assumption of stationarity within the analysis interval.

Granger causality, whether computed in the time domain or the frequency domain, assumes linear interactions by virtue of the auto-regressive model structure. This might lead to erroneous conclusions when nonlinear interactions occur, particularly in higher-level association cortices (Lin et al., 2006). For example, auditory neurons have been consistently found to perform coincidence detection of their dendritic inputs (Agmon-Snir et al., 1998; Peña et al., 2001), while posterior parietal cortex neurons perform gain modulation to guide motor behavior in response to visual stimuli (Salinas and Their, 2000; Cohen and Andersen, 2002). It has also been reported that Granger causality cannot detect inhibitory connections with the same accuracy as excitatory ones (Dahlhaus et al., 1997; Cadotte et al., 2008), and cannot discriminate between mono- and poly-synaptic connections.

5.2.4 **Generalized Linear Models**

The generalized linear model (GLM) is a generative model in wide use in many statistical problems. In the context of modeling population activity, GLM models the output of each neuron in terms of a conditional intensity function, $\lambda(t|F(t))$ of a stochastic point process, where $F(t)$ models all of the factors that influence neuron output (Okatan et al., 2005; Truccolo et al., 2005). For inhomogeneous Poisson processes, the conditional intensity of neuron i is expressed as

$$\lambda_i\left(t\,|F_i(t)\right) = \exp\left(\beta_i + \mathbf{k}_i \cdot \boldsymbol{\xi}^h + \sum_{j \in \pi_i} \boldsymbol{\alpha}_{ij} \cdot \mathbf{S}_j^h\right), \qquad (5.5)$$

where $F_i(t)$ is a linear sum of the neuron's background rate β_i, an extrinsic covariate $\boldsymbol{\xi}^h$ that the neuron independently encodes, and the firing history $\mathbf{S}_j^h$ of the pre-synaptic neurons forming the set π_i prior to time t (inclusive of neuron i to model self-inhibition caused by refractoriness). The parameters $\mathbf{k}_i$ and $\boldsymbol{\alpha}_{ij}$ model the neuron's receptive field and synaptic coupling with neuron j, respectively. While the GLM has been used as a generative model and not as a method to infer functional connectivity, optimizing its parameters to fit the observed data involves calculating the coupling parameters $\boldsymbol{\alpha}_{ij}$ and therefore it can be viewed as a tool to identify functional connectivity. Typically, $\boldsymbol{\alpha}_{ij}$ are calculated using maximum likelihood estimation techniques (Paninski et al., 2004).

GLMs have been successfully shown to fit data *in vivo* from the primary motor cortex during goal-directed 2D arm reach movements (Truccolo et al., 2005), as well as hippocampus place cell firing activity of rats in spatial navigation tasks (Barbieri et al., 2005). They have also been shown to fit population data from *in vitro* neural cultures such as the primate retinal ganglion cells (RGCs) (Pillow et al., 2008), a structure with well-characterized anatomical connectivity. Figure 5.2 shows a block diagram of the model for a pair of RGC neurons recorded in the study by Pillow et al. (2008), where an example of the estimated parameters for one ON cell is illustrated, including the estimated synaptic coupling parameters with the neighboring ON and OFF cells.

A critical assumption in the GLM is that the dependence of the firing rate on the external covariates is implicitly *linear*, with the possible inclusion of a nonlinearity, such as the exponential function in the Poisson case. There are many prime differences, however, between the structure of cortical networks observed *in vivo* and that of neuronal cultures observed *in vitro* such as the well-characterized spatial layout of the RGC system. More specifically, the linearity property may not always hold in many cortical areas where nonlinear mechanisms and non-Poisson output characteristics tend to predominate (Barbieri et al., 2001; Fujisawa et al., 2008; Cohen and Newsome, 2009; Dombeck et al., 2009). For example, neuronal responses in prefrontal areas have been shown to represent *discrete*—possibly binary—values (rather

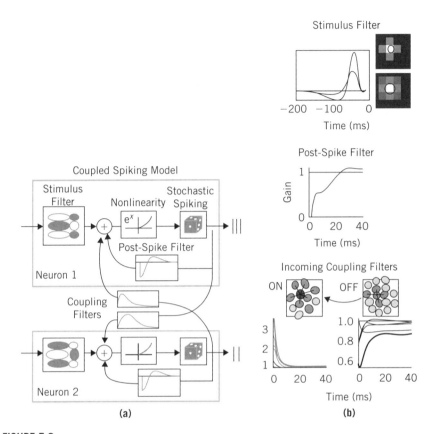

FIGURE 5.2

(a) GLM of a 2-neuron population used to fit spike train data from stimulus-driven retinal ganglion cell (RGC) responses *in vitro*. Each neuron input is processed by a stimulus filter $\mathbf{k}_i$, a post-spike filter $\boldsymbol{\alpha}_{ii}$, and coupling filters that capture other neurons' dependencies, $\boldsymbol{\alpha}_{ij}$. (b) Parameters of an example retinal ganglion ON cell model: temporal and spatial components of center and surround filter components, post-spike filter, and coupling filters from neighboring OFF and ON cells.
Source: Adapted from Pillow et al. (2008).

than a *continuum*) reflecting possible "decision-making" mechanisms reminiscent of sensorimotor integration (e.g., Gold and Shadlen, 2007). Many complex brain functions such as attention (Shadlen and Newsome, 1998; Paninski et al., 2007; Cohen and Maunsell, 2009) and motion perception are thought to be mediated by highly heterogeneous and nonlinear firing patterns of neurons in other areas such as area MT (Callaway, 1998; Yoshimura and Callaway, 2005; Churchland and Shenoy, 2007; Cohen and Newsome, 2009). Most important, the GLM assumes that the neuron responds to a covariate $\boldsymbol{\xi}^h$ that is known or can be explicitly measured. However, in many cortical areas (e.g., premotor), the intrinsic driving input to the observed population may not be explicitly measured, and therefore the exact functional form of the encoding mechanism—the mathematical relationship between extrinsic covariates, ξ^h's, and individual neurons' outputs, λ—remains a subject of intense debate (Moran and Schwartz, 1999; Carmena et al., 2003; Paninski et al., 2004; Wu et al., 2004).

From a computational standpoint, GLM fitting—like most maximum likelihood optimization techniques—is computationally intense and may tend to overfit the data because of the large number of parameters to be estimated. One potential remedy is Bayesian regularization techniques that attempt to maximize the posterior density instead of the likelihood, assuming a priori beliefs about receptive field parameters. Another remedy is to assume sparse connectivity and use known anatomical properties of the populations being measured, or to assume a specific functional form for the coupling between the neurons (Truccolo et al., 2005; Pillow et al., 2008), effectively reducing the number of parameters. Recent experimental evidence from selected brain areas seems to support this approach (Song et al., 2005; Histed et al., 2009).

5.3 GRAPHICAL MODELS

Graphical models are considered an elegant blend of probability and graph theory for capturing complex dependencies between random variables. They are a cornerstone of statistical inference and have been widely used in mathematics and machine learning applications such as bioinformatics, communication theory, statistical physics, signal and image processing, and information retrieval (Wainwright and Jordan, 2008). Their use in modeling neuroscience data has been surprisingly limited, but it is gaining strong attention in the computational and systems neuroscience communities.

A graphical model uses a *vertex* (or a node) to represent a neuron's state during a specific time interval, and an *edge* to connect the neuron to other neurons (vertices) in the observed population. Inferring graphs with undirected edges is therefore equivalent to discovering the functional connectivity between the neurons, whereas inferring graphs with directed edges is equivalent to finding their effective connectivity.

5.3.1 Effective Connectivity

Inferring causal links—that is, effective connectivity—between neurons is a challenge because an effect may have many causes, some of which may be unobservable, leading to the possibility of inferring spurious connections between neurons. For example, when neuron A excites neuron B and neuron B excites neuron C, A and B will have significant crosscorrelogram peaks that indicate a correctly inferred direct coupling between them. However, a crosscorrelogram of A and C will exhibit significant peaks that suggest A and C are coupled when in fact their coupling is mediated by B. This spurious coupling can be *explained away* by considering the effect of *both* A and B on C. The concept of "explaining away" (Wellman and Henrion, 1993) is a cornerstone in graphical models, particularly Bayesian networks (Pearl and Shafer, 1988), as it considers all possible individual and compound causes of the observed effect. Similar argument can be made when A excites both B and C, in which case explaining away the spurious coupling between B and C is achieved by considering the interaction between A and B as well as that between A and C.

In the case of time-varying stochastic processes with potentially many causes such as spike trains, Bayesian analysis provides a powerful framework because it treats all model quantities (observations, parameters, latent variables, etc.) as random variables causal links between these variables—now explicitly appearing as vertices in the graph—can therefore be represented as directed edges. In such a case, dynamic Bayesian networks (DBNs) (Eldawlatly et al., 2010) are well suited to inferring graphical models of spiking neurons because they explicitly model time dependencies between the variables (Murphy and Mian, 1999; Murphy, 2002). They have been used in a number of applications, for example, to infer transcriptional regulatory networks from gene expression data (Bernard and Hartemink, 2005; Dojer et al., 2006; Geier et al., 2007; Needham et al., 2007) and to infer connectivity between multiple brain areas from fMRI data (Zhang et al., 2006; Rajapakse and Zhou, 2007). They have also been shown to infer information transfer using coarse brain activity throughout the songbird auditory pathway and have agreed with anatomical findings (Smith et al., 2006).

A DBN represents sequential data in the form of a directed acyclic graph (DAG) (Heckerman, 1995; Pearl, 1988; Murphy, 2002). In the context of spiking cortical networks, a DBN representing spike train matrix $\mathbf{S}$ from N neurons is defined by $\mathbf{B} = <\mathbf{G}, \mathbf{P}>$, where $\mathbf{G}$ is a DAG and $\mathbf{P}$ is a set of conditional probabilities between the spike trains. Each graph $\mathbf{G}$ consists of a set of vertices $\mathbf{V}$ and edges $\mathbf{E}$. Each vertex in $\mathbf{V}$, denoted by $v_i^{(t)}$, corresponds to the state of neuron i at time t, denoted by $S_i(t)$. Each directed edge in $\mathbf{E}$, denoted by $v_i^{(t-\tau)} \rightarrow v_j^{(t)}$, indicates that $S_i(t-\tau)$ causes $S_j(t)$. Graphically, this means that $S_i(t-\tau)$ is a parent vertex of $S_j(t)$. When computing their joint density, the conditional dependence between the two variables is assumed to be time-invariant.

In a DBN, the firing state of any postsynaptic neuron i at time t, $S_i(t)$ is determined only by its parents' history, denoted by $\mathbf{S}_{\pi_i}(1:t-1)$, where π_i is the set of putative pre-synaptic neurons to neuron i and is independent of the status of any other neuron in the observed population. Thus, the joint probability $p(S_i(t),\ldots,S_N(t))$ is expressed in terms of the conditional probabilities

$$p(S_1(t),S_2(t),\ldots,S_N(t)) = \prod_{i=1}^{N} p\big(S_i(t)\big|\mathbf{S}_{\pi_i}(1:t-1)\big). \qquad (5.6)$$

In practice, the parents' history is considered up to a maximum Markov lag $\tau_{\max}$. Thus, the joint density can be rewritten as

$$p(S_1(t),S_2(t),\ldots,S_N(t)) = \prod_{i=1}^{N} p\big(S_i(t)\big|\mathbf{S}_{\pi_i}(t-\tau_{\max}:t-1)\big). \qquad (5.7)$$

Modeling time dependency allows DBN to capture relationships that appear at different time lags. This is useful when considerable latency between causal events results from aggregated synaptic delays. It also makes DBN capable of modeling *cycles* or *feedback loops*. Any loop that exists in the structure of the network is *unrolled* in the DBN representation, as illustrated in Figure 5.3.

Learning the structure of the DBN is the most important step because it permits inferring the network model that best fits the observed spike train

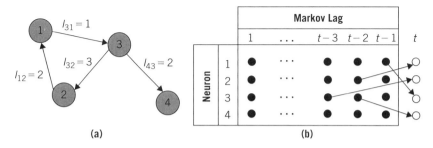

(a)

(b)

FIGURE 5.3

(a) Simple causal network where nodes represent neurons and edges represent connections. l_{ij} indicates the synaptic latency in units of time. (b) Corresponding DBN where black nodes represent the neuron firing state at a specific Markov lag and white nodes represent the present state.

data **D**. More precisely, this is expressed as

$$\mathbf{G}^* = \arg\max_{\mathbf{G}} (\log p(\mathbf{D}|\mathbf{G}) + \log p(\mathbf{G})) \qquad (5.8)$$

where $p(\mathbf{D}|\mathbf{G})$ is the likelihood of the data **D** given the structure **G**, and $p(\mathbf{G})$ is the prior probability of **G**. Structure learning transforms the inference problem from a maximum likelihood (ML) estimation to a maximum a posteriori (MAP) estimation by including a *prior* penalty term. If no prior information is available about the distribution of the expected graph, $p(\mathbf{G})$ is assumed to be uniform. Learning the structure of **G** is much more difficult compared to learning the parameters because once the structure is known, it is easy to learn the reduced set of parameters using maximum likelihood techniques (Stevenson et al., 2009).

There are many structure-learning algorithms. The most popular are score-based (Heckerman, 1995; Friedman et al., 1999; Hartemink et al., 2001). Figure 5.4 outlines the steps in a score-based approach. A scoring function is first defined by which a given structure is evaluated on a given data set, and then a search is performed through the space of all possible structures to find the one with the highest score. At each step of the search, only one modification is permitted in the structure that involves adding, removing, or reversing an edge. Score-based approaches are typically based on well-established statistical principles such as minimum description length (MDL) (Lam and Bacchus, 1994), Bayesian information criterion (BIC) (Schwarz, 1978), and Bayesian

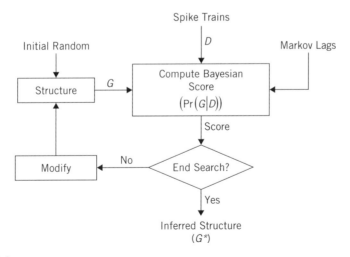

FIGURE 5.4

Block diagram of the DBN structure search algorithm.

Dirichlet equivalent (BDe) score (Heckerman et al., 1995). For instance, BIC approximates $p(\mathbf{D}|\mathbf{G})$ as

$$\log p(\mathbf{D}|\mathbf{G}) \approx \log p\left(\mathbf{D}\Big|\hat{\mathbf{P}},\mathbf{G}\right) - \frac{d}{2}\log Y, \qquad (5.9)$$

where $\hat{\mathbf{P}}$ is the maximum likelihood estimate of the parameters $\mathbf{P}$ given the structure, d is the number of parameters in $\mathbf{P}$, and Y is the number of samples. The first term measures how well the model fits the data, while the second penalizes the model's complexity. The search can be a greedy search or a simulated annealing search. The latter performs better as it avoids falling in local maxima (Kirkpatrick et al., 1983). Using the joint probability of the spike trains in the computation of the Bayesian score for structure inference ensures that the effect of the entire observed population is taken into account when searching for the best graph. This is a major advantage over pairwise methods such as the crosscorrelogram.

Graphical models such as DBNs provide a general framework for many well-known statistical models. For example, hidden Markov models (HMM) are a special case of DBNs where the state of the world is represented by a single discrete random variable that transits between K possible values, as opposed to a set of random variables in a factored and distributed graph in DBNs. As a result, the latter may have exponentially fewer parameters— thereby speeding up the inference process—than its corresponding HMM. State space models such as the Kalman filter and its many derivatives are also special cases of DBNs where the model is linear and Gaussian. GLMs are considered log linear equivalent models when the joint density belongs to the class of exponential families and the noise is non-Gaussian (McCulloch, 2000).

5.3.2 Graph-Based Functional Connectivity Inference

In addition to capturing multimodal, nonlinear, and non-Gaussian charac- teristics of joint population density, graphical representation of neuronal interactions is intuitive because it can be visualized easily and it allows the use of many established graph metrics to quantify certain features of the inferred graph that can be compared to imaging and anatomical neural data. For exam- ple, one may use single-vertex metrics such as the number of incoming and outgoing connections at each vertex, or pairwise vertex metrics such as the *path length* or the number of intermediate vertices between any pair of ver- tices, or metrics that quantify the entire graph structure such as the *clustering index* (Sporns et al., 2007).

Modeling temporal dependencies using graphical models such as DBNs, however, comes at a considerable computational cost. Vertices are arranged into columns that represent a particular time frame (as shown in Figure 5.3) because of the need to conduct the search in a high-dimensional graph space to find the best structure that explains the data. It is highly desirable to reduce the search space by eliminating unnecessary edges before effective connectivity is inferred. This can be done by computing some measure of *similarity* between spike trains and using it as undirected edges to connect the vertices in the graph. Pairwise Pearson correlation, mutual information, or any other suitable metric can serve this purpose. Another advantage of this approach is that, by restricting the number of vertices per neuron, the *state* of the interaction with other neurons can be time scale–independent, thereby reducing the effect of nonstationarity in the interaction between neurons that occurs over a multitude of time scales on the accuracy (Shadlen and Newsome, 1998; Averbeck et al., 2006; Schneidman, 2006).

The first step is to project the spike trains onto a "scale space." This space extracts information about changes in the firing pattern that are localized in both time and frequency. This is achieved by a wavelet transformation of each spike train using a Haar rectangular basis (Mallat, 1998). More precisely, the spike train at any given time scale k is expressed as

$$S_i^k(t) = W^{(k)} S_i(t) \qquad k = 0, 1, \ldots, K \qquad (5.10)$$

where $W^{(k)}$ is a lumped matrix representing the transform of size $2^{\log_2(L)-k+1} \times L$, L is the spike train length in bins, and K is the total number of time scales examined. Figure 5.5(a) outlines the steps of the algorithm.

The wavelet transformation extracts two important features from the spike train as shown in Figure 5.5(b). First, it extracts the probability of firing at time scale k indicated by the sequence $2^{-k/2} a_i^k(t)$ along the transform approximation path (left branch). This is proportional to the firing rate in bins of size 2^k. Second, it extracts the relative *change* in the probability of firing across successive bins indicated by the sequence $2^{-k/2} d_i^k(t)$ along the details path (right branch). Positive/negative values indicate a decrease/increase, respectively, in the probability of firing relative to the previous bin. Using this representation, small time scales account for fast-rising synaptic coupling while large time scales account for slow-decaying synapses. This information is important for determining the synchronization properties of neurons, ranging from weakly coupled oscillation states to persistent excitation states in recurrent networks when inputs are absent (Ermentrout, 2003).

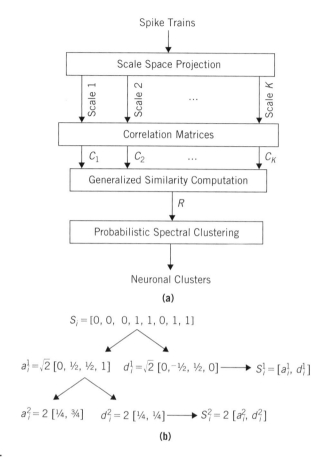

FIGURE 5.5

(a) Block diagram of the multiscale clustering algorithm for inferring functional connectivity. (b) Two-level scale space representation (using the Haar wavelet basis) of a sample spike train. The left branch represents the approximation path; the right branch represents the details path. At any time scale k, the wavelet coefficients of both the approximation and the detail are concatenated together to form S_i^k.

In the kth time scale, a similarity matrix is computed from the projections. In our case, the similarity measure is computed as the Pearson correlation expressed as

$$\sigma_{ij}^k = E[S_i^k S_j^k] \tag{5.11}$$

The full correlation matrix at time scale k, $\sum^{(k)} \in \mathscr{R}^{N \times N}$, is formed using Eq. (5.11) for all $i, j \in \{1, \dots, N\}$. Other measures of similarity, such as mutual information, can be used as well. The similarities computed across all time scales are fused by simultaneously diagonalizing the similarity matrices $\sum^{(k)}$ computed from Eq. (5.11) for $k = 0, 1, \ldots, K$. Specifically, this is achieved by forming a block diagonal matrix $\mathbf{R} \in \mathscr{R}^{(N \times K) \times (N \times K)}$ that has the kth block as the $N \times N_{\text{matrix}} \sum^{(k)}$. This matrix is diagonalized using singular value decomposition (SVD) as

$$
\mathbf{R} = \begin{bmatrix} \sum^{(0)} & & & \\ & \sum^{(1)} & & \\ & & \ddots & \\ & & & \sum^{(K)} \end{bmatrix} \cong \sum_{q=1}^{Q} \lambda_q \mathbf{u}_q \mathbf{u}_q^T \tag{5.12}
$$

where $\lambda_q / \mathbf{u}_q$ denote the eigenvalue/eigenvector pair associated with the qth dominant mode of the block diagonal matrix $\mathbf{R}$. The $N \times N$ similarity matrix is then formed by first vectorizing the K matrices $\sum^{(k)}$ to $N^2 \times 1$ vectors, concatenating the K vectors obtained in a matrix of dimension $N^2 \times K$, performing SVD on the resulting matrix and keeping the first Q modes in a matrix of size $N^2 \times Q$, and finally reshaping the $N^2 \times Q$ matrix to Q matrices of size $N \times N$. The *generalized similarity* matrix is formed by summing the Q principal matrices, weighted by the corresponding eigenvalues λ_q. Simultaneous diagonalization/fusion tells us how spike trains are similar in a *scale-free* space. This has the advantage of identifying functional connectivity independent of the length of the history interval governing the interaction between any neuron pair. Once generalized similarity measures have been computed, each neuron is represented as an *object* in a graph and connected by an edge whose weight w_{ij} equals the $(i, j)^{\text{th}}$ entry of the generalized similarity matrix. The outcome of this step is an undirected graph similar to the example shown in Figure 5.6.

The clustering of similar objects in the undirected graph can be done in many ways. Spectral clustering is one approach to identifying the clusters (Chung, 1994; Ding et al., 2001; Jin et al., 2005) and has been shown to perform better than some well-known clustering algorithms (Ng et al., 2001). It views

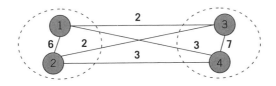

FIGURE 5.6

Undirected graph and its matrix representation as input to the spectral clustering algorithm. The dashed circles indicate the output of the clustering algorithm.

the problem as one of graph partitioning in which a *minimum cut* algorithm is applied to identify clusters with strongly similar objects. Specifically, if a_{ic} denotes the probability of the ith neuron being in the cth cluster, the minimum cut problem is solved by finding the set of probabilities $\{a_{ic}\}$ that maximize the following objective function (Jin et al., 2005):

$$O = \sum_{c=1}^{C} \frac{\sum_{i=1}^{N}\sum_{j=1}^{N} a_{ic}a_{jc}w_{ij}}{\sum_{i=1}^{N}\sum_{j=1}^{N} a_{ic}w_{ij}} \tag{5.13}$$

The final cluster memberships are derived from $c_j^* = \underset{c\in[1,...,C]}{\arg\max}\, a_{ic}$. Referring to the example in Figure 5.6, only edges of small weights are removed with this approach. The probabilistic spectral clustering algorithm overcomes many limitations when clustering data with arbitrary structures, since it is nonparametric and therefore can track a wide variety of cluster shapes. The probabilistic nature of the algorithm is useful in quantifying the degree of uncertainty in the membership of a given neuron in each cluster.

The clustering algorithm is useful in reducing the dimension of the neural space to a set of subspaces where statistical dependency is significant between neurons *within* a cluster and insignificant *across* clusters. Therefore, the two-step approach illustrated in Figure 5.7 constitutes an integrated framework for dealing with large-scale systems.

5.4 RESULTS

We tested the methods just discussed in analyzing both synthetic spike data and experimental spike data collected from awake, behaving rats in working memory tasks. Here we detail our most relevant findings.

5.4.1 Spiking Neural Model

We used a general formulation of the GLM to generate synthetic spike data. Specifically, the output of neuron i, $\lambda_i(t)$, was expressed as

$$\lambda_i\left(t|H_i(t),\xi(t),\theta_i\right) = f\left(\theta_i^b, h\left(\theta_i^h, H_i(t)\right), g\left(\theta_i^\xi, \xi(t)\right)\right), \tag{5.14}$$

This formulation includes the following set of parameters: $\theta_i = \{\theta_i^b, \theta_i^h, \theta_i^\xi\}$ is a vector in which θ_i^b expresses the neuron's background activity level, θ_i^h

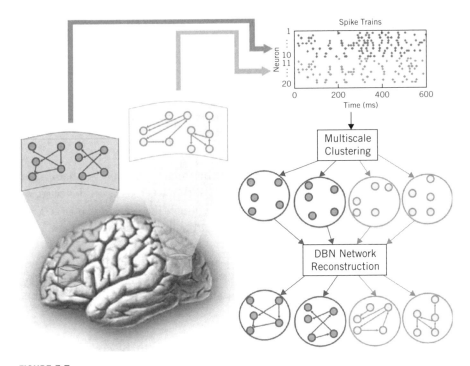

FIGURE 5.7

Schematic of the two-stage framework for inferring neuronal connectivity. Multiscale clustering infers functional connectivity and reduces the dimension of the neural space to a set of smaller subspaces (clusters). The dynamic Bayesian network (DBN) infers the effective connectivity within each subspace.

models the network interaction parameters, which are represented by α_{ij} in the GLM (Eq. (5.14)), and θ_i^{ξ} expresses the neuron's receptive field. The term $H_i(t)$ expresses the spiking history of the neuron itself and that of other neurons connected to it, while $\xi(t)$ expresses the external covariate (stimulus). One main difference is that the function $h(.)$ expresses a network term that models a time-varying relationship between the neuron and all other observable neurons that influence its firing. The model is dynamic in that the firing history term $H_i(t)$ is allowed to modify the coupling between the neurons θ_i^h. This represents a substantial difference from the static GLM, and it is more consistent with a large body of experimental studies on STDP in local cortical circuits (Song and Abbott, 2001; Caporale and Dan, 2008). The function $g(.)$ models the relationship between the external covariate and the receptive field of the neuron. The GLM in Eq. (5.5) is a special case of this form in which $f(.)$ is an exponential function, $h(.)$ is a binomial distribution with no dependence of α_{ij} on $H_i(t)$, and $g(.)$ is an inner product.

According to Eq. (5.5), the history term in the GLM takes the form $H_i(t) = \{S_j(t - M_{ij} : t)\}_{j \in \pi_i}$:

$$p(S_i(t) = 1) \cong \lambda_i(t \mid H_i(t), \xi(t), \theta_i)\, \Delta$$

$$= \exp\left(\beta_i + \mathbf{k}_i.\xi^h + \sum_{j \in \pi_i} \sum_{m=0}^{M_{ij}} \alpha_{ij}(m\Delta) S_j(t - m\Delta)\right) \Delta, \quad (5.15)$$

where M_{ij} is the number of history bins that relate the firing probability of neuron i at time t to activity from neuron j at time $(t - m\Delta)$. In this case, α_{ij} represents weight functions governing how much the occurrence of a spike from neuron j, going back M_{ij} bins in time, influences the probability of firing of neuron i at time t.

To mimic the influence of excitatory post-synaptic potential (EPSP) and inhibitory post-synaptic potential (IPSP), α_{ij} can be modeled as a decaying exponential function (Kuhlmann et al., 2002; Zhang and Carney, 2005; Sprekeler et al., 2007):

$$\alpha_{ij}^{\pm}(t) \begin{cases} 0, & \text{if } t < l_{ij}\Delta \\ \pm A_{ij}(t)\exp\left(-3000(t - l_{ij}\Delta)/M_{ij}\right), & \text{if } t \geq l_{ij}\Delta \end{cases} \quad (5.16)$$

where $+/-$ indicate excitatory/inhibitory interactions, $A_{ij}(t)$ models the time-dependent strength to account for plastic changes, and l_{ij} models the synaptic latency (in bins) associated with that connection.

Accounting for the external stimulus contribution using the explicit term as in Eq. (5.15) might be appropriate for early-stage sensory neurons that are directly *tuned* to the stimulus. Neurons in high-level association cortices, however, respond to the input stimulus in a much more complex and indirect manner. They are believed to dynamically shape their response characteristics using their connectivity to low-level sensory neurons through multiple, interleaved pathways with numerous feedback loops at different processing levels in the brain. An alternative way to generically model these complex interactions is

$$\lambda_i(t \mid H_i(\xi(t)), \theta_i) = f\left(\theta_i^b, h\left(\theta_i^h, H_i(\xi(t))\right)\right), \quad (5.17)$$

where the covariate $\xi(t)$ is now implicitly represented in the network term $h(.)$. This formulation attempts to generalize the relationship between the neuron's tuning characteristics and the external covariate to an arbitrary form. The functional relationship between the neuron's receptive field and the external covariate that it presumably encodes is therefore implicit. This leads to the following model of neural firing used in our simulations:

$$p(S_i(t) = 1) \cong \lambda_i (t \, | H_i(\xi(t)), \theta_i) \, \Delta = \left(\exp \left(\beta_i + \sum_{j \in \pi_i} \sum_{m=0}^{M_{ij}} \alpha_{ij}(m\Delta) S_j(t - m\Delta) \right) \right) \Delta,$$

(5.18)

This formulation may help address important questions regarding the putative role of correlation in encoding external covariates (Aertsen et al., 1989; Nirenberg and Latham, 2003; Averbeck et al., 2006; Pillow et al., 2008). Many experimental findings suggest that correlations are time scale–dependent (Kohn and Smith, 2005; Averbeck et al., 2006; Biederlack et al., 2006; Chacron and Bastian, 2008) and vary with stimuli (Gray et al., 1989; deCharms and Merzenich, 1996; Samonds et al., 2003; Kohn and Smith, 2005; Biederlack et al., 2006; Rocha et al., 2007; Chacron and Bastian, 2008). This leads to a more consistent formulation with the notion of *cell assemblies*, originally proposed by Hebb (1949), to underlie the filtering properties of the neural circuits in response to an external covariate.

5.4.2 Inferring Effective Connectivity

DBN inference accuracy can be quantified using the F-measure, defined as the harmonic mean of two quantities: recall and precision (Rijsbergen, 1979):

$$\text{Recall} = \frac{n}{n + \mu}, \quad \text{Precision} = \frac{n}{n + w}$$

$$F = \frac{2 \times \text{Recall} \times \text{Precision}}{\text{Recall} + \text{Precision}} = \frac{2n}{2n + \mu + w},$$

(5.19)

where n is the number of correctly inferred connections, μ is the number of missed connections, and w is the number of erroneously inferred connections. Thus, F is 0 if and only if $n = 0$ and 1 if and only if $\mu = 0$ and $w = 0$.

DBN performance was examined by varying the parameters of the model in Eq. (5.15). Initially, the synaptic latency l_{ij} for all neurons was fixed at 3 ms. DBN achieved more than 87% accuracy when applied to networks with a variable number of pre-synaptic connections for each neuron while fixing all other parameters as shown in Figure 5.8(a). It also achieved similar performance when applied to networks with variable history intervals M_{ij}, as illustrated in Figure 5.8(b), where all the interactions in a given population had the same M_{ij}. In both cases, the DBN Markov lag was set to match the population synaptic latency.

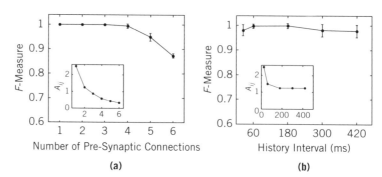

(a)   **(b)**

FIGURE 5.8

(a) DBN performance versus number of excitatory pre-synaptic connections per neuron; inset: connection strengths $A_{ij} = |\alpha_{ij}|$ for each choice of that number. A_{ij} was decreased as the number of pre-synaptic connections increased to keep the post-synaptic neuron firing rate steady and to avoid unstable network dynamics. (b) DBN performance versus firing history interval; inset: connection strengths $A_{ij} = |\alpha_{ij}|$ for each choice of this length that was decreased gradually as the history interval increased for stable dynamics.

In practice, the synaptic latency between any two neurons is unknown. We investigated performance when there is a mismatch between the DBN Markov lag (set by the user) and the true synaptic latency in the model. The synaptic weight functions shown in Figure 5.9(a) were used to mimic the EPSP/IPSP characteristic of the connections for different choices of M_{ij}. Given that, for relatively longer history intervals, the effect of a pre-synaptic spike on the firing probability of a post-synaptic neuron lasts longer, DBN can infer these connections even if the Markov lag is chosen to be larger than the true synaptic latency. On the other hand, we expected the performance to diminish for shorter history intervals. Figure 5.9(b) illustrates that when the Markov lag is set to a value *smaller* than the populations' synaptic latency, almost none of the connections were inferred. DBN performance was nearly perfect (close to unity) when the Markov lag matched the synaptic latency. The inference accuracy deteriorated slightly when the Markov lag was set larger than the true synaptic latency. The rate of deterioration was proportional to the history interval of the interactions—the larger the length of that interval, the more diminished the observed performance.

Functional cortical networks are more likely to have heterogeneous synaptic latencies that reflect their spatially diffuse structures and their local feedback loops (Averbeck et al., 2006; Brown and Hestrin, 2009). We used a *heterogeneity index* (HI), defined as the number of distinct synaptic latencies that exist in the network. Figure 5.9(c) demonstrates the accuracy for networks

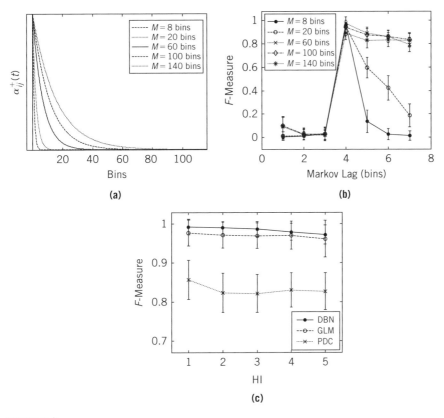

FIGURE 5.9

(a) Coupling function $\alpha_{ij}^{+}(t)$ for different history intervals M. (b) DBN performance versus Markov lag for variable history intervals and fixed synaptic latency of four bins. (c) DBN-, PDC-, and GLM-fit performance versus HI.

with different HIs. Each neuron received two random connections, one excitatory and one inhibitory. DBN accuracy was not significantly affected at high HI, as shown in Figure 5.9(c). DBN performance was also compared to two methods (reviewed in Section 5.2) that identify causal connections: namely, PDC, the frequency domain equivalent of Granger causality, and GLM fit. As can be seen, DBN outperformed PDC at all HIs. Careful examination of the networks inferred by PDC revealed that the major cause of diminished performance is the inability to detect most of the inhibitory connections. For the GLM fit, DBN performance was slightly superior. GLM performance was not surprising, given that it was tested on data obtained from the underlying generative model.

5.4.2.1 *Deciphering Mono-Synaptic Connectivity*

A main advantage of graphical models in structure learning is their ability to explain away spurious relationships between variables when the effect of the entire population is considered. This is most pronounced in the chain example illustrated in Figure 5.10(a). In this example, neurons A and B are the monosynaptic parents of neurons B and C, respectively. For simplicity, the synaptic latency of both connections was set to 1 bin. Based on this setting, a spurious monosynaptic connection between neurons A and C might be inferred at a 2-bin Markov lag. We assessed performance using 100 different network structures, 10 neurons each, with a fixed synaptic latency of 1 bin, and 2 presynaptic connections per neuron. Each of the 100 networks examined had a total of 40 chains with 3 neurons each.

Figure 5.10(b) shows that the number of spurious poly-synaptic connections per chain as a function of the Markov lag was insignificant when a Markov lag of 1 bin was used. This is not surprising given that the lag was shorter than the *superposition* of two synaptic latencies. Most interesting, the error was very small at a Markov lag of 2 bins, suggesting that the DBN tended to find the network structure with the least number of edges. Therefore, it would consider poly-synaptic connections $A \rightarrow C$ as redundant and eliminate them during the search, given that two monosynaptic connections had already been inferred. At a Markov lag range of 1 to 2, no polysynaptic connections were inferred since in this case the DBN found the lag (or a combination of lags) that best explained the data. This suggests that even if the

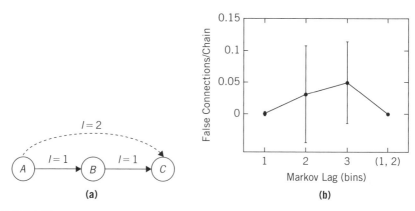

(a)   **(b)**

FIGURE 5.10

(a) Three-neuron chain, where the solid arrows represent true mono-synaptic connections and the dotted arrow represents a spurious second-order connection; l indicates the synaptic latency associated with each connection in bins. (b) Average number of spurious connections inferred per chain.

Markov lag were not correctly chosen, DBN could still restrict its inference to *true* monosynaptic connections provided that the lag range was chosen equal to or more than the maximum anticipated synaptic latency in the population (Eldawlatly et al., 2010).

5.4.2.2 *Unobserved Common Input and Scalability*

We further examined the performance of DBN in the case of unobserved inputs within the population. We randomly selected 6 neurons to be unobserved from 20-neuron populations. DBN was applied to the spike trains of the remaining 14 neurons. The inference accuracy in this case was 0.96 ± 0.03. We also looked at performance when some of the *observed* neurons were not elements of the functional network. These could be thought of as task-independent neurons such that their inclusion would be regarded as "noise." The inference accuracy obtained by applying the DBN to 100 networks of this type was 0.98 ± 0.07, which strongly indicates the ability of the method to tell true connections from spurious ones, as well as to identify the neural subspace where task information may reside.

DBN scalability was then assessed using larger populations. We examined performance with populations of 120 neurons partitioned into 12 equal clusters. Each neuron received 3 excitatory pre-synaptic connections from neurons belonging only to its own cluster. As illustrated in Figure 5.11, by limiting the search time to *one* minute (on a 64-bit, 2.33 GHz Dual Intel Xeon machine with 8 GB of memory), DBN was unable to infer the structure of these populations. However, its performance significantly improved when it was applied to arbitrarily selected subpopulations of 20 neurons, eventually reaching 100% accuracy with a fixed search time of one minute. This suggests that breaking large populations into smaller subpopulations facilitates

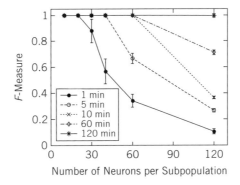

FIGURE 5.11

DBN performance versus number of neurons per subpopulation.

identification of connectivity with relatively short search times. Alternatively, when the search time was increased to two hours, as illustrated in Figure 5.11, DBN performance reached almost 100% without the need to subdivide the population into smaller clusters.

5.4.2.3 *Choosing the Markov Lag*

As noted earlier, DBN performance is highly dependent on the selection of the Markov lag; thus, some measure is needed to estimate this parameter in real data analysis. For each inferred connection, we computed the mean *influence score* (IS), defined as the average absolute value of the influence of a pre-synaptic neuron on a post-synaptic neuron *independent* of the output of other pre-synaptic neurons (Yu et al., 2004). This measure is computed with the conditional probabilities used in the inference step. The average of the absolute value of the IS over all inferred connections is maximized when the inferred structure matches the true one. Figure 5.12 shows the average IS at different Markov lags for the data analyzed in Figure 5.9(b). Comparing the results in both figures, it can be seen that for each choice of history interval length, the mean IS profile matches the F-measure profile obtained when exact knowledge of the network structures is available.

5.4.2.4 *Position-Specific Cell Assemblies in the Medial Prefrontal Cortex (mPFC)*

We used DBN to infer causal relationships between simultaneously recorded neurons in the medial prefrontal cortex (mPFC) of an awake, behaving rat during a working memory task (Fujisawa et al., 2008). This brain area is

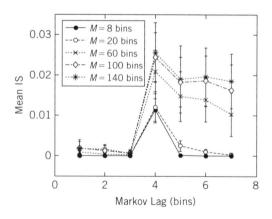

FIGURE 5.12

Mean IS versus Markov lag for variable history intervals for the same data analyzed in Figure 5.9(b). Note that the mean IS has a profile similar to that of the accuracy.

known to integrate multiple inputs during decision making to guide motor behavior. In this study, rats were trained to decide, based on an odor cue, whether to traverse the left arm or the right arm of a T-maze to be rewarded (Figure 5.13). A session consisted of 24 right-turn trials and 18 left-turn trials. Using crosscorrelogram features, recorded neurons were found to fire selectively in different regions of the maze and to exhibit some form of pairwise connectivity. These mPFC neurons were therefore argued to encode behavioral goal representation that is mediated by integrating a myriad of sensory inputs. In addition, inferring monosynaptic connectivity between some of the recorded neurons was hypothesized to form local dynamic circuits indicative of short-term plasticity.

Spike trains from 39 simultaneously recorded neurons in layer 2/3 of the mPFC were examined here for the existence of consistent networks using the graphical models discussed above. Spike trains were first down-sampled to 2000 Hz (0.5-ms bins). This permitted detecting connectivity with a synaptic

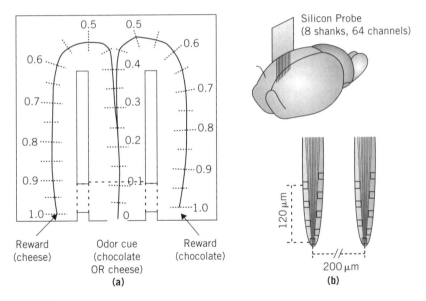

FIGURE 5.13

(a) Behavioral task: rats performed an odor-based matching-to-sample task in a T-maze. A cheese or chocolate odor cue was presented following a nose-poke in position 0, which signaled the availability of cheese or chocolate in the left or right goal area (position 1), respectively. (b) A 64 channel (8 shanks × 8 sites) silicon probe implanted in the mPFC with the dimensions shown was used to record neuronal activity. Spike trains from these neurons were obtained after spike sorting with custom-built MATLAB software.
Source: Adapted from Fujisawa et al. (2008).

latency of 0.5 ms or more. The length of each maze arm was normalized between 0 and 1 and divided into 5 intervals $(0 \to 0.2, 0.2 \to 0.4, 0.4 \to 0.6, 0.6 \to 0.8,$ and $0.8 \to 1)$. The spike train data corresponding to each interval were concatenated to form 5 position-specific data sets for each side (second row in Figure 5.14(a)). Each position-specific data set was subsequently divided into 8 smaller data sets, formed by randomly selecting 6 subtrials and grouping them (third row in Figure 5.14(a)). A jittered version of each of the 8 data sets was formed in which each spike was randomly jittered

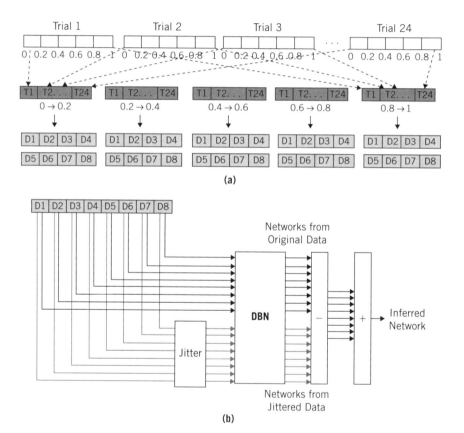

FIGURE 5.14

(a) Partitioning of spike trains to obtain position-specific data sets for 24 right-turn trials (similar analysis for 18 left-turn trials). (b) Network model fitting each data set, obtained by eliminating the common connections in networks from the original and jittered data. In the subtraction step, the network inferred from the jittered data set is subtracted from the network inferred from the corresponding original data set. The resulting networks are added together to obtain one inferred network with the frequency of occurrence of each connection as an indicator of its consistency across data sets.

around its position in a uniform range $[-5, +5]$ ms. This kept the firing rate of each neuron the same but destroyed information about time-dependent correlation across neurons. Figure 5.14(b) illustrates the steps of the analysis where connections obtained from the jittered data sets were eliminated from the networks obtained from the original data sets.

For each position-specific data set and its corresponding jittered version, we identified connections with synaptic latency in the range $[0.5, 3]$ ms with a step of 0.5 ms. Figure 5.15 illustrates the networks obtained for each maze position interval for the left and right data sets. Only connections inferred in at least 3 out of the 8 data sets examined are shown.

We then looked at the similarity between the inferred networks for different maze positions in Figure 5.16(a) using the F-measure (Eq. (5.19)). Figure 5.16(b) shows that the similarity decreases with increasing separation between maze positions, suggesting a graded transition between network states as the rat navigates through the maze. In addition, we found that networks corresponding to left-arm data sets have more connections (17.8 ± 5.3) than those corresponding to right-arm data sets (12.6 ± 7). This may be attributed to the difference in firing rate of each neuron for left and right trials, where 62% of the analyzed neurons were more active during left trials than in right trials.

5.4.3 Identifying Functional Connectivity

We analyzed spike trains obtained from the synthetic network model shown in Figure 5.17. The objective was to cluster neurons having statistically dependent spike trains with heterogeneous history intervals within each cluster. Here the true structure of the network was known, and therefore the clustering performance could be quantified as

$$\text{Clustering Accuracy} = \frac{1}{N} \sum_{c=1}^{C} r_c \tag{5.20}$$

where N is the population size, C is the total number of clusters, and r_c is the number of neurons correctly identified as members of cluster c.

Initially, the history interval length M_{ij} was set to 120 bins for all interactions and the bin width of the spike trains Δ was set to 3 ms. These choices correspond to an interaction length of $M_{ij}\Delta = 360$ ms. Figure 5.17(c) shows the similarity matrix obtained by considering time scales up to 360 ms. Neurons in the same cluster are observed to share significantly greater similarity than those across clusters. The average clustering accuracy in Figure 5.17(d)

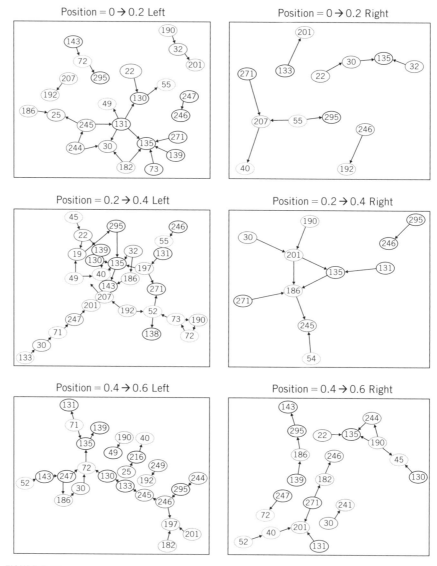

FIGURE 5.15

(*Continued*)

(solid curve) illustrates a ~96% accuracy at time scales in the range of ~200 ms to ~1.5 sec. Most important, the performance peaks around 360 ms, consistent with the history interval chosen in the generative model. The same result was observed when the history interval of interaction was varied across clusters, as shown in Figure 5.17(e).

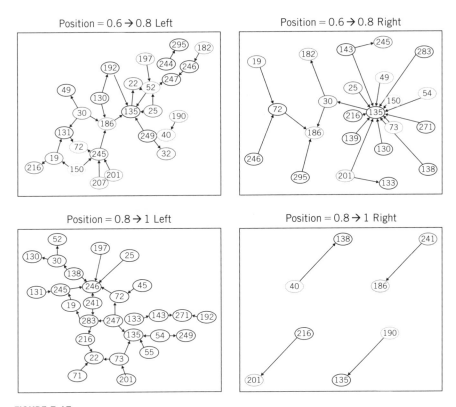

FIGURE 5.15

Networks inferred from left and right data for different maze position intervals. Only connections appearing in at least three out of the eight data sets of each position interval are shown. A clear dissimilarity between network models of the left and right trials can be seen.

The effect of each computational step in the algorithm on clustering accuracy was assessed to determine its contribution to accuracy. When the similarity matrix was computed without the scale space projection step (Eq. (5.10)), the average accuracy was only ∼51%. When the k-means algorithm was used to perform the clustering step with the inclusion of the scale space projection (Jain et al., 2000), Figure 5.17(d) shows that clustering accuracy deteriorated to 61%. We therefore concluded that the combination of both steps was important to achieving optimal performance.

Figure 5.18 demonstrates that the performance of the crosscorrelogram and the JPSTH is significantly inferior to the multiscale clustering algorithm. When variable history of interaction occurred *within* a cluster, an average accuracy of 97% was obtained.

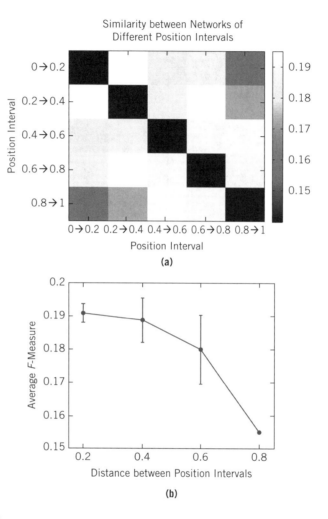

FIGURE 5.16

(a) Similarity between the inferred networks at different maze positions computed using the *F*-Measure averaged across both sides. On-diagonal entries are eliminated. (b) Average similarity between the inferred networks versus distance between maze position intervals. Values were extracted from the matrix shown in (a). It can be seen that networks become increasingly dissimilar as a function of the distance between position intervals, suggesting a differential population encoding mechanism that depends on the behavioral goal.

5.4.3.1 *Scalability*

We tested the scalability and adaptation of the algorithm when dealing with larger populations. We simulated populations of 120 neurons partitioned into four equal clusters. Each cluster was formed as a 2D grid of neurons, as illustrated in Figure 5.19(a). To mimic plastic changes in connectivity, the strength

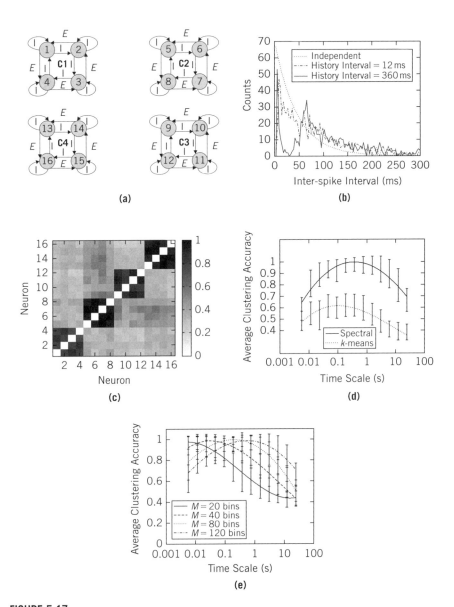

FIGURE 5.17

(a) Population of 16 neurons with 4 independent clusters; I = inhibition, E = excitation. (b) ISI histograms for an independent Poisson neuron and two dependent neurons with 12-ms and 360-ms interaction history intervals. (c) Fused correlation matrix using time scales up to 360 ms for the population in (a) with 360-ms interaction history intervals. (d) Average clustering accuracy versus time scale obtained using probabilistic spectral clustering and k-means clustering for the population in (a) with 360-ms interaction history intervals. (e) Clustering accuracy versus time scale for networks with different history intervals. Within each network, all interactions had the same history interval.

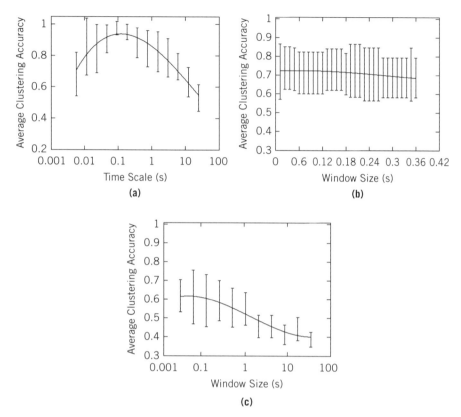

FIGURE 5.18

(a) Average clustering accuracy over all trials versus time scale for populations with variable history intervals within each cluster. (b) Average clustering accuracy using crosscorrelograms for different window sizes. (c) Average clustering accuracy using JPSTH for different window sizes.

of the connections between clusters 1 and 2 was manually varied by increasing the amplitude of the weight functions $A_{ij}(\underline{t})$ in Eq. (5.16). Figure 5.19(b) demonstrates that when *across-cluster* connections got stronger, the clustering accuracy reached its maximum when three clusters, not four, were assigned to the data. This suggests that the algorithm is capable of detecting *merging* clusters, which is particularly important when tracking plastic changes in neuronal connectivity that may be reminiscent of learning and memory formation.

5.4.3.2 *Choosing the Time Scale*
Much like choosing the Markov lag in DBN analysis, choosing the *depth* of scale space projection (i.e., the maximum level of time scale decomposition)

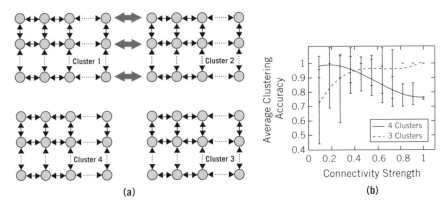

FIGURE 5.19

(a) Population of 120 neurons with 4 clusters. Bidirectional connections refer to an excitation connection in one direction and an inhibition in the opposite direction. Black connections represent *within-cluster* connections; gray connections represent *across-cluster* connections that were strengthened over time. (b) Average clustering accuracy for variable across-cluster connectivity strengths (clusters 1 and 2).

is important in the analysis. This parameter is unknown in a real data analysis situation. To overcome this limitation, a *clustering validity index*—the Dunn index (DI) (Bezdek and Pal, 1998)—was used:

$$DI = \min_{\substack{1 \le i \le N, 1 \le j \le N \\ i \ne j}} d\left(c_i, c_j\right) \Big/ \max_{1 \le k \le N} diam\left(c_k\right) \tag{5.21}$$

where $d(c_i, c_j)$ is the average distance between spike trains of pairwise neurons in clusters i and $j(i \ne j)$, $diam(c_k)$ is the maximum pairwise distance between spike trains of neurons in cluster k, and N is the total number of clusters. The distance between the spike trains of any pair of neurons i and j was computed as $1 - \mathbf{R}(i,j)$, where $\mathbf{R}$ is the augmented similarity matrix in Eq. (5.12). This index measures the relative *dispersion* in the clusters and therefore is maximized when the across-cluster distance (the numerator) is large and the within-cluster distance (the denominator) is small. Thus, it can identify sets of clusters that are compactly separated. Figure 5.20 shows the average DI as a function of the depth of the scale space projection for the data sets analyzed in Figure 5.17(e). It can be seen that for each choice of history interval M, the DI is maximized around the same time scale where clustering accuracy peaks, suggesting that dispersion of the clusters in the feature space is an adequate measure of confidence in the results.

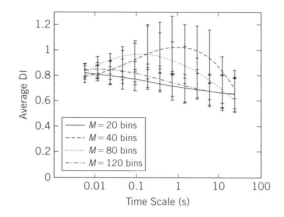

FIGURE 5.20

Average DI over all trials versus time scale for the network in Figure 5.17(a) with different history intervals.

5.5 DISCUSSION AND FUTURE DIRECTIONS

Devising methods to identify neuronal connectivity from simultaneously recorded spike trains is crucial to understanding how neurons *collectively and dynamically* respond to external stimuli or how they give rise to an observed behavior. We reviewed a number of methods that have been proposed and briefly discussed their utility as well as their limitations in analyzing population activity. In particular, we showed that model-based techniques are generally well suited for analyzing small populations of a few tens of neurons but significantly suffer from an "explosion" in the number of parameters as the population size increases, eventually overfitting the data. Non-model-based techniques, on the other hand, are suited for small population sizes but tend to fail in large populations, as they do not consider the joint influence of the constituents, making them prone to spurious inferences.

We emphasized graph theory as a way to circumvent many of the afore-mentioned limitations. We provided an elegant mix of both model-based and non-model-based methods in the context of graphical models. In particular, the multiscale clustering algorithm for inferring functional connectivity falls in the category of non-model-based techniques, as it attempts to find disjoint subspaces with arbitrary structures in the entire neural space using unlabeled data. The graphical-model-based approach exemplified in the use of DBNs then finds the intrinsic structure of the system underlying each sub-space by transforming the inference problem into a structure learning problem for which many efficient search algorithms exist. This two-step approach therefore combines the advantages of model-based and non-model-based approaches to decipher the complex connectivity patterns that may exist in large populations.

From a system identification standpoint, the multiscale clustering step helps to alleviate uncertainty about the *temporal* dynamics of the system that are governed by a time scale–dependent interaction between elements. Thus, it takes into consideration the effect of pre-synaptic cells with the hard-to-measure synaptic latencies on post-synaptic cell firing probability. The DBN, on the other hand, helps to alleviate uncertainty about the *spatial* dynamics of the system that are governed by the specific element labels. It does so by integrating immediate evidence with long-term knowledge (prior information) and uses it to *explain away* unlikely causes of the observed effects. In doing so, it searches for the best graphical model that fits the data. Taken together, both steps are essential to the capture of uncertainties about dynamic interactions between neurons in large complex networks.

Nevertheless, the methods have a number of limitations. For example, identifying effective connectivity using the DBN approach requires supervision. More work is needed to devise algorithms for estimating some parameters during the analysis such as the best time scale for the clustering step and the Markov lag for the DBN. We devised metrics such as the Dunn Index and the Mean Influence Score to suggest near-optimal values for these parameters. The metrics, however, require running the algorithms using a range of values for each parameter and then choosing the value that maximizes it. This is a computationally intense task. Future work should provide more systematic ways of estimating parameters from the data, perhaps using maximum likelihood, prior to running the algorithms.

The methods presented in this chapter were designed specifically to analyze spike train data. An important factor to consider that may affect their performance is the accuracy of spike sorting (Harris et al., 2000; Bar-Gad et al., 2001). This was discussed in detail in Chapter 2. Although we did not assess the effect of spike sorting errors on accuracy, we suspect that errors above a certain level may affect accuracy when, for example, resolving spike overlaps. A detailed study is needed to address how sensitive these methods are to spike detection and sorting deficiencies.

Another area that might directly benefit from the development of functional and effective connectivity algorithms is neural decoding. In a Bayesian sense, knowledge of the stimulus prior as well as the connectivity between observed neurons may improve the ability to decode spike trains (Barbieri et al., 2005; Pillow et al., 2005; Wu et al., 2005; Aghagolzadeh et al., 2009). One way to think about this is to consider factorizing the likelihood function—the joint neural response given the stimulus—as a product of marginal densities. Each density would express an independent neural response to the stimulus *given* its pre-synaptic inputs, which would be inferred by the connectivity algorithm. Several questions arise in this case:

- Would we expect to see distinct network connectivity patterns for different stimulus parameter values?
- How consistent would these networks be across multiple repeated trials?
- Does information about connectivity improve decoding performance?

Recently, we began to address these questions both in simulations and in experiments. In a simulation study, our results suggested that distinct and consistent network connectivity patterns may arise during goal-directed 2D arm reach movements (Aghagolzadeh et al., 2009). When these connectivity patterns were used to estimate the parameters of a Bayesian decoder, the decoding performance was significantly improved compared to the case in which information about network connectivity was absent (Aghagolzadeh et al., 2009). Experimental results by other groups using the activity of retinal ganglion cells *in vitro* to decode visual scenes agree with our findings—namely, that better scene reconstruction was obtained when information about connectivity was taken into account (Pillow et al., 2008). *In vivo*, the preliminary results of our analysis of neural population responses to whisker stimulation in the somatosensory cortex of anesthetized rats perfectly agree with these hypotheses (Eldawlatly and Oweiss, 2010).

Lastly, quantifying neural plasticity can greatly benefit from the ability to systematically identify network connectivity. For example, learning and memory formation have long been hypothesized to accompany changes in network connectivity (Schnupp and Kacelnik, 2002; Wickens et al., 2003; Gharbawie and Whishaw, 2006; Baeg et al., 2007; Dupret et al., 2007; Fusi et al., 2007; Feldman, 2009), as well as recovery after spinal cord injury, traumatic brain injury, and stroke (Dancause et al., 2005; Gharbawie and Whishaw, 2006; Girgis et al., 2007; Winship and Murphy, 2008). We showed that the methods presented here can track plastic changes using quantitatively observable changes in the shape and number of neuronal clusters (Eldawlatly et al., 2010). An ultimate goal is to validate these methods with anatomical and imaging data. If the internal structure of the probed neural system were to be identified, multiple avenues for controlling its behavior would become feasible. Our ongoing work is geared in this direction.

Acknowledgments

This work was supported by the National Institute of Neurological Disorders and Stroke, NINDS grant number NS054148.

References

Abeles, M., Gerstein, G.L., 1988. Detecting spatiotemporal firing patterns among simultaneously recorded single neurons. J. Neurophysiol. 60, 909–924.

Aertsen, A.M., Gerstein, G.L., Habib, M.K., Palm, G., 1989. Dynamics of neuronal firing correlation: modulation of "effective connectivity." J. Neurophysiol. 61, 900–917.

Aghagolzadeh, M., Eldawlatly, S., Oweiss, K., 2009. Identifying functional connectivity of motor neuronal ensembles improves the performance of population decoders. In: Proc. of 4th Int. IEEE EMBS Conf. on Neural Engineering, pp. 534–537.

Aghagolzadeh, M., Eldawlatly, S., Oweiss, K., 2010. Synergestic coding by cortical neural ensembles. IEEE Trans. Inf. Theory 56, 875–889.

Agmon-Snir, H., Carr, C.E., Rinzel, J., 1998. The role of dendrites in auditory coincidence detection. Nature 393, 268–272.

Averbeck, B.B., Latham, P.E., Pouget, A., 2006. Neural correlations, population coding and computation. Nat. Rev. Neurosci. 7, 358–366.

Baeg, E.H., Kim, Y.B., Kim, J., Ghim, J.W., Kim, J.J., Jung, M.W., 2007. Learning-induced enduring changes in functional connectivity among prefrontal cortical neurons. J. Neurosci. 27, 909.

Barbieri, R., Quirk, M.C., Frank, L.M., Wilson, M.A., Brown, E.N., 2001. Construction and analysis of non-Poisson stimulus-response models of neural spiking activity. J. Neurosci. Methods 105, 25–37.

Barbieri, R., Wilson, M.A., Frank, L.M., Brown, E.N., 2005. An analysis of hippocampal spatiotemporal representations using a Bayesian algorithm for neural spike train decoding. IEEE Trans. Neural Syst. Rehabil. Eng. 13, 131–136.

Bar-Gad, I., Ritov, Y., Vaadia, E., Bergman, H., 2001. Failure in identification of overlapping spikes from multiple neuron activity causes artificial correlations. J. Neurosci. Methods 107, 1–13.

Bernard, A., Hartemink, A.J., 2005. Informative structure priors: joint learning of dynamic regulatory networks from multiple types of data. Pac. Symp. Biocomput. 10, 459–470.

Bettencourt, L.M.A., Stephens, G.J., Ham, M.I., Gross, G.W., 2007. Functional structure of cortical neuronal networks grown in vitro. Phys. Rev. E Stat. Nonlin. Soft Matter Phys. 75, 021915.

Bezdek, J.C., Pal, N.R., 1998. Some new indexes of cluster validity. IEEE Trans. Syst. Man. Cybern B 28, 301–315.

Bi, GQ., Poo, M.M., 1998. Synaptic modifications in cultured hippocampal neurons: dependence on spike timing, synaptic strength, and postsynaptic cell type. J. Neurosci. 18, 10464–10472.

Biederlack, J., Castelo-Branco, M., Neuenschwander, S., Wheeler, D.W., Singer, W., Nikolić, D., 2006. Brightness induction: rate enhancement and neuronal synchronization as complementary codes. Neuron 52, 1073–1083.

Bressler, S.L., 1995. Large-scale cortical networks and cognition. Brain Res. 20, 288–304.

Brown, E.N., Kass, R.E., Mitra, P.P., 2004. Multiple neural spike train data analysis: state-of-the-art and future challenges. Nat. Neurosci. 7, 456–461.

Brown, S.P., Hestrin, S., 2009. Intracortical circuits of pyramidal neurons reflect their long-range axonal targets. Nature 457, 1133–1136.

Buzsaki, G., 2004. Large-scale recording of neuronal ensembles. Nat. Neurosci. 7, 446–451.

Caclin, A., Fonlupt, P., 2006. Functional and effective connectivity in an fMRI study of an auditory-related task. Eur. J. Neurosci. 23, 2531–2537.

Cadotte, A.J., DeMarse, T.B., He, P., Ding, M., 2008. Causal measures of structure and plasticity in simulated and living neural networks. PLoS ONE 3, e3355.

Callaway, E.M., 1998. Local circuits in primary visual cortex of the macaque monkey. Annu. Rev. Neurosci. 21, 47–74.

Caporale, N., Dan, Y., 2008. Spike timing dependent plasticity: a hebbian learning rule. Annu. Rev. Neurosci. 31, 25–46.

Carmena, J.M., Lebedev, M.A., Crist, R.E., O'Doherty, J.E., Santucci, D.M., Dimitrov, D.F., et al., 2003. Learning to control a brain-machine interface for reaching and grasping by primates. PLoS Biol. 1, e2.

Catani, M., fftyche, D.H., 2005. The rises and falls of disconnection syndromes. Brain 128, 2224–2239.

Chacron, M.J., Bastian, J., 2008. Population coding by electrosensory neurons. J. Neurophysiol. 99, 1825–1835.

Chapin, J.K., Moxon, K.A., Markowitz, R.S., Nicolelis, M.A., 1999. Real-time control of a robot arm using simultaneously recorded neurons in the motor cortex. Nat. Neurosci. 2, 664–670.

Chung, F.R.K., 1994. Spectral Graph Theory, vol. 92. American Mathematical Society, Providence, RI.

Churchland, M.M., Shenoy, K.V., 2007. Temporal complexity and heterogeneity of single-neuron activity in premotor and motor cortex. J. Neurophysiol. 97, 4235.

Cohen, M.R., Maunsell, J.H., 2009. Attention improves performance primarily by reducing interneuronal correlations. Nat. Neurosci. 12, 1594–1600.

Cohen, M.R., Newsome, W.T., 2009. Estimates of the contribution of single neurons to perception depend on timescale and noise correlation. J. Neurosci. 29, 6635–6648.

Cohen, Y.E., Andersen, R.A., 2002. A common reference frame for movement plans in the posterior parietal cortex. Nat. Rev. Neurosci. 3, 553–562.

Cover, T.M., Thomas, J.A., 2006. Elements of Information Theory. Wiley-Interscience, New York.

Csicsvari, J., Henze, D.A., Jamieson, B., Harris, K.D., Sirota, A., Barthó, P., et al., 2003. Massively parallel recording of unit and local field potentials with silicon-based electrodes. J. Neurophysiol. 90, 1314–1323.

Dahlhaus, R., Eichler, M., Sandkühler, J., 1997. Identification of synaptic connections in neural ensembles by graphical models. J. Neurosci. Methods 77, 93–107.

Dan, Y., Poo, M.M., 2004. Spike timing-dependent plasticity of neural circuits. Neuron 44, 23–30.

Dancause, N., Barbay, S., Frost, S.B., Plautz, E.J., Chen, D., Zoubina, E.V., et al., 2005. Extensive cortical rewiring after brain injury. J. Neurosci. 25, 10167–10179.

de la Rocha, J., Doiron, B., Shea-Brown, E., Josić, K., Reyes, A., 2007. Correlation between neural spike trains increases with firing rate. Nature 448, 802–806.

deCharms, R.C., Merzenich, M.M., 1996. Primary cortical representation of sounds by the coordination of action-potential timing. Nature 381, 610–613.

Destexhe, A., Marder, E., 2004. Plasticity in single neuron and circuit computations. Nature 431, 789–795.

Ding, C., He, X., Zha, H., Gu, M., Simon, H., 2001. A min-max cut algorithm for graph partitioning and data clustering. In: Proceedings of the 2001 IEEE International Conference on Data Mining. pp. 107–114. IEEE Computer Society, Washington, DC.

Dojer, N., Gambin, A., Mizera, A., Wilczyński, B., Tiuryn, J., 2006. Applying dynamic Bayesian networks to perturbed gene expression data. BMC Bioinformatics 7, 249.

Dombeck, D.A., Graziano, M.S., Tank, D.W., 2009. Functional clustering of neurons in motor cortex determined by cellular resolution imaging in awake behaving mice. J. Neurosci. 29, 13751–13760.

Dupret, D., Fabre, A., Döbrössy, M.D., Panatier, A., Rodríguez, J.J., Lamarque, S., et al., 2007. Spatial learning depends on both the addition and removal of new hippocampal neurons. PLoS Biol. 5, e214.

Dzakpasu, R., Zochowski, M., 2005. Discriminating different types of synchrony in neural systems. Physica D 208, 115–122.

Eldawlatly, S., Jin, R., Oweiss, K.G., 2009. Identifying functional connectivity in large-scale neural ensemble recordings: a multiscale data mining approach. Neural Comput. 21, 450–477.

Eldawlatly, S., Oweiss, K., 2010. Millisecond precision reflects the collective dynamics of population codes in the rat somatosensory cortex. In preparation.

Eldawlatly, S., Zhou, Y., Jin, R., Oweiss, K.G., 2010. On the use of dynamic Bayesian networks in reconstructing functional neuronal networks from spike train ensembles. Neural Comput. 22, 158–189.

Ermentrout, B., 2003. Dynamical consequences of fast-rising, slow-decaying synapses in neuronal networks. Neural Comput. 15, 2483–2522.

Feldman, D.E., 2009. Synaptic mechanisms for plasticity in neocortex. Annu. Rev. Neurosci. 32, 33–55.

Franaszczuk, P.J., Bergey, G.K. and Kaminski, M.J., 1994. Analysis of mesial temporal seizure onset and propagation using the directed transfer function method. Electroenceph. Clin. Neurophysiol. 91, 413–427.

Frerking, M., Schulte, J., Wiebe, S.P., Stäubli, U., 2005. Spike timing in CA3 pyramidal cells during behavior: implications for synaptic transmission. J. Neurophysiol. 94, 1528–1540.

Friedman, N., Nachman, I., Peer, D., 1999. Learning bayesian network structure from massive datasets: the sparse candidate algorithm. In Proc. of the 15th Conference on Uncertainty in Artificial Intelligence (UAI) (pp. 206–215). San Francisco: Morgan Kaufmann.

Friston, K.J., 1994. Functional and effective connectivity in neuroimaging: a synthesis. Hum. Brain Mapp. 2, 56–78.

Fujisawa, S., Amarasingham, A., Harrison, M.T., Buzsáki, G., 2008. Behavior-dependent short-term assembly dynamics in the medial prefrontal cortex. Nat. Neurosci. 11, 823–833.

Fusi, S., Asaad, W.F., Miller, E.K., Wang, X.J., 2007. A neural circuit model of flexible sensorimotor mapping: learning and forgetting on multiple timescales. Neuron 54, 319–333.

Geier, F., Timmer, J., Fleck, C., 2007. Reconstructing gene-regulatory networks from time series, knock-out data, and prior knowledge. BMC Syst. Biol. 1, 11.

Gerstein, G.L., 2000. Cross correlation measures of unresolved multi-neuron recordings. J. Neurosci. Methods 100, 41–51.

Gerstein, G.L., Perkel, D.H., 1969. Simultaneously recorded trains of action potentials: analysis and functional interpretation. Science 164, 828–830.

Gharbawie, O.A., Whishaw, I.Q., 2006. Parallel stages of learning and recovery of skilled reaching after motor cortex stroke: "oppositions" organize normal and compensatory movements. Behav. Brain Res. 175, 249–262.

Girgis, J., Merrett, D., Kirkland, S., Metz, G.A., Verge, V., Fouad, K., 2007. Reaching training in rats with spinal cord injury promotes plasticity and task specific recovery. Brain 130, 2993.

Gold, J.I., Shadlen, M.N., 2007. The neural basis of decision making. Annu. Rev. Neurosci. 30, 535–574.

Gourévitch, B., Eggermont, J.J., 2007. Evaluating information transfer between auditory cortical neurons. J. Neurophysiol. 97, 2533–2543.

Granger, C.W.J., 1969. Investigating causal relations by econometric models and cross-spectral methods. Econometrica 37, 424–438.

Gray, C.M., König, P., Engel, A.K., Singer, W., 1989. Oscillatory responses in cat visual cortex exhibit inter-columnar synchronization which reflects global stimulus properties. Nature 338, 334–337.

Greicius, M.D., Krasnow, B., Reiss, A.L., Menon, V., 2003. Functional connectivity in the resting brain: a network analysis of the default mode hypothesis. PNAS 100, 253–258.

Grün, S., Diesmann, M., Aertsen, A., 2002. Unitary events in multiple single-neuron spiking activity: II. Nonstationary data. Neural Comput. 14, 81–119.

Harris, K.D., Henze, D.A., Csicsvari, J., Hirase, H., Buzsáki, G., 2000. Accuracy of tetrode spike separation as determined by simultaneous intracellular and extracellular measurements. J. Neurophysiol. 84, 401–414.

Hartemink, A.J., Gifford, D.K., Jaakkola, T.S., Young, R.A., 2001. Using graphical models and genomic expression data to statistically validate models of genetic regulatory networks. Pac. Symp. Biocomput., 422–433, Hawaii.

Hebb, D.O., 1949. The Organization of Behavior; a Neuropsychological Theory. Wiley, New York.

Heckerman, D., 1995. A tutorial on learning with bayesian networks. Technical Report MSR-TR-95-06, Microsoft Research.

Heckerman, D., Geiger, D., Chickering, D.M., 1995. Learning bayesian networks: the combination of knowledge and statistical data. Mach. Learn. 9, 197–243.

Histed, M.H., Bonin, V., Reid, R.C., 2009. Direct activation of sparse, distributed populations of cortical neurons by electrical microstimulation. Neuron 63, 508–522.

Hochberg, L.R., Serruya, M.D., Friehs, G.M., Mukand, J.A., Saleh, M., Caplan, A.H., et al., 2006. Neuronal ensemble control of prosthetic devices by a human with tetraplegia. Nature 442, 164–171.

Jain, A.K., Duin, R.P.W., Mao, J., 2000. Statistical pattern recognition: a review. IEEE Trans. Pattern Anal Mach. Intell. 22, 4–37.

Jarvis, M.R., Mitra, P.P., 2001. Sampling properties of the spectrum and coherency of sequences of action potentials. Neural Comput. 13, 717–749.

Jin, R., Ding, C., Kang, F., 2005. A probabilistic approach for optimizing spectral clustering. In: Y. Weiss, B. Schölkopf, and J. Platt (Eds.), Advances in neural information processing systems, 18. Cambridge, MA: MIT Press.

Kamiński, M., Ding, M., Truccolo, W.A., Bressler, S.L., 2001. Evaluating causal relations in neural systems: granger causality, directed transfer function and statistical assessment of significance. Biol. Cybern. 85, 145–157.

Kirkpatrick, S., Gelatt Jr., C.D., Vecchi, M.P., 1983. Optimization by simulated annealing. Science 220, 671–680.

Kohn, A., Smith, M.A., 2005. Stimulus dependence of neuronal correlation in primary visual cortex of the macaque. J. Neurosci. 25, 3661–3673.

Kuhlmann, L., Burkitt, A.N., Paolini, A., Clark, G.M., 2002. Summation of spatiotemporal input patterns in leaky integrate-and-fire neurons: application to neurons in the cochlear nucleus receiving converging auditory nerve fiber input. J. Comput. Neurosci. 12, 55–73.

Lam, W., Bacchus, F., 1994. Learning bayesian belief networks: an approach based on mdl principle. Comput. Intell. 10, 269–293.

Lin, L., Osan, R., Tsien, J.Z., 2006. Organizing principles of real-time memory encoding: neural clique assemblies and universal neural codes. Trends Neurosci. 29, 48–57.

Liu, Y., Liang, M., Zhou, Y., He, Y., Hao, Y., Song, M., et al., 2008. Disrupted small-world networks in schizophrenia. Brain 131, 945–961.

Logothetis, N.K., 2003. MR imaging in the non-human primate: studies of function and of dynamic connectivity. Curr. Opin. Neurobiol. 13, 630–642.

Logothetis, N.K., 2008. What we can do and what we cannot do with fMRI. Nature 453, 869–878.

Maass, W., Natschläger, T., 2000. A model for fast analog computation based on unreliable synapses. Neural Comput. 12, 1679–1704.

Mallat, S., 1998. A Wavelet Tour of Signal Processing, second ed. Kluwer Acedemic Publishers Norwood, MA.

Markram, H., Lübke, J., Frotscher, M., Sakmann, B., 1997. Regulation of synaptic efficacy by coincidence of postsynaptic APs and EPSPs. Science 275, 213–215.

McCulloch, C., 2000. Generalized linear models. J. Am. Stat. Assoc. 95, 1320–1324.

Moran, D.W., Schwartz, A.B., 1999. Motor cortical representation of speed and direction during reaching. J. Neurophysiol. 82, 2676–2692.

Murphy, K., 2002. Dynamic Bayesian Networks: Representation, Inference and Learning, PhD thesis. Computer Science Division, University of California, Berkeley, CA.

Murphy, K., Mian, S., 1999. Modelling Gene Expression Data Using Dynamic Bayesian Networks. Computer Science Division, University of California, Berkeley, CA.

Needham, C.J., Bradford, J.R., Bulpitt, A.J., Westhead, D.R., 2007. A primer on learning in Bayesian networks for computational biology. PLoS Comput. Biol. 3, e129.

Ng, A., Jordan, M., and Weiss, A., 2001. On spectral clustering: Analysis and an algorithm. In T.G. Dietterich, S. Becker, and Z. Ghahramani (Eds.), Advances in neural information processing systems, 14. Cambridge, MA: MIT Press.

Nicolelis, M.A.L., 1999. Methods for Neural Ensemble Recordings. CRC Press, Boca Raton, FL.

Nirenberg, S., Latham, P.E., 2003. Decoding neuronal spike trains: how important are correlations? Proc. Natl. Acad. Sci. 100, 7348–7353.

Normann, R.A., Maynard, E.M., Rousche, P.J., Warren, D.J., 1999. A neural interface for a cortical vision prosthesis. Vision Res. 39, 2577–2587.

Okatan, M., Wilson, M.A., Brown, E.N., 2005. Analyzing functional connectivity using a network likelihood model of ensemble neural spiking activity. Neural Comput. 17, 1927–1961.

Paninski, L., Fellows, M.R., Hatsopoulos, N.G., Donoghue, J.P., 2004. Spatiotemporal tuning of motor cortical neurons for hand position and velocity. J. Neurophysiol. 91, 515–532.

Paninski, L., Pillow, J., Lewi, J., 2007. Statistical models for neural encoding, decoding, and optimal stimulus design. Prog. Brain Res. 165, 493.

Paninski, L., Pillow, J.W., Simoncelli, E.P., 2004. Maximum likelihood estimation of a stochastic integrate-and-fire neural encoding model. Neural Comput. 16, 2533–2561.

Pearl, J., 1988. Probabilistic Reasoning in Intelligent Systems. Morgan Kaufmann, San Francisco, CA.

Pearl, J., Shafer, G., 1988. Probabilistic Reasoning in Intelligent Systems: Networks of Plausible Inference. Morgan Kaufmann, San Mateo, CA.

Peña, J.L., Viete, S., Funabiki, K., Saberi, K., Konishi, M., 2001. Cochlear and neural delays for coincidence detection in owls. J. Neurosci. 21, 9455–9459.

Perkel, D.H., Gerstein, G.L., Moore, G.P., 1967. Neuronal spike trains and stochastic point processes. II. Simultaneous spike trains. J. Biophys. 7, 419–440.

Pillow, J.W., Paninski, L., Uzzell, V.J., Simoncelli, E.P., Chichilnisky, E.J., 2005. Prediction and decoding of retinal ganglion cell responses with a probabilistic spiking model. J. Neurosci. 25, 11003–11013.

Pillow, J.W., Shlens, J., Paninski, L., Sher, A., Litke, A.M., Chichilnisky, E.J., et al., 2008. Spatiotemporal correlations and visual signalling in a complete neuronal population. Nature 454, 995–999.

Quiroga, R.Q., Reddy, L., Koch, C., Fried, I., 2007. Decoding visual inputs from multiple neurons in the human temporal lobe. J. Neurophysiol. 98, 1997–2007.

Rajapakse, J.C., Zhou, J., 2007. Learning effective brain connectivity with dynamic Bayesian networks. Neuroimage 37, 749–760.

Rijsbergen, C.J. v., 1979. Information Retrieval. Buttersworth, London.

Roberts, P.D., Bell, C.C., 2002. Spike-timing dependent synaptic plasticity in biological systems. Biol. Cybern. 87, 392–403.

Roudi, Y., Nirenberg, S., Latham, P.E., 2009. Pairwise maximum entropy models for studying large biological systems: when they can work and when they can't. PLoS Comput. Biol. 5, e1000380.

Salinas, E., Thier, P., 2000. Gain modulation: a major computational principle of the central nervous system. Neuron 27, 15–21.

Sameshima, K., Baccalá, L.A., 1999. Using partial directed coherence to describe neuronal ensemble interactions. J. Neurosci. Methods 94, 93–103.

Samonds, J.M., Allison, J.D., Brown, H.A., Bonds, A.B., 2003. Cooperation between area 17 neuron pairs enhances fine discrimination of orientation. J. Neurosci. 23, 416.

Schneidman, E., Berry 2nd., MJ., Segev, R., Bialek, W., 2006. Weak pairwise correlations imply strongly correlated network states in a neural population. Nature 440, 1007.

Schnupp, J.W.H., Kacelnik, O., 2002. Cortical plasticity: learning from cortical reorganisation. Curr. Biol. 12, R144–R146.

Schwarz, G., 1978. Estimating the dimension of a model. Ann. Stat. 6, 461–464.

Serruya, M.D., Hatsopoulos, N.G., Paninski, L., Fellows, M.R., Donoghue, J.P., 2002. Brain-machine interface: instant neural control of a movement signal. Nature 416, 141–142.

Shadlen, M.N., Newsome, W.T., 1998. The variable discharge of cortical neurons: implications for connectivity, computation, and information coding. J. Neurosci. 18, 3870–3896.

Smith, V.A., Yu, J., Smulders, T.V., Hartemink, A.J., Jarvis, E.D., 2006. Computational inference of neural information flow networks. PLoS Comput. Biol. 2, 1436–1449.

Smith, W.S., Fetz, E.E., 2009. Synaptic interactions between forelimb-related motor cortex neurons in behaving primates. J. Neurophysiol. 102, 1026–1039.

Song, S., Abbott, L.F., 2001. Cortical development and remapping through spike timing-dependent plasticity. Neuron 32, 339–350.

Song, S., Sjöström, P.J., Reigl, M., Nelson, S., Chklovskii, D.B., 2005. Highly nonrandom features of synaptic connectivity in local cortical circuits. PLoS Biol. 3, e68.

Sporns, O., Honey, C.J., Kötter, R., 2007. Identification and classification of hubs in brain networks. PLoS ONE 2, e1049.

Sprekeler, H., Michaelis, C., Wiskott, L., 2007. Slowness: an objective for spike-timing-dependent plasticity? PLoS Comput. Biol. 3, e112. doi:10.1371/journal.pcbi.0030112.

Stevenson, I.H., Rebesco, J.M., Hatsopoulos, N.G., Haga, Z., Miller, L.E., Körding, K.P., 2009. Bayesian inference of functional connectivity and network structure from spikes. IEEE Trans. Neural Syst. Rehabil. Eng. 17, 203–213.

Stosiek, C., Garaschuk, O., Holthoff, K., Konnerth, A., 2003. In vivo two-photon calcium imaging of neuronal networks. Proc. Natl. Acad. Sci. 100, 7319.

Supekar, K., Menon, V., Rubin, D., Musen, M., Greicius, M.D., 2008. Network analysis of intrinsic functional brain connectivity in Alzheimer's disease. PLoS Comput. Biol. 4, e1000100.

Tang, A., Jackson, D., Hobbs, J., Chen, W., Smith, J.L., Patel, H., et al., 2008. A maximum entropy model applied to spatial and temporal correlations from cortical networks in vitro. J. Neurosci. 28, 505–518.

Truccolo, W., Eden, U.T., Fellows, M.R., Donoghue, J.P., Brown, E.N., 2005. A point process framework for relating neural spiking activity to spiking history, neural ensemble, and extrinsic covariate effects. J. Neurophysiol. 93, 1074–1089.

Tsao, D.Y., Schweers, N., Moeller, S., Freiwald, W.A., 2008. Patches of face-selective cortex in the macaque frontal lobe. Nat. Neurosci. 11, 877–879.

Tsodyks, M.V., Markram, H., 1997. The neural code between neocortical pyramidal neurons depends on neurotransmitter release probability. PNAS 94, 719–723.

Varela, F., Lachaux, J.P., Rodriguez, E., Martinerie, J., 2001. The brainweb: phase synchronization and large-scale integration. Nat. Rev. Neurosci. 2, 229–239.

Verhagen, L., Dijkerman, H.C., Grol, M.J., Toni, I., 2008. Perceptuo-motor interactions during prehension movements. J. Neurosci. 28, 4726–4735.

Wagenaar, D.A., Pine, J., Potter, S.M., 2006. Searching for plasticity in dissociated cortical cultures on multi-electrode arrays. J. Negat. Results Biomed. 5, 16.

Wainwright, M.J., Jordan, M., 2008. Graphical models, exponential families, and variational inference. Found. Trends Mach. Learn. 1, 1–305.

Wallach, A., Eytan, D., Marom, S., Meir, R., 2008. Selective adaptation in networks of heterogeneous populations: model, simulation, and experiment. PLoS Comput. Biol 4(2), e29.

Wellman, M., Henrion, M., 1993. Explaining'explaining away. IEEE Trans. Pattern Anal. Mach. Intell. 15, 287–292.

Wickens, J.R., Reynolds, J.N., Hyland, B.I., 2003. Neural mechanisms of reward-related motor learning. Curr. Opin. Neurobiol. 13, 685–690.

Winship, I.R., Murphy, T.H., 2008. In vivo calcium imaging reveals functional rewiring of single somatosensory neurons after stroke. J. Neurosci. 28, 6592–6606.

Wise, K.D., Anderson, D.J., Hetke, J.F., Kipke, D.R., Najafi, K., 2004. Wireless implantable microsystems: high-density electronic interfaces to the nervous system. Proc. IEEE 92 (1), 76–97.

Wu, W., Black, M.J., Mumford, D., Gao, Y., Bienenstock, E., Donoghue, J.P., 2004. Modeling and decoding motor cortical activity using a switching Kalman filter. Biomed. Eng. IEEE Trans. 51, 933–942.

Wu, W., Gao, Y., Bienenstock, E., Donoghue, J.P., Black, M.J., 2005. Bayesian population decoding of motor cortical activity using a Kalman filter. Neural Comput. 18, 80–118.

Xu, T., Yu, X., Perlik, A.J., Tobin, W.F., Zweig, J.A., Tennant, K., et al., 2009. Rapid formation and selective stabilization of synapses for enduring motor memories. Nature 462, 915–919.

Yamada, S., Matsumoto, K., Nakashima, M., Shiono, S., 1996. Information theoretic analysis of action potential trains. II. Analysis of correlation among n neurons to deduce connection structure. J. Neurosci. Methods 66, 35–45.

Yoshimura, Y., Callaway, E.M., 2005. Fine-scale specificity of cortical networks depends on inhibitory cell type and connectivity. Nat. Neurosci. 8, 1552–1559.

Yu, J., Smith, V.A., Wang, P.P., Hartemink, A.J., Jarvis, E.D., 2004. Advances to Bayesian network inference for generating causal networks from observational biological data. Bioinformatics 20, 3594–3603.

Zhang, L., Samaras, D., Alia-Klein, N., Volkow, N., Goldstein, R., 2006. Modeling neuronal interactivity using dynamic Bayesian networks. In: Y. Weiss, B. Schölkopf, and J. Platt (Eds.), Advances in neural information processing systems, 18. Cambridge, MA: MIT Press.

Zhang, X., Carney, L.H., 2005. Response properties of an integrate-and-fire model that receives subthreshold inputs. Neural Comput. 17, 2571–2601.

State Space Modeling of Neural Spike Train and Behavioral Data

Zhe Chen, Riccardo Barbieri, Emery N. Brown

Neuroscience Statistics Research Laboratory, Department of Brain and Cognitive Sciences, Massachusetts Institute of Technology, Cambridge, Massachusetts; Harvard-MIT Division of Health Sciences and Technology, Massachusetts Institute of Technology, Cambridge, Massachusetts; Massachusetts General Hospital, Harvard Medical School, Boston, Massachusetts

6.1 INTRODUCTION

State space modeling is an established framework for analyzing stochastic and deterministic dynamical systems that are measured or observed through a stochastic process. This highly flexible paradigm has been successfully applied in engineering, statistics, computer science, and economics to solve a broad range of dynamical systems problems (Mendel, 1995; Kitagawa and Gersh, 1996; Shephard and Pitt, 1997; Kim and Nelson, 1999; Durbin and Koopman, 2001). Other terms used to describe state space modeling are hidden Markov models (Cappé et al., 2005) and latent process models (Fahrmeir and Tutz, 2001). The most-studied state space modeling tool is the Kalman filter, which defines an optimal algorithm for analyzing linear Gaussian systems measured with Gaussian error (Mendel, 1995).

The revolution in neuroscience recording technologies in the last 20 years has provided many novel ways to study the dynamic activity of the brain and central nervous system. These technologies include multi-electrode recording arrays (Wilson and McNaughton, 1993, 1994), functional magnetic imaging (Kwong et al., 1992; Ogawa et al., 1992), electroencephalography and magnetoencephalography (Hämäläinen et al., 1993), diffuse optical tomography (Boas et al., 2001), calcium imaging (Denk et al., 1990), and behavioral data (Wirth et al., 2003). Because a fundamental feature of many neuroscience data analysis problems is that the underlying neural system is dynamic and is observed indirectly through measurements from one or a combination of these different recording modalities,

Statistical Signal Processing for Neuroscience and Neurotechnology. DOI: 10.1016/B978-0-12-375027-3.00006-5

the state space paradigm provides an ideal framework for developing statistical tools to analyze neural data.

Neural spiking activity recorded from single or multiple electrodes is one of the principal types of data recorded in neurophysiological experiments. Because neural spike trains are time series of action potentials (i.e., all-or-nothing responses), point process theory has been shown to provide an accurate framework for modeling the stochastic structure of single and multiple neural spike trains (Brown et al., 2003, 2004; Brown, 2005). An important objective of these experiments is to study how neurons represent and transmit information in their ensemble spiking activity. In behavioral learning experiments, a frequently recorded type of data is time series of binary responses that track whether a subject responds correctly or incorrectly to a task across series of multiple presentations or trials (Jog et al., 1999; Wirth et al., 2003; Barnes et al., 2005; Law et al., 2005). Each binary measurement may be considered an observation of a Bernoulli random variable. The objective of the analysis is to determine how this Bernoulli probability changes as a function of repeated attempts at the task.

Both the neural spiking activity recorded from neurophysiological experiments and the binary responses recorded in learning experiments may be viewed as time series of 0–1 responses. Moreover, because the underlying systems being studied in neurophysiological and learning experiments are dynamic, these processes readily lend themselves to analysis using the state space paradigm. Point process measurements and binary observations are types of observation models not commonly studied in state space modeling. However, they are appropriate for two types of neuroscience data.

In this chapter, we discuss the application of the state space paradigm to the analysis of neural spike trains and behavioral learning data. We first define point process and binary observation models and review the framework for state space filtering and state space smoothing. To illustrate the state space paradigm we discuss four data analysis problems:

- Ensemble neural spike train decoding (Brown et al., 1998; Barbieri et al., 2004)

- Analysis of neural receptive field plasticity in one and two dimensions (Brown et al., 2001; Ergun et al., 2007)

- Dynamic analysis of individual and population behavioral learning (Smith and Brown, 2003; Smith et al., 2004)

- Estimation of cortical UP/DOWN states (Chen et al., 2009)

For each problem we describe the observation model, the state space model, and the estimation algorithm, and apply the paradigm in the analysis of actual data.

6.2 STATE SPACE MODELING PARADIGM

The state space model consists of two equations: the state equation and the observation equation. The state equation defines the temporal evolution of the state process as a dynamical system. The states can be described either in continuous time by a deterministic or stochastic differential equation, or in discrete time by a deterministic or stochastic difference equation. If the current state depends only on the previous state, the state space model is termed Markovian. The observation equation defines how the state is observed or measured. The observation model is generally stochastic, and the values of the observation process can be either continuous or discrete. In the analyses we present, the observation processes are discrete.

6.2.1 Notation

Because our objective is to analyze neural spike train and behavioral data, we assume that the observations are recorded from either a point process or a binary process. To define the point process we assume that we have an observation interval $(0, T]$ and that we divide it into subintervals of width Δ. We take x_t to be the value of the state process and y_t to be the observation recorded in the interval $[t, t + \Delta)$. A point process is a binary $(0 - 1)$ stochastic process defined in continuous time characterized by its conditional intensity function (CIF), $\lambda(t|x_t, H_t)$, where H_t represents the relevant history of the point process up to time t (Daley and Vere-Jones, 2002; Brown, 2005). The CIF defines the probability of observing an event from the point process in the interval $[t, t + \Delta)$ as $\Pr(y_t = 1|x_t, H_t) = \lambda(t|x_t, H_t)\Delta + o(\Delta)$. That is, as Δ becomes small, the probability of more than one event in the interval $[t, t + \Delta)$ tends to 0. The CIF is a history-dependent generalization of the rate function of a Poisson process. The point process models are used to analyze neural spiking activity.

To define the discrete binary process we assume that in the interval $(0, T]$ there is a set of lattice points $t = 1, 2, \ldots, T$. We let y_t denote the discrete binary observation and x_t denote the state process at time t. The discrete binary process is a $(0 - 1)$ stochastic process recorded in discrete time defined by the probability of observing an event as $\Pr(y_t = 1|x_t) = p_t$, for $t = 1, \ldots, T$ (Fahrmeir and Tutz, 2001; Smith and Brown, 2003). We use the discrete binary process to model the binary responses recorded in learning experiments. For both the point process and the discrete binary process, we take $y_{0:t} = (y_1, \ldots, y_t)$ to be the data recorded in $(0,t]$ for $t = 1, \ldots, T$.

6.2.2 Recursive Form of Bayes' Rule

The objective of state space modeling is to compute the optimal estimate of the state given the observed data. There are two important cases that are typically considered. To state them we consider $y_{0:s}$ for $s = 1, \ldots, T$. If we take $s = t$ and we wish to determine the best estimate of the state x_t given the observation $y_{0:t}$, then we seek a solution to the filtering problem; if we take $s = T$ and we wish to determine the best estimate of the state x_t given the observation $y_{0:T}$, then we seek a solution to the smoothing problem.

To develop the state space analysis paradigm and make formal statements on the filtering and smoothing problems, we first derive a recursive form of Bayes' rule. The objective for the filtering problem is to compute at each time t the probability density of the state given the data. Because of the state space structure, this posterior density can be expressed in a recursive form as

$$
\begin{aligned}
p(x_t \mid y_{0:t}) &= \frac{p(x_t, y_{0:t})}{p(y_{0:t})} \\
&= \frac{p(x_t, y_{0:t} \mid y_{0:t-1}) p(y_{0:t-1})}{p(y_t \mid y_{0:t-1}) p(y_{0:t-1})} \\
&= \frac{p(x_t \mid y_{0:t-1}) p(y_{0:t} \mid x_t, y_{0:t-1})}{p(y_t \mid y_{0:t-1})} \\
&= \frac{p(x_t \mid y_{0:t-1}) p(y_t \mid x_t, y_{0:t-1})}{p(y_t \mid y_{0:t-1})}
\end{aligned}
\tag{6.1}
$$

where the Chapman-Kolmogorov, or one-step prediction, equation is

$$
p(x_t \mid y_{0:t-1}) = \int p(x_t \mid x_{t-1}) p(x_{t-1} \mid y_{0:t-1}) dx_{t-1}
\tag{6.2}
$$

and the normalizing constant is

$$
p(y_t \mid y_{0:t-1}) = \int p(x_t \mid y_{0:t-1}) p(y_t \mid x_t, y_{0:t-1}) dx_t
\tag{6.3}
$$

where $p(y_t \mid x_t, y_{0:t-1})$ is the observation equation and $p(x_t \mid x_{t-1})$ is the state equation in the state space model. The posterior probability density (Eq. (6.1)) and the Chapman-Kolmogorov equation (Eq. (6.2)) constitute the

recursive form of Bayes' rule. Equations (6.1) and (6.2) provide the fundamental relations to develop our state space analyses for point process and
binary observations. These two equations are used to derive the Kalman
filter (Haykin, 2002), conduct time series analyses with smoothness priors
(Kitagawa and Gersh, 1996; Durbin and Koopman, 2001), design dynamic
generalized linear models (Fahmeir and Tutz, 2001), and build particle filters
(Doucet et al., 2001).

To solve the smoothing problem, we are interested in computing the posterior density of the state x_t given all of the observed data $y_{0:T}$. We can express
the smoothing probability density in terms of the filtering probability density
as follows:

$$
\begin{aligned}
p(x_t|y_{0:T}) &= \int p(x_t, x_{t+1}|y_{0:T})dx_{t+1} \\
&= \int p(x_{t+1}|y_{0:T})p(x_t|x_{t+1}, y_{0:T})dx_{t+1} \\
&= \int p(x_{t+1}|y_{0:T})p(x_t|x_{t+1}, y_{0:t})dx_{t+1}. \qquad (6.4) \\
&= p(x_t|y_{0:t}) \int p(x_{t+1}|y_{0:T})p(x_{t+1}|x_t, y_{0:t})dx_{t+1} \\
&= p(x_t|y_{0:t}) \int \frac{p(x_{t+1}|y_{0:T})p(x_{t+1}|x_t)}{p(x_{t+1}|y_{0:T})}dx_{t+1}
\end{aligned}
$$

If the observation equation is linear and Gaussian and the state equation is
linear and Gaussian, or if Gaussian approximations have been used to estimate
the filtering probability density, then the solution to the smoothing probability
density is Gaussian and its mean and variance can be computed recursively
by the fixed-interval smoothing algorithm (Fahrmeir and Tutz, 2001).

6.2.3 Classes of Filtering and Smoothing Problems

Applying Eqs. (6.1), (6.2), and (6.4) to actual data analyses requires that a
statistical model be defined for each problem. This model will have a set of
parameters θ, which must be estimated from either current or previous data.
We can use θ to parameterize the state equation, the observation equation, or
both. For both the filtering and smoothing problems, θ can be either known
or estimated and either static or dynamic. We then define our applications of
the state space paradigm in terms of different classes of filtering and smoothing problems by expressing Eqs. (6.1), (6.2), and (6.4) as functions of θ.
The three classes of filtering and smoothing problems we consider are state
space filtering with a known or estimated static parameter θ, $p(x_t|y_{0:t}, \theta)$; state

space filtering with an estimated dynamic parameter θ_t, $p(x_t|y_{0:t}, \theta_t)$; and state space smoothing with an estimated static parameter θ, $p(x_t|y_{0:T}, \theta)$. A third problem is one of prediction, in which we consider $p(x_t|y_{0:T}, \theta)$ where $t > T$. This problem arises naturally in the design of algorithms for brain–computer interfaces (Serruya et al., 2002; Taylor et al., 2002; Musallam et al., 2004; Hochberg et al., 2006; Santhanam et al., 2006).

6.3 APPLICATIONS OF THE STATE SPACE PARADIGM IN NEUROSCIENCE DATA ANALYSIS

The following sections discuss how the state space paradigm is applied in neuroscientific data analysis.

6.3.1 Neural Spike Train Decoding and Point Process Filter Algorithms

The development of multiple electrode arrays allows neuroscientists to record the simultaneous spiking activity of large numbers of neurons. For example, the technical capability to record the spiking activity of large numbers of hippocampal place cells—between 20 and 100—along with a rat's position in its environment has made feasible a formal quantitative study of how rats encode spatial information in short-term memory and use it for navigation. Investigation of these questions requires statistical methods to analyze how the animal's position in the environment (the continuous stimulus) is represented in the firing patterns of place cells (multidimensional point processes).

The mathematical techniques used in neuroscience to study how spike train firing patterns represent biological signals and external stimuli are termed neural decoding algorithms (Rieke et al., 1997; Brown et al., 1998; Barbieri et al., 2004; Eden et al., 2004; Wu et al., 2006, 2009). Many data analyses of population coding have used decoding algorithms. Examples include position representation by ensembles of rat hippocampal neurons (Brown et al., 1998; Zhang et al., 1998), velocity encoding by fly H1 neurons (Bialek et al., 1991), velocity and position encoding by M1 neurons (Shoham et al., 2005; Truccolo et al., 2005), and natural scene representations by catLGN neurons (Stanley et al., 1999). How to develop optimal strategies for constructing and testing decoding algorithms is an important question in computational neuroscience. The decoding analysis proceeds typically in two stages: encoding and decoding. In the encoding stage, neural spiking activity is characterized as a function of the biological signal (Figure 6.1(a)). In the decoding stage, the relation is inverted and the signal is estimated from the spiking activity of

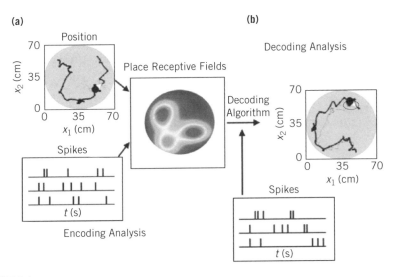

FIGURE 6.1

Decoding of position from ensemble rat neural spiking activity. (a) Encoding analysis: The relation between the biological stimulus (trajectory of the rat in the environment, solid black line in the Position panel) and spiking activity (Spikes panel) is estimated as place receptive fields for three neurons. (b) Decoding analysis: The estimated place receptive fields are used in a Bayesian decoding algorithm to compute the predicted position (thin black line) of the rat in the environment from new spiking activity of the neural ensemble recording during the decoding stage. The predicted position is compared to the observed position (thick black line) during the decoding stage. The oval outline defines a 95% confidence region centered at that location. Please see this figure in color at the companion web site: www.elsevierdirect.com/companions/9780123750273

Source: Brown et al. (2004), *Nature Neuroscience,* reprinted with permission.

the neurons (Figure 6.1(b)). The state space framework provides a systematic means of formulating the decoding problem.

To illustrate the application of a state space model to the neural spike trains decoding problem, we assume that we observe the simultaneous neural spiking activity of C hippocampal neurons of a rat foraging in an open circular environment (Brown et al., 1998; Barbieri et al., 2004). During the decoding interval $(0, T^d]$, we record the spiking activity of C neurons. For $c = 1, \ldots, C$ and $t \in (0, T^d]$, let y_t^c be 1 if there is a spike in the interval $[t, t + \Delta)$ from neuron c and 0 otherwise; let $y_t = \{y_t^1, \ldots, y_t^C\}$ be the set of observations in $[t, t + \Delta)$ from the C neurons. Also, let $y_{0:t}^c = \{y_1^c, y_2^c, \ldots, y_t^c\}$ be the spiking activity from neuron c and $y_{0:t} = \{y_{0:t}^1, \ldots, y_{0:t}^C\}$ be the ensemble spiking activity in $(0, t]$. The state is the animal's position in two dimensions $x_t = (x_{1,t}, x_{2,t})$. The CIF for neuron c is $\lambda(x_t | y_{0:t-1}, \theta^c)$, where θ^c is a parameter that depends on the equation being used to model the place receptive field.

For a small Δ, under the assumption that the spiking activity of the neurons in the ensemble is conditionally independent, the joint probability density or observation model for the interval $[t, t + \Delta)$ is as follows (Brown et al., 1998; Smith and Brown, 2003):

$$p(y_t|x_t, y_{0:t-1}, \theta) = \prod_{c=1}^{C} p(y_t^c|x_t, y_{0:t-1}, \theta^c)$$

(6.5)

$$= \exp\left[\sum_{c=1}^{C} y_t^c \log \lambda^c(x_t|y_{0:t-1}, \theta^c) - \sum_{c=1}^{C} \lambda^c(x_t|y_{0:t-1}, \theta^c)\Delta\right]$$

where $\theta = \{\theta^1, \ldots, \theta^C\}$. Conditional independence means that neurons are only related through the dynamics of the state variable.

Analysis of the decoding problem without satisfying this assumption requires a network likelihood model, such as the one used in Okatan et al. (2005) to model dependence among the neurons not related to the state variable. We do not consider this problem. Here, the state equation for x_t is the first-order auto-regressive model

$$x_t = Fx_{t-1} + v_t$$

(6.6)

where v_t is independent two-dimensional Gaussian noise with zero mean and covariance matrix Σ_v. The CIF model of the receptive field of neuron c is either the Gaussian function (Brown et al., 1998):

$$\lambda(x_t, \theta^c) = \exp\left\{\alpha^c - \frac{1}{2}(x_t - \mu^c)(W^c)^{-1}(x_t - \mu^c)\right\}$$

(6.7)

where $\theta^c = (\alpha^c, \mu^c, W^c)$, or the log linear function defined by

$$\lambda(x_t, \theta^c) = \exp\left\{\sum_{j=1}^{J} \theta_j^c z_j(x_t)\right\}$$

(6.8)

where the z_j are Zernike polynomials—that is, orthogonal polynomials on a unit disk (Barbieri et al., 2004).

To evaluate the system in Eqs. (6.1) and (6.2), as related to the model in Eqs. (6.5) and (6.6), we apply the *maximum a posteriori* (MAP) derivation of the Kalman filter and approximate $p(x_t|y_{1:t})$ as Gaussian probability densities by recursively computing their means (modes) and negative inverse of the Hessian matrices (covariance matrices) (Mendel, 1995; Brown et al., 1998). To initiate the recursive algorithm, let the notation $t|j$ denote the expectation of the state variable at t given the responses up to time j. We assume that the mean $x_{t-1|t-1}$ and covariance matrix $W_{t-1|t-1}$ have been estimated at time $t - 1$. That is, we take $p(x_{t-1}|y_{0:t-1})$, the posterior probability density at time $t - 1$, to be the Gaussian probability density with mean $x_{t-1|t-1}$ and covariance matrix $W_{t-1|t-1}$. The next step is to compute $p(x_t|y_{0:t-1})$, the one-step prediction probability density at time t. This is the probability density of the predicted position at t given the spiking activity in $(0, t - 1]$. It follows from standard properties of integrals of Gaussian functions that the mean and covariance matrix are respectively defined as

$$x_{t|t-1} = E(x_t|y_{0:t}) = \mu_x + Fx_{t-1|t-1} \tag{6.9}$$

$$W_{t|t-1} = Var(x_t|y_{0:t}) = FW_{t-1|t-1}F' + RW_\varepsilon \tag{6.10}$$

which correspond, respectively, to the one-step prediction estimate and the one-step prediction variance. Assuming a Gaussian approximation, we x have

$$p(x_t|y_{0:t-1}) = (2\pi)^{-1} \left|W_{t|t-1}\right|^{-\frac{1}{2}} \exp\left\{-\frac{1}{2}(x_t - x_{t|t-1})'W_{t|t-1}^{-1}(x_t - x_{t|t-1})\right\}. \tag{6.11}$$

Substituting Eqs. (6.5) and (6.11) into Eq. (6.1), and neglecting the denominator and other terms that do not depend on position, we obtain an explicit expression for the probability density of position at time t given the spiking activity in $(0, t]$:

$$p(x_t|y_{0:t}) \propto \exp\left\{-\frac{1}{2}(x_t - x_{t|t-1})'W_{t|t-1}^{-1}(x_t - x_{t|t-1})\right\}$$

$$\times \exp\left[\sum_{c=1}^{C} y_t^c \log \lambda^c(x_t|y_{0:t-1}) - \sum_{c=1}^{C} \lambda^c(x_t|y_{0:t-1})\Delta\right]. \tag{6.12}$$

To evaluate Eq. (6.12), we further assume a Gaussian approximation by deriving a quadratic approximation to $\log p(x_t|y_{0:t})$. Thus, we rewrite Eq. (6.1) as

$$p(x_t|y_{0:t}) \propto p(x_t|y_{0:t-1})p(y_t|x_t). \tag{6.13}$$

Given $p(x_t|y_{0:t})$, the MAP estimate of x_t is

$$\hat{x}_t = \arg\ \max p(x_t|y_{0:t}). \tag{6.14}$$

Simply stated, the MAP estimate is the mode of the posterior density. Given a posterior probability density $p(x_t|y_{0:t})$ and $\hat{x}_t$, an approximate MAP estimate of x_t, we can expand the log posterior probability density in a Taylor series around $\hat{x}_t$ to obtain

$$\log p(x_t|y_{0:t}) = \log p(\hat{x}_t|y_{0:t}) + \nabla \log p(\hat{x}_t|y_{0:t})(x_t - \hat{x}_t)$$
$$+ \frac{1}{2}(x_t - \hat{x}_t)'\nabla^2 \log p(\hat{x}_t|y_{0:t})(x_t - \hat{x}_t) + \cdots$$
$$\approx \log p(\hat{x}_t|y_{0:t}) + \frac{1}{2}(x_t - \hat{x}_t)'\nabla^2 \log p(\hat{x}_t|y_{0:t})(x_t - \hat{x}_t)$$

because $\nabla \log p(\hat{x}_t|y_{0:t}) = 0$ by definition of the MAP estimate. Hence,

$$p(x_t|y_{0:t}) \approx p(\hat{x}_t|y_{0:t})\exp\left\{\frac{1}{2}(x_t - \hat{x}_t)'\nabla^2 \log p(\hat{x}_t|y_{0:t})(x_t - \hat{x}_t)\right\}. \tag{6.15}$$

gives the Gaussian approximation to a posterior of probability density defined as

$$p(x_t|y_{0:t}) \approx N(\hat{x}_t, W_{t|t}) \tag{6.16}$$

where $\hat{x}_t$ is the posterior mode (the MAP estimate of x_t) in Eq. (6.15) and

$$W_{t|t} = -\left[\nabla^2 \log p(\hat{x}_t|y_{0:t})\right]^{-1}. \tag{6.17}$$

In our example, from Eq. (6.12), the log posterior $\log p(x_t|y_{0:t})$ is

$$\log p(x_t|y_{0:t}) \propto -\frac{1}{2}(x_t - x_{t|t-1})'W_{t|t-1}(x_t - x_{t|t-1})$$
$$+ \sum_{c=1}^{C} y_{t-1:t}^c \log \lambda^c(x_t|y_{0:t-1}) - \sum_{c=1}^{C} \lambda^c(x_t|y_{0:t-1})\Delta. \tag{6.18}$$

The gradient and the Hessian matrix required for the quadratic approximation are given respectively by

$$\nabla \log p(x_t|y_{0:t}) = -W_{t|t-1}^{-1}(x_t - x_{t|t-1})$$

$$+ \sum_{c=1}^{C} \nabla \log \lambda^c(x_t|y_{0:t-1}) \left[y_{t-1:t}^c - \lambda^c(x_t|y_{0:t-1})\Delta \right]$$

(6.19)

$$\nabla^2 \log p(x_t|y_{0:t}) = -W_{t|t-1}^{-1} + \sum_{c=1}^{C} \nabla^2 \log \lambda^c(x_t|y_{0:t-1}) \left[y_t^c - \lambda^c(x_t|y_{0:t-1})\Delta \right]$$

$$- \nabla \log \lambda^c(x_t|y_{0:t-1}) \left[\nabla \lambda^c(x_t|y_{0:t-1})\Delta \right]'.$$

(6.20)

Solving for x_t in Eq. (6.19) yields the posterior mode, and taking the negative inverse of the Hessian from Eq. (6.20) yields the covariance matrix. We term the resulting recursive algorithm the point process filter algorithm (Brown et al., 1998; Barbieri et al., 2004) and summarize it as follows:

- One-step prediction:

$$x_{t|t-1} = \mu_x + \hat{F} x_{t-1|t-1},$$

(6.21)

- One-step prediction variance:

$$W_{t|t-1} = \hat{F} W_{t-1|t-1} \hat{F}' + R \hat{W}_\varepsilon$$

(6.22)

- Posterior mode:

$$x_{t|t} = x_{t|t-1} + W_{t|t-1} \sum_{c=1}^{C} \nabla \log \lambda^c(x_{t|t}|\hat{\theta}_j^c) \left[y_t^c - \lambda^c(x_{t|t}|\hat{\theta}_j^c)\Delta \right]$$

(6.23)

- Posterior variance:

$$W_{t|t}^{-1} = \left[W_{t|t-1}^{-1} - \sum_{c=1}^{C} \left[\nabla^2 \log \lambda^c(x_{t|t}|\hat{\theta}_j^c)[y_t^c - \lambda^c(x_{t|t}|\hat{\theta}_j^c)\Delta] \right. \right.$$

$$\left. \left. - \nabla \log \lambda_c(x_{t|t}|\hat{\theta}_j^c)[\nabla \lambda_c(x_{t|t}|\hat{\theta}_j^c)\Delta]' \right] \right]$$

(6.24)

for $t = 1, \ldots, T$, where $\lambda^c (x_{t|t}|\hat{\theta}_j^c)$ is the conditional intensity (rate) function for either the spatial Gaussian or Zernike model for neuron c; $\hat{\theta}_j^c$ is the associated model parameter for either the spatial Gaussian ($j = G$) or Zernike ($j = Z$) model estimated from the encoding analysis; $\hat{F}$ and $\hat{W}_\varepsilon$ are, respectively, the estimates of the transition matrix and the white noise covariance matrix for the $AR(1)$ model from the encoding analysis; and R is a scale factor.

The two-dimensional position and the spiking activity of 32 (34) neurons recorded from two rats, animal 1 (2) for 23 (25) minutes, were analyzed. The encoding analysis was performed on the first 13 (15) minutes of ensemble neural spiking activity from animals 1 (2), during which the parameters of the Gaussian and Zernike place receptive field models, as well as the position model, were estimated for each neuron for both animals. The decoding algorithm was applied to the last 10 minutes for each. An example of one minute of the application of the decoding algorithm comparing the Gaussian and Zernike models is shown in Figure 6.2(a). The median error for each of the 15-second segments is smaller for the Zernike model than for the Gaussian model. In general, the Zernike model gave more accurate decoding results than the Gaussian model, as illustrated by the box plots of the error distributions in Figure 6.2(b). The median error for the Zernike (Gaussian) model was 5.9 and (7.9) for animal 1, and 5.5 and (7.7) for animal 2. For animal 1 the error distribution of the Zernike model lies to the left of the error distribution for the Gaussian model. Although for animal 2, the error distribution of the Zernike model lies to the left of the error distribution of the Gaussian model up to the 75th percentile, the Zernike model error distribution has a larger right tail. The analyses from both models demonstrate that the hippocampus maintains a dynamic representation of the animal's position in its environment as encoded in the ensemble spiking activity.

6.3.2 Neural Receptive Field Plasticity and Instantaneous Steepest Descent Filtering

The receptive fields of neurons are dynamic. That is, their responses to relevant stimuli change with experience. Experience-dependent change or plasticity has been documented in a number of brain regions. In the rat hippocampus, the pyramidal neurons in the CA1 region have spatial receptive fields. As a rat executes a behavioral task, a given CA1 neuron fires only in a restricted region of the experimental environment, termed the cell's spatial or place receptive field. Place fields change in a reliable manner as the animal executes its task. When the experimental environment is a linear track, these spatial receptive fields have been shown to migrate and skew in the direction opposite the cell's preferred direction of firing relative to

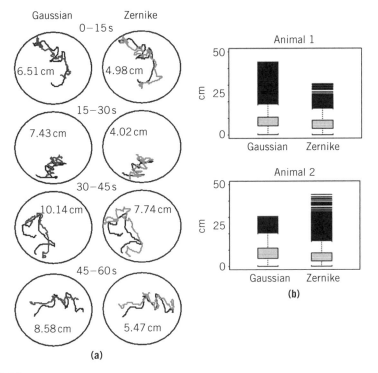

(a)

(b)

FIGURE 6.2

Decoding comparison and box plots of the error distributions for the 10 minutes of the decoding analysis. (a) Trajectory decoding comparison between two encoding models from 60s-recordings of one animal. (b) Box plot summaries of the decoding error (defined as |True Position—Estimated Position|) distributions for the spatial Gaussian model and the optimal Zernike polynomial. The lower border of the box is the 25th percentile of the distribution and the upper border is the 75th percentile. The white bar within the box is the median of distribution. The distance between the 25th and 75th percentiles is the interquartile range (IQR). The lower (upper) whisker is at 1.5 × IQR below (above) the 25th (75th) percentile. All the black bars below (above) the lower (upper) whiskers are far outliers. For reference, less than 0.35% of the observations from a Gaussian distribution would lie beyond the 75th percentile plus 1.5 × IQR. Please see this figure in color at the companion web site: www.elsevierdirect.com/companions/9780123750273

Source: Barbieri et al. (2004), *Neural Computation*, Reprinted with permission, Copyright © 2004 MIT Press.

the animal's movement, and to increase in scale and maximum firing rate (Frank et al., 2004).

As an illustration of this behavior, an actual place cell spike train recorded from a rat running back and forth for 1200 s on a 300-cm U-shaped track is shown in Figure 6.3. To display all of the experimental data on a

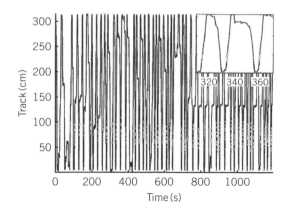

FIGURE 6.3

Place specific firing dynamics of an actual CA1 place cell recorded from a rat running back and forth on a 300-cm U-shaped track for 1200 s. The track was linearized to display the entire experiment in a single graph. The vertical lines show the animal's position and the light gray dots indicate the time at which a spike was recorded. The inset is an enlargement of the display from 320 to 360 s to show the cell's unidirectional firing—that is, spiking only when the animal runs from the bottom to the top of the track. Please see this figure in color at the companion web site: www.elsevierdirect.com/companions/9780123750273

Source: Brown et al. (2001). Proceedings of National Academy of Sciences, Reprinted with permission, copyright © 2001 National Academy of Science, U.S.A.

single graph, we represent the track linearly. The actual trajectory is irregular because the animal stops and starts several times and, in two instances (50 s and 650 s), turns around shortly after initiating its run. On several of the upward passes, particularly in the latter part of the experiment, the animal slows as it approaches the curve in the U-shaped track at approximately 150 cm. The strong place-specific firing of the neuron is readily visible because the spiking activity occurs almost exclusively between 50 and 100 cm. The spiking activity of the neuron is entirely unidirectional as the cell discharges only as the animal runs up and not down the track (Figure 6.3, inset).

Brown et al. (2001) proposed a state space model to analyze these data, in which the observation equation is

$$\Pr(y_t | \lambda(x_t, \theta_t)) = \exp(y_t \log \lambda(x_t, \theta_t) \Delta - \lambda(x_t, \theta_t) \Delta) \qquad (6.25)$$

and the state equation is

$$\theta_t = I\theta_{t-1} \qquad (6.26)$$

where I is an identity matrix. The CIF models the receptive field of the neuron as a Gaussian function:

$$\lambda(x_t,\theta_t) = \exp\left(\alpha_t - \frac{(x_t - \mu_t)^2}{2\sigma_t^2}\right) \tag{6.27}$$

and $\theta_t = (\alpha_t,\mu_t,\sigma_t^2)$.

The joint probability density of a neural spike train may be written as

$$p(y_{0:T}) = \exp\left\{\int_0^T \log \lambda(u|H_u)\,dy(u) - \int_0^T \lambda(u|H_u)\,du\right\} \tag{6.28}$$

If the probability density in Eq. (6.28) depends on estimation of an unknown q-dimensional parameter θ, then the logarithm of Eq. (6.28) viewed as a function of θ given $y_{0:T}$ is the sample path of the log likelihood defined as

$$L(\theta|y_{0:T}) = \int_0^T \ell_u(\theta)\,du \tag{6.29}$$

where $\ell_t(\theta)$ is the integrand in Eq. (6.29) or the "instantaneous" log likelihood defined as

$$\ell_t(\theta)dt = \log[\lambda(t|H_t,\theta)]y_t - \lambda(t|H_t,\theta). \tag{6.30}$$

Heuristically, Eq. (6.30) measures the instantaneous accrual of "information" from the spike train about the parameter θ. We use it as the criterion function in our point process adaptive filter algorithm.

To derive an adaptive point process filter algorithm we assume that the q-dimensional parameter θ in the instantaneous log likelihood. Eq. (6.30) is time-varying. The interval $(0,T]$ is divided into equal bins with bin width Δ. The parameter estimates are updated at every time bin indexed by t. A standard prescription for constructing an adaptive filter algorithm to estimate a time-varying parameter is to use the instantaneous steepest descent (Haykin, 2002):

$$\hat{\theta}_t = \hat{\theta}_{t-1} - \varepsilon \nabla \ell_t(\hat{\theta}_{t-1}) \tag{6.31}$$

which, by a rearrangement of terms, gives the *instantaneous steepest descent* adaptive filter algorithm for point process observations (Brown et al., 2001):

$$\hat{\theta}_t = \hat{\theta}_{t-1} - \varepsilon \nabla \log \lambda(t|H_t,\hat{\theta}_{t-1})[y_t - \lambda(t|H_t,\hat{\theta}_{t-1})\Delta]. \qquad (6.32)$$

Namely, the parameter update $\hat{\theta}_t$ at t is the previous parameter estimate $\hat{\theta}_{t-1}$ plus a dynamic gain coefficient, $-\varepsilon \nabla \log \lambda(t|H_t,\hat{\theta}_{t-1})$, multiplied by an innovation or error signal $[y_t - \lambda(t|H_t,\hat{\theta}_{t-1})\Delta]$. The error signal provides the new information coming from the spike train and is defined by comparing the predicted probability of a spike, $\lambda(t|\hat{\theta}_{t-1})\Delta$, at t with y_t, which is 1 if a spike is observed in $((t-1),t]$ and 0 otherwise. How much the new information is weighted depends on the magnitude of the dynamic gain coefficient. The parallel between the error signal in Eq. (6.32) and that in standard recursive estimation algorithms suggests that the instantaneous log likelihood is a reasonable criterion function for adaptive filters with point process observations. The instantaneous steepest descent algorithm can also be derived as a special case of the point process filter by setting $W_{t|t-1} = \varepsilon$ at each update.

We used the instantaneous steepest descent algorithm to track the evolution of the place field from the neural spiking activity in the rat CA1 neuron shown in Figure 6.3. The activity was recorded for 1200 seconds of the animal running back and forth on the U-shaped track. The evolution of the individual parameters across the experiment is shown in Figure 6.4. During the 20 minutes of the experiment, the center of the place field moved from 85 cm to 65 cm, the scale of the field increased from 10 cm to 16 cm, and the amplitude of the field increased from 1 spike/sec to 27 spikes/sec. The evolution of the place field can be seen in Figure 6.5, which shows its shape at four different time points. The displacement of the field and its growth in height and scale are evident. The maximum likelihood estimate of the field (the dashed line in the figure), computed using a static model, overestimated the field's shape and location during the early part of the experiment, and underestimated them during the latter part of the experiment.

These results illustrate that even in an environment that is familiar to the animal the place receptive fields change their representations of the space. The instantaneous steepest descent algorithm allows us to track those changes on a millisecond-to-millisecond time scale.

6.3.3 Tracking Spatial Receptive Field and Particle Filtering

The challenge of using the Bayes-Chapman-Kolmogorov equations is in evaluating Eqs. (6.1) and (6.2) for a generic set of observation and state equations.

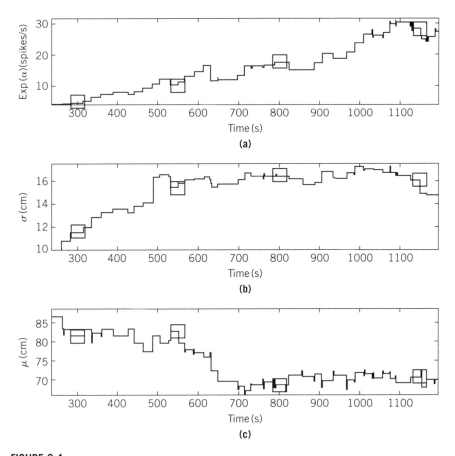

FIGURE 6.4

Adaptive filter estimates of the trajectories. (a) Maximum spike rate, $\exp(\alpha)$. (b) Place field scale, σ. (c) Place field center. Adaptive estimates were updated at 1-ms intervals. The squares in the panels at 300, 550, 800, and 1150 s are the times at which the place fields are displayed in Figure 6.5. The growth of the maximum spike rate (a), the variability of the place field scale (b), and the migration of the place field center (c) are all readily visible. *Source*: Brown et al. (2001), *Proceedings of National Academy of Sciences*, Reprinted with permission, copyright © 2001 National Academy of Science, U.S.A.

As noted earlier, there are four computational approaches to performing Bayesian calculations: analytically, Gaussian or other approximation, exact numerical integration, and Monte Carlo. As an alternative to the point process filter algorithms, we can develop sequential Monte Carlo (SMC) or particle filter algorithms for evaluating the Bayes-Chapman-Kolmogorov system. The SMC algorithm is a recursive Monte Carlo procedure for simulating Eqs. (6.1) and (6.2). When the Monte Carlo samples or particles used in the algorithm

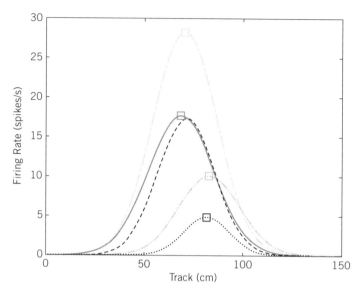

FIGURE 6.5

Estimated place fields at times 300 (dotted line), 550 (spike above dotted line), 800 (solid line), and 1150 (highest spike) sec. As in Figure 6.4, the black dashed line is the maximum likelihood estimate (MLE) of the place field obtained by using all the spikes in the experiment. The MLE ignores the temporal evolution of the place field. Please see this figure in color at the companion web site: www.elsevierdirect.com/companions/ 9780123750273

Source: Brown et al. (2001), *Proceedings of National Academy of Sciences,* Reprinted with permission, copyright © 2001 National Academy of Science, U.S.A.

are properly placed, weighted, and propagated, this system can be estimated sequentially through time. Moreover, it can be shown that as the number of particles increases, the output of the algorithm approaches an exact evaluation of these equations (Doucet et al., 2001; Cappé et al., 2005, 2007).

To define the SMC algorithm, we let π denote the proposal probability density, which defines the probability density from which we sample the particles at each time t of the algorithm, and we let $p(\theta_0)$ define the probability density for the state at time 0. We let $\hat{\theta}_t^{(i)}$ denote the value of particle i at t, and we let $w_t^{(i)}$ denote the weight of particle i at t for $i = 1,\ldots,M$, where M is the total number of particles sampled at each update of the algorithm. The SMC algorithm is therefore given by

Step 1. Set $t = 0$ and, for $i = 1,\ldots,M$ particles, draw the initial states $\theta_0^{(i)}$ from the $p(\theta_0)$ and set $w_0^{(i)} = M^{-1}$ for all i. Set $t = 1$.

Step 2. For $i = 1,\ldots,M$, sample $\hat{\theta}_t^{(i)}$ from $\pi(\theta_t | \theta_{0:t-1}^{(i)}, y_{1:t})$ and set $\hat{\theta}_{0:t}^{(i)} = (\theta_{0:t-1}^{(i)}, \hat{\theta}_t^{(i)})$. Evaluate the importance weights:

$$w_t^{(i)} = w_{t-1}^{(i)} \frac{p\left(y_t | \hat{\theta}_t^{(i)}\right) p\left(\hat{\theta}_t^{(i)} | \theta_{0:t-1}^{(i)}\right)}{\pi\left(\hat{\theta}_t^{(i)} | \theta_{0:t-1}^{(i)}, y_{1:t}\right)}$$

Normalize the importance weights:

$$\tilde{w}_t^{(i)} = w_t^{(i)} \left[\sum_{j=1}^{M} w_t^{(j)} \right]^{-1}$$

Step 3. Resample with replacement M particles $\theta_{0:t}^{(i)}$; $i = 1,\ldots,M$ from the set $(\hat{\theta}_{0:t}^{(i)}; i = 1,\ldots,M)$ with selection probability defined by the normalized importance weights $(\tilde{w}_t^{(i)})$, and using the sampling-resampling method (SIR) with added jitter (Djuric, 2001), to obtain samples approximately distributed according to $p(\theta_{0:t}^{(i)} | y_{0:t})$. For $i = 1,\ldots,M$, set $w_t^{(i)} = \tilde{w}_t^{(i)} = M^{-1}$ to obtain the Monte Carlo probability density

$$p(\theta_{0:t}^{(i)} | y_{1:t}) \approx M^{-1} \sum_{i=1}^{M} \delta_{(\theta_{0:t}^{(i)})}(d\theta_{0:t})$$

where $\delta_{(\theta_{0:t}^{(i)})}$ is the Dirac delta function indicating a point mass at $\theta_{0:t}^{(i)}$.

Step 4. Compute any summary statistics of interest as

$$E[g_t(\theta_{0:t})] = \int g_t(\theta_{0:t}) p(\theta_{0:t} | y_{1:t}) d\theta_{0:t} \approx M^{-1} \sum_{i=1}^{M} g_t(\theta_{0:t}^{(i)})$$

where, for example, $g(\theta) = \theta$ gives the estimate of the conditional mean $\mu_{t|t}$, and $g_t(\theta_t) = \theta_t \theta_t' - \mu_{t|t} \mu_{t|t}'$ gives the estimate of the conditional variance. Set $t = t + 1$. If $t \leq T$, return to step 2; otherwise stop.

The SMC methods have been successfully applied to the problem of movement decoding from neural spike activity recorded in motor cortex (Brockwell et al., 2004; Shoham et al., 2005; Brockwell et al., 2007). In particular, they have been applied to the problem of tracking the two-dimensional spatial receptive field of a rat hippocampal neuron. The specific filter used in this analysis combines both SMC and point process filter (PPF) techniques, and has been referred to as the SMC-PPF algorithm (Ergun et al., 2007). Here the PPF algorithm is used to directly compute the proposal density.

We illustrate the algorithm as applied in an actual experiment by analyzing a spike train recorded from a single pyramidal neuron in the CA1 region of the rat hippocampus, recorded while the animal foraged for 900 s in a 70-cm-diameter open circular environment. As illustrated in Section 6.3.2 hippocampal neurons have well-defined spatial receptive fields with known dynamic properties, particularly in one-dimensional linear environments. Because the dynamics of these receptive fields have been less well studied in two-dimensional environments, we used the SMC-PPF algorithm to analyze the temporal evolution of a hippocampal neuron's receptive field. We model the two-dimensional spatial receptive field as an exponentiated, linear combination of Zernike polynomials and rewrite Eq. (6.8) as

$$\lambda(t|x_t,\theta_t) = \exp\left\{\sum_{\ell=0}^{L}\sum_{m=-\ell}^{\ell}\theta_t^{\ell,m}z_\ell^m(x_t)\right\} \tag{6.33}$$

where z_l^m is the mth component of the ℓth-order Zernike polynomial, $\theta^{\ell,m}$ is the associated coefficient, and $x_t = [x_{1,t},x_{2,t}]$ is the position of the animal at time t. The Zernike polynomials form an orthogonal basis whose support is restricted to the circular environment. Following Barbieri et al. (2004), the trade-off between model flexibility and computational complexity is balanced by choosing $L = 3$, yielding a total of 16 coefficients of which 10 are nonzero. The CIF defines the spatial receptive field of the neuron. In order to track the temporal evolution of the receptive field, it suffices to track the temporal evolution of $\theta_t = [\theta_t^{0,0} \quad \theta_t^{1,-1} \quad \cdots \quad \theta_t^{3,3}]$, which is the vector of the 10 nonzero Zernike coefficients. It is assumed that the state space model for the Zernike coefficients is a Gaussian random-walk model:

$$\theta_t = \theta_{t-1} + \eta_t \tag{6.34}$$

with zero mean and variance Σ_η. In this analysis, a total of 10,000 particles was used, and the Zernike coefficients were updated at $\Delta = 33$ ms. Initial $Q = I_{10\times10}\sigma^2$ was chosen as the diagonal covariance matrix, where $\sigma^2 = 1 \times 10^{-3}$, and θ_0, the maximum likelihood estimate of the parameter computed for this neuron from the preceding 500 s in this experiment, was used as the initial guess of the Zernike coefficients.

The results from this analysis are presented in Figure 6.6. To display all the position and spike train data, the experimental data are subdivided into nine consecutive 100-s epochs (parts (a) through (i) of the figure). In each panel, the upper plot shows the trajectory of the animal during the 100-s interval, with the spiking activity of the neuron superimposed as light gray asterisks on the trajectory to show the location of the animal at the time each spike was fired. The lower plot in each panel shows the intensity function computed by averaging the means of the estimated SMC-PPF posterior densities for each of the 10 Zernike coefficients in that 100-s segment. The spiking activity was concentrated initially in the upper left area of the environment

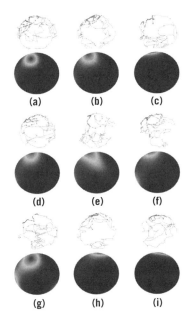

FIGURE 6.6

Tracking the temporal evolution of a 2D place field. Results are presented subdividing the 900-s experiment into nine consecutive 100 s of exploration in a circular environment 70 cm in diameter (from (a) to (i)). For each subplot, the upper panel shows the trajectory of the animal (dark gray curves) during the 100-s interval, with the spiking activity of the neuron superimposed on the trajectory (light gray asterisks). The lower panel of each subplot shows the 2D spatial receptive field computed from the average of the ten Zernike coefficients estimated in each 100-s segment. Please see this figure in color at the companion web site: www.elsevierdirect.com/companions/9780123750273
Source: Ergun et al. (2007), *IEEE Transactions on Biomedical Engineering*, Reprinted with permission, copyright © 2007 IEEE.

(Figure 6.6(a)–(b), upper panel) and the shape of the receptive field was more symmetric, with its higher-intensity peak located in the upper left area (Figure 6.6(a)–(b), lower panel).

Between 201 and 300 s, the spiking activity concentrated around the upper rim of the environment (Figure 6.6(c), upper panel); between 301 to 400 s, it reemerged in the upper left area (Figure 6.6(d), upper panel); then, between 401 to 500 s, it spread into the center of the environment (Figure 6.6(e), upper panel). From 501 to 600 s, the spiking activity was once again concentrated on the upper rim of the environment (Figure 6.6(f), upper panel). The change in the location of the spiking activity is captured nicely by the dynamic of the receptive fields between 201 and 600 s estimated from the SMC-PPF (Figure 6.6(c)–(f), lower panels). The spiking activity showed a similar pattern from 601 to 700 s (Figure 6.6(g)–(i), upper panels), and the SMC-PPF estimates of the receptive fields (Figure 6.6(g)–(i), lower panels) once again tracked this evolution. During the 900-s recording session of this experiment, the center of this spatial receptive field migrated 50 cm.

These results show that the SMC-PPF algorithm tracked the dynamics of this neuron's spatial receptive field; they also illustrate that the temporal evolution of the spatial receptive fields of hippocampal pyramidal neurons observed in linear environments can be observed in open circular environments as well.

6.3.4 Dynamic Analysis of Behaviorial Learning Experiments and the Expectation-Maximization Algorithm

Learning experiments in neuroscience can be analyzed using a state space model with a discrete observation process. A learning experiment consists of a sequence of trials in which a subject executes a task correctly or incorrectly. That is, we have a sequence of learning trials with binary responses and we let y_t denote the response in trial t, where $y_t = 1$ is a correct response (Figure 6.7, medium gray dots at top of the panel) and $y_t = 0$ is an incorrect response (Figure 6.7, dark gray dots at the top of the panel).

We let p_t denote the probability of a correct response t. At trial t, the observation model defines the probability of observing y_t given the value of the cognitive state process x_t. The observation model is the Bernoulli probability mass function:

$$\Pr(y_t|p_t,x_t) = p_t^{y_t}(1-p_t)^{1-y_t} \tag{6.35}$$

where p_t is defined by the logistic equation

$$p_t = [1 + \exp(\mu + x_t)]^{-1} \exp(\mu + x_t) \tag{6.36}$$

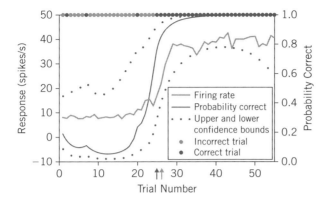

FIGURE 6.7

Sequence of correct responses (dark gray dots) and incorrect responses (medium gray dots) from a monkey executing a location-scene association task. Learning curve (solid dark gray curve) and 90% confidence intervals (dark gray dotted curves) are shown. The medium gray curve is the estimated firing rate of a hippocampal neuron recorded simultaneously as the animal executed the task. The dark gray horizontal line is the probability of a correct response by chance. Please see this figure in color at the companion web site: www.elsevierdirect.com/companions/9780123750273
Source: Wirth et al. (2003), *Science*, Reprinted with permission, Copyright © 2003 AAAS.

and μ is determined by the probability of a correct response by chance in the absence of learning or experience. We define the unobservable cognitive state process as a random walk:

$$x_t = x_{t-1} + \varepsilon_t \qquad (6.37)$$

where the ε_t are independent Gaussian random variables with mean 0 and variance σ_ε^2. In this analysis the objective is to estimate the learning curve— that is, the probability of a correct response as a function of trial number given the entire sequence of correct and incorrect responses in the experiment. Unlike the other examples we have discussed, this is a state space smoothing problem. In other words, we wish to estimate the most likely values of the learning curve at each time given all of the data in the experiment.

In the learning example, we let $x = x_{1:T}$ be the vector of the unobservable or hidden cognitive state process; x_0 and σ_ε^2 are parameters. We use the expectation-maximization (EM) algorithm to estimate them by maximum likelihood (Dempster et al., 1977). The EM algorithm is a well-known procedure for performing maximum likelihood estimation when there is an unobservable process or missing observations. We used it to estimate state space models from point process observations with linear Gaussian state processes (Smith and Brown, 2003). The current EM algorithm is a special case of the same algorithm in Smith and Brown (2003).

We state first the general formulation of the EM algorithm. If y is the observed data, x is the missing or unobservable data, and θ is the parameter to be estimated, then the EM algorithm maximizes the likelihood $p(y|\theta)$ by iterating between an expectation step (E-step) and a Maximization step (M-step). To do so, we define the complete data likelihood $p(y,x|\theta)$. The general formulation is as follows. Given an estimate of the parameter θ at iteration ℓ we compute the in step $\ell + 1$ as follows:

$$\text{E-step:} \quad Q(\theta|\theta^{(\ell)}) = E[\log p(y,x|\theta)\,||\,y,\theta^{(\ell)}] \tag{6.38}$$

which is the expectation of the complete data log likelihood.

$$\text{M-step:} \quad \theta^{(\ell+1)} = \max_{\theta} Q(\theta|\theta^{(\ell)}). \tag{6.39}$$

The algorithm iterates between the E-step in Eq. (6.38) and the M-step in Eq. (6.39). The maximum likelihood estimate of θ is $\theta_{ML} = \lim \theta^{(\ell)}{}_{\ell \to \infty} = \theta^{(\infty)}$.

The use of the EM algorithm to analyze our state space learning model here requires computing the complete data likelihood, which is the joint probability density of $y_{1:T}$ and x. The EM algorithm allows computing the maximum likelihood estimate of x_0 and σ_ε^2 and hence, the learning curve. The complete data likelihood is

$$p(y_{1:T},x|\sigma_\varepsilon^2,x_0) = \prod_{t=1}^{T} p_t^{y_t}(1-p_t)^{1-y_t} p(x|\sigma_\varepsilon^2,x_0) \tag{6.40}$$

$$= p(y_{1:T}|x)p(x|\sigma_\varepsilon^2,x_0)$$

where the first term on the right-hand side is defined by the Bernoulli probability mass function in Eq. (6.35), and second term is the joint probability density of the cognitive state process defined by the Gaussian model in Eq. (6.37). At iteration $(\ell + 1)$ of the algorithm, we compute in the E-step the expectation of the complete data log likelihood given the responses $y_{1:T}$ across the T trials and $\sigma_\varepsilon^{2(\ell)}$, as well as the parameter estimate from iteration ℓ, which is defined as follows:

E-Step:

$$Q(\sigma_\varepsilon^2 | \sigma_\varepsilon^{2(\ell)})$$

$$= E\left[\log p(y_{1:T}, x | \sigma_\varepsilon^2) \Big|\Big| y_{1:T}, \sigma_\varepsilon^{2(\ell)}, x_0\right]$$

$$= E\left[\sum_{t=1}^{T} n_t(\mu + x_t) - \log[1 + \exp(\mu + x_t)] \Big|\Big| y_{1:T}, \sigma_\varepsilon^{2(\ell)}, x_0\right].$$

$$(6.41)$$

$$+ E\left[\sum_{t=1}^{T} -\frac{1}{2}\frac{(x_t - x_{t-1})^2}{\sigma_\varepsilon^2} - \frac{T}{2}\log 2\pi - \frac{T}{2}\log \sigma_\varepsilon^2 \Big|\Big| y_{1:T}, \sigma_\varepsilon^{2(\ell)}, x_0\right]$$

Expanding the right-hand side of Eq. (6.41), we see that calculating the expected value of the complete data log likelihood requires computing the expected value of the state variable $E[x_t || y_{1:T}, \sigma_\varepsilon^{2(\ell)}, x_0]$, and the covariances $E[x_t^2 || y_{1:T}, \sigma_\varepsilon^{2(\ell)}, x_0]$ and $E[x_t x_{t+1} || y_{1:T}, \sigma_\varepsilon^{2(\ell)}, x_0]$. We denote them as

$$x_{t|T} \equiv E[x_t || y_{1:T}, \sigma_\varepsilon^{2(\ell)}, x_0] \qquad (6.42)$$

$$W_{t|T} \equiv E[x_t^2 || y_{1:T}, \sigma_\varepsilon^{2(\ell)}, x_0] \qquad (6.43)$$

$$W_{t,t+1|T} \equiv E[x_t x_{t+1} || y_{1:T}, \sigma_\varepsilon^{2(\ell)}, x_0] \qquad (6.44)$$

for $t = 1, \ldots, T$. To compute these quantities efficiently we decompose the E-step into three parts: a nonlinear recursive filter algorithm to compute $x_{t|t}$, a fixed-interval smoothing algorithm to estimate $x_{t|T}$, and a state space covariance algorithm to estimate $W_{t|T}$ and $W_{t,t-1|T}$. The specific algorithms for our model are

Filter Algorithm.   Given $\sigma_\varepsilon^{2(\ell)}$, we first compute recursively the state variable, $x_{t|t}$, and its variance, $\sigma_{t|t}^2$. We accomplish this using the filter algorithm (Smith and Brown, 2003):

- One-step prediction:

$$x_{t|t-1} = x_{t-1|t-1} \qquad (6.45)$$

- One-step prediction variance:

$$\sigma_{t|t-1}^2 = \sigma_{t-1|t-1}^2 + \sigma_\varepsilon^{2(\ell)} \qquad (6.46)$$

- Posterior mode:

$$x_{t|t} = x_{t|t-1} + \sigma_{t|t-1}^2 [y_t - p_{t|t}] \tag{6.47}$$

- Posterior variance:

$$\sigma_{t|t}^2 = \left[(\sigma_{t|t-1}^2)^{-1} + p_{t|t}(1 - p_{t|t}) \right]^{-1} \tag{6.48}$$

where $p_{t|t}$ is computed from Eq. (6.36) evaluated at $x_{t|t}$ for $t = 1, \ldots, T$ and $j = t$. The initial condition is $x_0 = 0$ and $\sigma_{0|0}^2 = \sigma_\varepsilon^{2(\ell)}$. The algorithm is nonlinear because $x_{t|t}$ appears on both left- and right-hand sides of Eq. (6.47). The derivation of this algorithm for the arbitrary point process observation model and the linear state space model is given in Smith and Brown (2003).

Fixed-Interval Smoothing Algorithm. Given the sequence of posterior mode estimates $x_{t|t}$ (Eq. (6.47)) and the variance $\sigma_{t|t}^2$ (Eq. (6.48)), we use the fixed-interval smoothing algorithm (Shumway and Stoffer, 1982; Mendel, 1995; Brown et al., 1998; Smith and Brown, 2003) to compute $x_{t|T}$ and $\sigma_{t|T}^2$. The smoothing algorithm is formulated as

$$x_{t|T} = x_{t|t} + A_t(x_{t+1|T} - x_{t+1|t}) \tag{6.49}$$

$$A_t = \sigma_{t|t}^2 (\sigma_{t+1|t}^2)^{-1}$$

$$\sigma_{t|T}^2 = \sigma_{t|t}^2 + A_t^2 (\sigma_{t+1|T}^2 - \sigma_{t+1|t}^2) \tag{6.50}$$

for $t = T - 1, \ldots, 1$ and initial conditions $x_{T|T}$ and $\sigma_{T|T}^2$.

State Space Covariance Algorithm. The covariance estimate, $\sigma_{t,u|T}$, can be computed from the state space covariance algorithm (de Jong and MacKinnon, 1988) and is given as

$$\sigma_{t,u|T} = A_t \, \sigma_{t+1,u|T} \tag{6.51}$$

for $1 \leq t \leq u \leq T$.

It follows that the expectations required in Eq. (6.42) are given by Eq. (6.49) and that the covariance terms required for Eqs. (6.43) and (6.44) of the E-step in the EM algorithm are

$$W_{t,t-1|T} = \sigma_{t-1,t|T} + x_{t|T} x_{t-1|T} \tag{6.52}$$

and

$$W_{t|T} = \sigma_{t|T}^2 + x_{t|T}^2. \tag{6.53}$$

In the M-step we maximize the expected value of the complete data log likelihood in Eq. (6.41) with respect to $\sigma_\varepsilon^{2(\ell+1)}$, which yields

$$\text{M-Step:} \quad \sigma_\varepsilon^{2(\ell+1)} = T^{-1}\left(2\left(\sum_{t=1}^T W_{t|T} - \sum_{t=1}^T W_{t-1,t|T}\right) - (W_{T|T} - W_{0|T})\right).$$

$$(6.54)$$

The algorithm iterates between the E-step (Eq. (6.41)) and the M-step (Eq. (6.54)). The maximum likelihood estimate of σ_ε^2 is $\sigma_\varepsilon^{2(\infty)}$. The convergence criteria for the algorithm are those used in Smith and Brown (2003). The filter algorithm (Eqs. (6.45) through (6.48)), evaluated at the maximum likelihood estimate of σ_ε^2, together with Eq. (6.36), give the filter algorithm estimate of the learning curve, whereas the fixed-interval smoothing algorithm evaluated at a maximum likelihood estimate of σ_ε^2 (Eqs. (6.49) through (6.50)), together with Eq. (6.36), give the maximum likelihood (empirical Bayes) or smoothing algorithm estimate of the learning curve (Smith et al., 2004).

We illustrate the algorithm applied to an actual learning experiment by analyzing the sequence of correct and incorrect responses of a macaque monkey in a location–scene association task, described in detail in Wirth et al. (2003). In this task, each trial started when the monkey fixated on a centrally presented cue on a computer screen. The animal was then presented with four identical targets (north, south, east, and west) superimposed on a novel visual scene. After the scene disappeared, the targets remained on the screen during a delay period. At the end of the delay period, the fixation point disappeared, cueing the animal to make an eye movement to one of the four targets. For each scene, only one target was rewarded and the positions of rewarded locations were counterbalanced across all new scenes. Two to three novel scenes were typically learned simultaneously, and trials of novel scenes were interspersed with trials in which two to four well-learned scenes were presented. Because there were four locations, the monkey could choose as a response; the probability of a correct response occurring by chance was 0.25. To characterize learning, the correct and incorrect responses as a function of trial number were recorded for each scene (Figure 6.7). The objective of the study was to track learning as a function of trial number and to relate it to changes in the activity of simultaneously recorded hippocampal neurons (Wirth et al., 2003).

Figure 6.7 shows the correct responses (medium gray dots) and incorrect responses (dark gray dots) responses given by the monkey across 55 trials for one novel scene. The sequence of correct and incorrect responses and the learning curve (Figure 6.7, dark gray curve) show that the animal's performance changed dramatically after approximately trial 23. The 90% confidence intervals for the learning curve were computed using the change-of-variable formula in Smith et al. (2004). With the criterion of the trial being that beyond which performance better than chance was certain with a probability of at least 0.95 for the balance of the experiment (Smith et al., 2004), the learning trial was identified as trial 23 (Figure 6.7, dark gray arrow). A dynamic estimate of the spike rate function (Figure 6.7, medium gray curve) for a hippocampal neuron was computed using the instantaneous steepest descent algorithm (Eq.(6.32)). The trial in which the spiking activity of the neurons was identified as being different from baseline was trial 24 (Figure 6.7, medium gray arrow). The change in the spike rate function for this neuron occurred almost simultaneously with the change in the monkey's performance of the task, which suggests that the neuron's activity mirrored the learning process.

These results illustrate how the state space smoothing algorithm for binary responses can be used along with the instantaneous steepest descent algorithm to relate changes in behavior on a learning task to changes in the activity of hippocampal neurons.

6.3.5 Markov Chain Monte Carlo Methods and the Analysis of Cortical UP/DOWN States

Neuronal UP and DOWN states, the periodic fluctuations between increased and decreased spiking activity of a neuronal population, are fundamental features of cortical circuits (Battaglia et al., 2004; Haider et al, 2007; Ji and Wilson, 2007). Developing quantitative descriptions of UP/DOWN state dynamics is important for understanding how these circuits represent and transmit information. Given multiunit spike train recordings, the objective of the analysis is to estimate the states, the location, and the number of state transitions, and parameters associated with the sojourn time probability densities of the states. This is a joint state and parameter estimation problem, in which the state is discrete and the parameter is static. For this purpose, we develop an EM algorithm.

The E-step of the EM algorithm requires computing the distribution of the missing data or hidden states conditional on the observed data and the current parameter estimate (Eq. (6.38)). In the applications of this algorithm presented earlier, we used Gaussian approximations to compute these conditional distributions. In recent years many new techniques have been developed in Bayesian analyses to compute conditional or posterior distributions. Therefore, in principle, any Bayesian method used to compute posterior distributions may be used to evaluate the conditional distributions in the E-step of an EM algorithm. Markov chain Monte Carlo method (MCMC) techniques are now among those principally used to conduct Bayesian analyses (Spall, 2003; Gelman et al., 2004; Robert and Casella, 2004). In our analysis of cortical UP/DOWN states, we use the MCMC method to evaluate the conditional distributions in the E-step because it provides an efficient way to determine a critical unknown in this problem: the number of transitions between the UP and DOWN states. These algorithms are termed MCEM algorithms (Chan and Ledolter, 1995; Tanner, 1996; McLachlan and Krishnan, 2008).

When posterior distributions cannot be computed exactly, numerical methods are not feasible because an analytic approximation may be inaccurate. MCMC methods can often be constructed to simulate posterior distributions by appropriately structured random sampling. Instead of drawing independent samples from the posterior distribution directly, the MCMC algorithm constructs a Markov chain of sample draws whose equilibrium distribution comes near to the posterior distribution desired. That is, sample $i + 1$ is drawn conditional on sample i following a Markov chain rule. The key step in designing an MCMC algorithm is the choice of proposal density that governs the transition between sample draws. The Markov chain theory states that, given an arbitrary initial value, the chain will converge to the equilibrium point provided that the chain is run for a sufficiently long period of time. Various statistical tests have been developed to assess convergence (Brooks and Gelman, 1998; Gelman et al., 2004).

In most applications of MCMC algorithms, the dimension of the posterior distribution to be estimated is fixed. However, there are problems for which the dimension of the posterior distribution is unknown and must be determined as part of the analysis. The study of cortical UP/DOWN states presents such challenges because the unknown dimension is indeed the number of state transitions. For this purpose, we use reversible jump MCMC (RJM-CMC), for which the proposal density generates moves reversibly among parameter spaces of different dimensions (Green, 1995). Analysis with an RJMCMC method yields estimates of the states and a probability density over the dimension of the states.

To formulate the RJMCMC, we consider the transition from a current state space X to another state space X', where $\dim(X) \neq \dim(X')$. Let $R^{(i)}(X \to X'|\theta)$ be the proposal density associated with the proposal X' for the move type i, and let q_i be the probability of choosing the move type (i). The proposal density for the move from a current state X to X' is thus given by

$$R(X \to X'|\theta) = \sum_i q_i R^{(i)}(X \to X'|\theta). \qquad (6.55)$$

The acceptance probability for the move $X \rightarrow X'$ is given by $A = \min\{1, B\}$, where

$$B = \frac{p(y, X'|\theta) R(X' \rightarrow X|\theta)}{p(y, X|\theta) R(X \rightarrow X'|\theta)} |J| = \frac{p(y|X', \theta) p(X'|\theta) R(X' \rightarrow X|\theta)}{p(y|X, \theta) p(X|\theta) R(X \rightarrow X'|\theta)} |J|.$$

$$(6.56)$$

$p(y, X|\theta)$ is an unnormalized posterior density or, in our case, a complete data likelihood, and $|J|$ is the determinant of the Jacobian of a mapping from X to X'. If there is no change in dimension, then $|J| = 1$.

For the current UP/DOWN estimation problem, we model the state as binary (0–1) being either DOWN or UP, and we assume that a Markov or semi-Markov model governs the transitions between the states. To estimate the UP/DOWN states from multiunit spiking activity, we develop a continuous-time Markov discrete state space model with time-varying state transition with point process observations. Let X_t be the state of the continuous-time Markov process at time t, which is 0 for the DOWN state and 1 for the UP state. Let $\xi_k(j, i) = \Pr(X_{k-1} = j, X_k = i)$ be the transition probability from state j to state i at the kth sojourn, $F(t|\theta_j)$ be the cumulative distribution function of the sojourn time for the DOWN states for $j = 0$ and for the UP states for $j = 1$. We let τ_k be the length of the kth sojourn time, $v_k = \sum_{i=0}^{k-1} \tau_i$ for $k = 1, \ldots, n$, where n is the total number of state jumps or transitions. We model the sojourn time distributions for the UP and DOWN states, respectively, as exponential and log normal. It is assumed that, conditional on the state X_t, the recorded C tetrodes of multiunit spiking activity are conditionally independent. We model the CIF for the multiunit activity on tetrode c for $c = 1, \ldots, C$ by a generalized linear model:

$$\log \lambda(t|\theta^c, X_t) = \mu^c + \alpha^c X_t + \int_0^t \exp(-\beta^c(t - u)) dy(u) \qquad (6.57)$$

where $\theta^c = (\mu^c, \alpha^c, \beta^c)$. The complete data likelihood function is

$$p(y, X_{0:T}|\theta) = p(y|X_{0:T}, \theta) p(X_{0:T}|\theta)$$

$$= \exp\left(\sum_{c=1}^{C} \sum_{k=1}^{n} \int_{v_{k-1}}^{v_k} [\log \lambda^c(t|X_t, \theta^c) dy_t^c - \lambda^c(t|X_t, \theta^c) dt] \right)$$

$$(6.58)$$

$$\times \prod_{i,j} [1 - F(\tau_k|\theta_j)]^{I_k(j,j)} [F(\tau_k|\theta_j)]^{I_k(j,i)}$$

where the first term on the right-hand side of Eq. (6.58) defines the multiunit activity conditional on the states, and the second term defines the probability of state transitions between the neighboring sojourns in which $I_k(j,i)$ is the indicator function for the transition from state j to state i at the kth sojourn.

To illustrate this method, the above model was applied to the analysis of multiunit spiking recorded on $C = 8$ tetrodes in the primary somatosensory cortex of a rat. A single time series of approximately 15.7 minutes of multiunit spiking activity from the 8 tetrodes was constructed by concatenating multiunit activity recorded from 11 slow-wave sleep (SWS) periods with an average length of 85.7 s and with a standard deviation of 35.8 s. An MCEM algorithm using RJMCMC with seven categories of moves at each E-step was implemented to estimate the state transitions, to estimate $\xi_k(j,i)$ and to estimate n. The parameter θ was computed at the M-step. The details of this EM algorithm are given in Chen et al. (2009).

A 5-second segment of the data is shown in Figure 6.8. In the analysis we compared the MCEM method with a simpler hidden Markov model (HMM) also fit with an EM algorithm and an empirical, threshold-based method (Ji and Wilson, 2007). In the 5-second segment the state transitions identified by the HMM (Figure 6.8(b)) agreed most of the time with those identified by the threshold method (Figure 6.8(a)). In general, the MCEM method estimated less frequent state transitions than did the other two (Figure 6.8(c)) because the frame occurrence rates for the threshold, HMM and MCEM methods were respectively 95, 103, and 82 min^{-1}. The MCEM analysis gave fewer short sojourns because it allowed merges of neighboring sojourns during the RJMCMC procedure. The frame rate from our MCEM analysis was still greater than the rate of 43.5 min^{-1} reported in the visual cortex recordings (Ji and Wilson, 2007).

These results show that it is possible to develop stochastic models and model fitting algorithms to analyze cortical UP/DOWN states, and that different model formulations can reach different conclusions regarding the number of state transitions. As more experimental evidence accrues on the properties of cortical UP/DOWN states, the future challenge will be to determine which model most accurately describes the underlying neurophysiology.

6.3.6 State Space Smoothing, Dynamic Parameter Estimation, and Analysis of Population Learning

In population learning studies, across-subject response differences are an important source of variance that must be characterized to accurately identify features of the learning process that are common to the population. Although

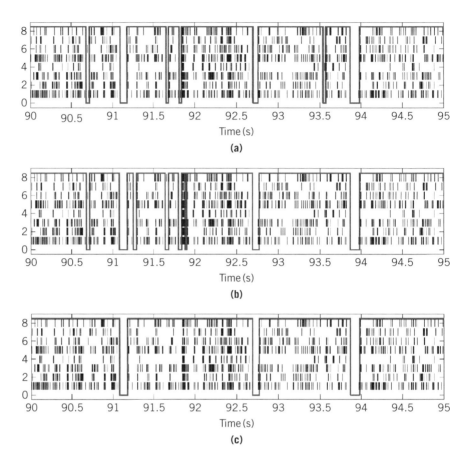

FIGURE 6.8

Multi-unit activity spike trains of 8 tetrodes recorded from the primary somatosensory cortex of one rat. (a) A selected 5-s segment of the MUA spike trains during SWS and its UP/DOWN state classification via the threshold-based method (segmented by the thick solid line). (b) The hidden state estimation result obtained from the discrete-time HMM (used as the initial state for the continuous-time RJMCMC sampler). (c) The hidden state estimation obtained from the MCEM algorithm. In this example, the MCEM algorithm merged several neighboring sojourns that were estimated differently from the HMM. *Source*: Chen et al. (2009), *Neural Computation*, reprinted with permission, Copyright © 2009 MIT Press.

learning is a dynamic process, current population analyses do not use dynamic estimation methods, nor do they compute both population and individual learning curves; rather, they use learning criteria that are less than optimal. We present a *state space random effects* (SSRE) model to estimate population and individual learning curves, ideal observer curves, and learning trials, and to make dynamic assessments of learning between two populations and within the same population that avoid multiple hypothesis tests.

To define the model, we assume that J subjects participate in a learning experiment with T trials, where we index the trials by t for $t = 1,\ldots,T$ and the subjects by j for $j = 1,\ldots,J$. To define the observation equation we let y_j^k denote the response in trial t, from subject j, where $y_t^j = 1$ is a correct response and $y_t^j = 0$ is an incorrect response. We let p_t^j denote the probability of a correct response t from subject j. We assume that the probability of a correct response in trial t from subject j is governed by an unobservable learning state process x_t, which characterizes the dynamics of learning as a function of trial number. At trial t, for subject j, the observation model defines the probability of observing y_t^j, (i.e., either a correct or an incorrect response) given the value of the state process x_t. The observation model can be expressed as the Bernoulli probability mass function

$$Pr(y_t^j|p_t^j,x_t) = (p_t^j)^{n_t^j}(1 - p_t^j)^{1-n_t^j} \tag{6.59}$$

where p_t^j is defined by the logistic function

$$p_t^j = \exp(\mu + \beta^j x_t)[1 + \exp(\mu + \beta^j x_t)]^{-1}. \tag{6.60}$$

The parameter μ in Eq. (6.60) is determined by the probability of a correct response by chance in the absence of learning or experience, and β^j is the learning modulation parameter for subject j. We define the random effects component of our state space model by assuming that the modulation parameters β^j are independent Gaussian random variables with mean β_0 and variance $\sigma_\beta^2 I_{J\times J}$, where $I_{J\times J}$ is a $J \times J$ identity matrix. Therefore, we define the probability of a correct response for the population as

$$p_t = [1 + \exp(\mu + \beta_0 x_t)]^{-1} \exp(\mu + \beta_0 x_t). \tag{6.61}$$

We define the unobservable learning state process as a random walk:

$$x_t = x_{t-1} + \varepsilon_t \tag{6.62}$$

where the ε_t are independent Gaussian random variables with mean 0 and variance σ_ε^2.

We construct a nonlinear recursive filtering algorithm, a fixed-interval smoothing algorithm, and a covariance smoothing algorithm to evaluate these expectations, as in Smith and Brown (2003) and Smith et al. (2004). To do so, we first construct the augmented state space model (Jones, 1993) in order to include the random effects component of the model in the state equation. The augmented state space model is

$$\beta_t^* = \beta_{t-1}^* + \varepsilon_t^* \tag{6.63}$$

where $\beta_t^* = (x_t, \beta_t^1, \beta_t^2, \beta_t^3, \ldots, \beta_t^J)$ and $\varepsilon_t^* = (\varepsilon_t, 0, \ldots, 0)$. The stochastic properties of x_t are defined by Eq. (6.63), whereas the stochastic properties of β come from the assumption that the modulation parameters are jointly Gaussian-distributed with mean β_0 and covariance $\sigma_\beta^2 I_{J \times J}$. Our representation of the random effects in the state space model ensures that the stochastic properties of these parameters remain constant as the filter and smoothing algorithms evolve across trials (Jones, 1993). In summary, the algorithms are structured as follows.

Filter Algorithm.   Given $\theta^{(\ell)}$, we can first compute recursively the state variable, $\beta_{t|t}^*$, and its variance, $W_{t|t}$. This is accomplished by using the following vector-valued nonlinear filter algorithm for the augmented state space model (Eden et al., 2004):

$$\beta_{t|t-1}^* = \beta_{t-1|t-1}^* \tag{6.64}$$

$$W_{t|t-1} = W_{t-1|t-1} + W_{\beta^*} \tag{6.65}$$

$$\beta_{t|t}^* = \beta_{t|t-1}^* + W_{t|t-1} F_t \tag{6.66}$$

$$W_{t|t} = \left[W_{t|t-1}^{-1} - G_t \right]^{-1} \tag{6.67}$$

$t = 1, \ldots, T$, where W_{β^*} is the $(J+1) \times (J+1)$ diagonal covariance matrix whose $\{1,1\}$ element is $\sigma_\varepsilon^{2(\ell)}$ and whose remaining elements are 0, and where F_t is the $(J+1) \times 1$ vector whose elements are

$$F_t = \begin{bmatrix} \sum_{j=1}^{J} \beta_t^j (y_t^j - p_t^j) \\ x_t (y_t^1 - p_t^1) \\ \vdots \\ x_t (y_t^J - p_t^J) \end{bmatrix} \tag{6.68}$$

and G_t is the $(J+1) \times (J+1)$ matrix whose elements are

$$
G_{t,i,m} = \begin{cases} -\sum_{j=1}^{J} (\beta_t^j)^2 p_t^j (1 - p_t^j) & i = m = 1 \\[2ex] -x_t \beta_t^j p_t^j (1 - p_t^j) + (y_t^j - p_t^j) & \begin{aligned} & i = 1, \ m = 2, \ldots, J+1; \\ & i = 2, \ldots, J+1, \ m = 1 \end{aligned} \\[2ex] -x_t^2 p_t^j (1 - p_t^j) & i = m = 2, \ldots, J \\[2ex] 0 & \text{otherwise.} \end{cases}
$$

$$(6.69)$$

The initial conditions are $\beta_0^{*(\ell)} = (x_0^{(\ell)}, \beta_0^{(\ell)})$ and

$$
W_0 = \begin{bmatrix} \sigma_\varepsilon^{2(\ell)} & 0 \\ 0 & \sigma_\beta^{2(\ell)} I_{J \times J} \end{bmatrix}.
$$

$$(6.70)$$

In these analyses we take $x_0^\ell = 0$.

The parameters are estimated using an EM algorithm as described in Section 6.3.4. The details of the algorithm are given in Smith et al. (2005).

To illustrate an application of the methods just described, we analyzed the learning behavior of four groups of rats in a set-shift experiment (Figure 6.9). Each animal executed 80 trials of a binary choice task. One group was treated with MK801 (treatment group), an NMDA antagonist used as a pharmacological model for schizophrenia; the other group received a placebo consisting of only the vehicle (vehicle group) in which the MK801 was dissolved. The animals in each group were further subdivided based on whether the reward arm on the previous day had been light or dark. The trial responses of the four groups are shown in Figure 6.9 as dark gray and light gray marks corresponding, respectively, to correct and incorrect responses across the 80 trials. In Figure 6.9, parts (a) and (b) (Figure 6.9, parts (c) and (d)) are the responses from the vehicle (treatment) group. Figure 6.9(a) (6.9(c)) shows the vehicle (treatment) animals rewarded for the light reward arm on the previous day, and Figure 6.9(b) (6.9(d)) shows the vehicle (treatment) animals rewarded for the dark reward arm on the previous day. Further subgroups were defined as vehicle light (6 animals), vehicle dark (7 animals), treatment light (3), and treatment dark (6).

Learning trials were identified using the criterion in Smith et al. (2004). The objective of the analysis was to estimate, for each animal, the individual

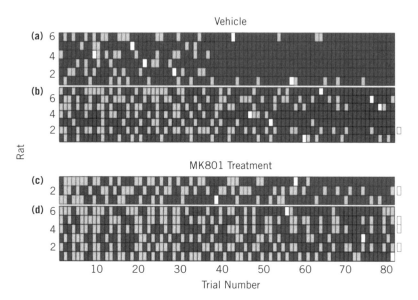

FIGURE 6.9

Behavior learning with multiple subjects. Responses of the 13 rats in the vehicle group ((a) and (b)) and 9 rats in the MK801 treatment group ((c) and (d)) during the 80 trials of set 2. Dark gray and light gray squares indicate correct and incorrect responses, respectively. Each row gives the responses of a different animal and each column represents a different trial. Both groups were trained to discriminate one of the categories of the brightness dimension in set 1. Parts (a) and (b) are the respective responses of the 6 (7) vehicle rats that were rewarded for entering the lighter (darker) arm in set 1. Parts (c) and (d) are the respective responses of the 3 (6) treatment rats that were rewarded for entering the lighter (darker) arm in set 1. White squares indicate the 8-CCR (consecutive correct response) learning trial estimate of the learning trial for each individual. Rats not achieving the 8-CCR criterion were assigned a learning trial of 80.
Source: Smith et al. (2005), *Journal of Neurophysiology*. Reprinted with permission, Copyright © 2005, American Physiological Association.

learning curve and to estimate the population learning curves for each of the four groups and to characterize the differences in learning between treatment and vehicle and between light and dark exposure on the previous day (Figure 6.10). The analysis with the population learning model shows that the two treatment groups had impaired learning (trials of 29 and 44) relative to the two vehicle groups (11 and 30). It also shows that the light exposure groups had more rapid learning (trials 11 and 29) compared with trials 30 and 44 for the dark exposure groups.

These results are consistent with the expected effect of MK801 on task acquisition and with expected behaviors of rats exposed to light and dark environments. Furthermore, they demonstrate the benefits of a hierarchical

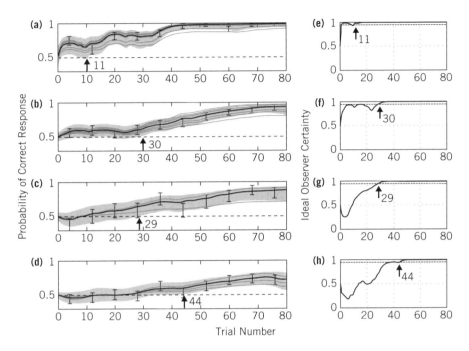

FIGURE 6.10

State space random effects (SSRE) estimation of population and individual learning. Population learning curves (dark gray) for the vehicle light (a), vehicle dark (b), treatment light (c), and treatment dark (d) subgroups and the associated 90% confidence intervals (light gray shaded area). Medium gray curves are the individual SSRE learning curves in each subgroup. Black error bars are the learning curve estimates from the 8-TB method $\pm$ SE. Arrows indicate the respective ideal observer (IO(0.95)) learning trials. Ideal observer curves (dark gray) for the vehicle light (e), vehicle dark (f), treatment light (g), and treatment dark subgroups (h). All of the ideal observer curves (dark gray) start at 0.5, the probability of a correct response by chance. The trial where an ideal observer curve crosses and stays above 0.95 (dashed horizontal line) is the IO(0.95) learning trial for that group. Please see this figure in color at the companion web site: www.elsevierdirect.com/companions/9780123750273

Source: Smith et al. (2005), *Journal of Neurophysiology,* Reprinted with permission, Copyright © 2005, American Physiological Association.

model for simultaneously characterizing individual and population learning dynamically with a single analysis.

6.4 DISCUSSION

An important objective of neuroscience data analysis is the development of statistical techniques to characterize the dynamic features inherent in most neuroscience data. State space models provide a tractable and flexible statistical framework for dynamic modeling. The estimation strategies chosen depend on the state space and observational model characteristics. We reviewed the state space paradigm with point process and binary observation models by discussing its application to the analysis of neural spike train and behavioral data. We considered the problems of neural spike train decoding, characterizing neural plasticity, analyses of behavior learning experiments, and the estimation of cortical neuronal UP/DOWN states. We conducted simultaneous state and parameter estimation using maximum likelihood methods implemented with EM algorithms in which Gaussian approximations and Monte Carlo methods were used to carry out the E-steps.

We reviewed only a small number of state space modeling applications in neuroscience. State space modeling has been applied to a variety of neural data analyses, including

- Analysis of between-trial and within-trial neural spiking dynamics (Czanner et al., 2008)

- Analysis of time-varying high-order correlations among simultaneously recorded neural spike trains (Shimazaki et al., 2009)

- Estimation of spike rate functions (Smith et al., 2010)

- Design of algorithms for control of neural prosthetic devices (Wu et al., 2006; Srinivasan et al., 2006, 2007; Hatsopoulos and Donoghue, 2009)

- Solving inverse problems for either functional imaging (Penny et al., 2005) or multi-channel EEG or MEG recordings (Barton et al., 2009; Lamus et al., 2010)

State space models have also been used to analyze calcium imaging (Vogelstein et al., 2009), to decode biophysical recordings (Huys and Paninski, 2009; Paninski, 2009), and to conduct causality analyses (Wong and Ozaki, 2007).

There are several important future directions in which the state space modeling paradigm can be extended. First, maximum likelihood algorithms have obvious Bayesian extensions. Extensions for the analysis of behavioral data can be easily implemented using standard software (Smith et al., 2007). In general, the development of computationally efficient estimation algorithms for the analysis of large-scale neuroscience data (Long et al., 2006; Ahmadian et al., 2009; Koyama and Paninski, 2009; Paninski et al., 2009; Koyama et al., 2010a,b) is another important direction. Second, the observation models can be extended to include mixed (continuous, point process, and binary) observation models for multimodal neural data that are recorded on the same or different time scales. For example, a state and parameter estimation algorithm was recently developed for cognitive state estimation from simultaneously recorded continuous and binary measures of performance (Prerau et al., 2008; Prerau et al., 2009). Finally, the development of state space models to handle state switching (Kim and Nelson, 1999; Ghahramani and Hinton, 2000; Wu et al., 2004; Srinvasan et al., 2007; Fox et al., 2009) and hierarchical state space models (Ghahramani, 1998; Friston, 2008) for neuroscience applications is another important direction. All of these can be treated within the unifying framework of graphical models (Jordan, 1997; Airoldi, 2007).

Acknowledgments

This work was supported by NIH grants R01-DA015644, R01-MH071847, and R01-HL084502, and by NIH Director Pioneer Award DP1-OD003646.

References

Ahmadian, Y., Pillow, J., Paninski, L., 2009. Efficient Markov chain Monte Carlo methods for decoding neural spike trains. Submitted.

Airoldi, E.M., 2007. Getting started in probabilistic graphical models. PLoS Comput. Biol. 3 (12), e252.

Barbieri, R., Frank, L.M., Nguyen, D.P., Quirk, M.C., Solo, V., Wilson, M.A., et al., 2004. Dynamic analyses of information encoding by neural ensembles. Neural Comput. 16, 277–307.

Barton, M.J., Robinson, P.A., Kumar, S., Galka, A., Durrant-Whyte, H.F., Guivant, J., et al., 2009. Evaluating the performance of Kalman-filter-based EEG source localization. IEEE Trans. Biomed. Eng. 56, 122–136.

Barnes, T.D., Kubota, Y., Hu, D., Jin, D.Z.Z., Graybiel, A.M., 2005. Activity of striatal neurons reflects dynamic encoding and recoding of procedural memories. Nature 437, 1158–1161.

Battaglia, F.P., Sutherland, G.R., McNaughton, B.L., 2004. Hippocampal sharp wave bursts coincide with neocortical "up-state" transitions. Learn. Mem. 11, 697–704.

Bialek, W., Rieke, F., de Ruyter van Steveninck, R.R., Warland, D., 1991. Reading a neural code. Science 252, 1854–1857.

Boas, D.A., Brooks, D.H., Miller, E.L., DiMarzio, C.A., Kilmer, M., Gaudette, R.J., et al., 2001. Imaging the body with diffuse optical tomography. IEEE Signal Process. Mag. 18, 57–75.

Brockwell, A.E., Rojas, A.L., Kass, R.E., 2004. Recursive Bayesian decoding of motor cortical signals by particle filtering. J. Neurophysiol. 91, 1899–1907.

Brockwell, A.E., Kass, R.E., Schwartz, A.B., 2007. Statistical signal processing and the motor cortex. Proc. IEEE 95, 881–898.

Brooks, S., Gelman, A., 1998. General methods for monitoring convergence of iterative simulations. J. Comput. Graph. Stat. 7, 434–455.

Brown, E.N., 2005. The theory of point processes for neural systems. In: Chow, C.C., Gutkin, B., Hansel, D., Meunier, C., Dalibard, J. (Eds.), Methods and Models in Neurophysics. Elsevier, Paris, pp. 691–726.

Brown, E.N., Barbieri, R., Eden, U.T., Frank, L.M., 2003. Likelihood methods for neural data analysis. In: Feng, J. (Ed.), Computational Neuroscience: A Comprehensive Approach. CRC Press, London, pp. 253–286.

Brown, E.N., Barbieri, R., Ventura, V., Kass, R.E., Frank, L.M., 2002. The time-rescaling theorem and its application to neural spike data analysis. Neural Comput. 14, 325–346.

Brown, E.N., Frank, L.M., Tang, D., Quirk, M.C., Wilson, M.A., 1998. A statistical paradigm for neural spike train decoding applied to position prediction from ensemble firing patterns of rat hippocampal place cells. J. Neurosci. 18, 7411–7425.

Brown, E.N., Mitra, P.P., Kass, R.E., 2004. Multiple neural spike train data analysis: state-of-the-art and future challenges. Nat. Neurosci. 7, 456–461.

Brown, E.N., Nguyen, D.P., Frank, L.M., Wilson, M.A., Solo, V., 2001. An analysis of neural receptive field plasticity by point process adaptive filtering. Proc. Natl. Acad. Sci. 98, 12261–12266.

Cappé, O., Moulines, E., Rydén, T., 2005. Inference in Hidden Markov models. Springer, Berlin.

Cappé, O., Godsill, S.J., Moulines, E., 2007. An overview of existing methods and recent advances in sequential Monte Carlo. Proc. IEEE 95 (5), 899–924.

Carlin, B.P., Polson, N.G., Stoffer, D.S., 1992. A Monte Carlo approach to nonnormal and nonlinear state-space modeling. J. Am. Stat. Assoc. 87, 493–500.

Chan, K.S., Ledolter, J., 1995. Monte Carlo EM estimation for time series models involving counts. J. Am. Stat. Assoc. 90, 242–252.

Chen, Z., Vijayan, S., Barbieri, R., Wilson, M.A., Brown, E.N., 2009. Discrete- and continuous-time probabilistic models and algorithms for inferring neuronal UP and DOWN states. Neural Comput. 21, 1797–1862.

Czanner, G., Eden, U.T., Wirth, S., Yanike, M., Suzuki, W.A., Brown, E.N., 2008. Analysis of between-trial and within-trial neural spiking dynamics. J. Neurophysiol. 99, 2672–2693.

Daley, D., Vere-Jones, D., 2002. An Introduction to the Theory of Point Processes, Vol. 1: Elementary Theory and Methods. Springer, New York.

deJong, P., MacKinnon, M.J., 1988. Covariances for smoothed estimates in state space models. Biometrika 75, 601–602.

Dempster, A., Laird, N., Rubin, D.B., 1977. Maximum likelihood from incomplete data via the EM algorithm. J. R. Stat. Soc. B39, 1–38.

Denk, W., Strickler, J.H., Webb, W.W., 1990. Two-photon laser scanning fluorescence microscopy. Science 248, 73–76.

Djuric, P., 2001. Sequential estimation of signals under model uncertainty. In: Sequential Monte Carlo Methods in Practice. Springer, New York.

Doucet, A., de Freitas, N., Gordon, N., 2001. Sequential Monte Carlo Methods in Practice. Springer, New York.

Durbin, J., Koopman, S.J., 2001. Time Series Analysis by State Space Methods. Oxford Univ. Press.

Eden, U.T., Frank, L.M., Barbieri, R., Solo, V., Brown, E.N., 2004. Dynamic analyses of neural encoding by point process adaptive filtering. Neural Comput. 16, 971–998.

Eden, U.T., Brown, E.N., 2008. Continuous-time filters for state estimation from point process models of neural data. Stat. Sin. 18, 1293–1310.

Ergun, A., Barbieri, R., Eden, U.T., Wilson, M.A., Brown, E.N., 2007. Construction of point process adaptive filter algorithms for neural systems using sequential Monte Carlo methods. IEEE Trans. Biomed. Eng. 54 (3), 419–428.

Fahrmeir, L., Tutz, G., 2001. Multivariate Statistical Modelling Based on Generalized Linear Models, second ed. Springer, New York.

Fox, E.L., Sudderth, E.B., Jordan, M.I., Willsky, A.S., 2009. Nonparametric Bayesian learning of switching linear dynamical systems. In: Advances in Neural Information Processing Systems 22. MIT Press, Cambridge, MA.

Frank, L.M., Stanley, G.B., Brown, E.N., 2004. Hippocampal plasticity across multiple days of exposure to novel environments. J. Neurosci. 24, 7681–7689.

Friston, K., 2008. Hierarchical models in the brain. PLoS Comput. Biol. 4 (11), e1000211. doi:10.1371/journal.pcbi.1000211.

Gelman, A., Carlin, J.B., Stern, H.S., Rubin, D.B., 2004. Bayesian Data Analysis, second ed. Chapman & Hall/CRC, Boca Raton, FL.

Ghahramani, Z., 1998. Learning dynamic Bayesian networks. In: Giles, C.L., Gori, M. (Eds.), Adaptive Processing of Sequences and Data Structures. Springer-Verlag, Berlin, pp. 168–197.

Ghahramani, Z., Hinton, G.E., 2000. Variational learning for switching state-space models. Neural Comput. 12, 831–864.

Gilks, W.R., Richardson, S., Spiegelhalter, D.J., 1996. Markov Chain Monte Carlo in Practice. Chappman & Hall/CRC Press, Boca Raton, FL.

Green, P.J., 1995. Reversible jump Markov chain Monte Carlo computation and Bayesian model determination. Biometrika 82, 711–732.

Haider, B., Duque, A., Hasentaub, A.R., Yu, Y., McCormick, D.A., 2007. Enhancement of visual responsiveness by spontaneous local network activity in vivo. J. Neurophysiol. 97, 4186–4202.

Hämäläinen, M., Hari, R., Ilmoniemi, R., Knuutila, J., Lounasmaa, O., 1993. Magnetoencephalography–theory, instrumentation, and applications to noninvasive studies of the working human brain. Rev. Mod. Phys. 65, 1–93.

Hatsopoulos, N.G., Donoghue, J.P., 2009. The science of neural interface systems. Annu. Rev. Neurosci. 32, 249–266.

Haykin, S., 2002. Adaptive Filter Theory, fourth ed. Prentice Hall, Upper Saddle River, NJ.

Hochberg, L.R., Serruya, M.D., Friehs, G.M., Mukand, J.A., Saleh, M., Caplan, A.H., et al., 2006. Neuronal ensemble control of prosthetic devices by a human with tetraplegia. Nature 442, 164–171.

Huys, Q.J.M., Paninski, L., 2009. Smoothing of, and parameter estimation from, noisy biophysical recordings. PLoS Comput. Biol. 5 (5), e1000379.

Ji, D., Wilson, M.A., 2007. Coordinated memory replay in the visual cortex and hippocampus during sleep. Nature Neurosci. 10, 100–107.

Jog, M.S., Kubota, Y., Connolly, C.I., Hillegaart, V., Graybiel, A.M., 1999. Building neural representations of habits. Science 286, 1745–1749.

Jones, R.H., 1993. Longitudinal Data with Serial Correlation: A State-Space Approach. Chapman & Hall, New York.

Jordan, M.I. (Ed.), 1997. Learning in Graphical Models. MIT Press, Cambridge, MA.

Kim, C.-J., Nelson, C.R., 1999. State-Space Models with Regime Switching: Classical and Gibbs-Sampling Approaches with Applications. MIT Press, Cambridge, MA.

Kitagawa, G., Gersh, W., 1996. Smoothness Priors Analysis of Time Series. Springer, New York.

Koyama, S., Paninski, L., 2009. Efficient computation of the maximum a posteriori path and parameter estimation in integrate-and-fire and more general state-space models. J. Comput. Neurosci. (in press).

Koyama, S., Eden, U.T., Brown, E.N., Kass, R.E., 2010a. Bayesian decoding of neural spike trains. Ann. Inst. Stat. Math. 62, 37–59.

Koyama, S., Perez-Bolde, L.C., Shalizi, C.R., Kass, R.E., 2010b. Approximate methods for state-space models. J. Amer. Stat. Assoc. 105, 170–180.

Kwong, K.K., Belliveau, J.W., Chesler, D.A., Goldberg, I.E., Weisskoff, R.M., Poncelet, B.P., et al., 1992. Dynamic magnetic resonance imaging of human brain activity during primary sensory stimulation. Proc. Natl. Acad. Sci. U.S.A. 89, 5675–5679.

Lamus, C. Hamalainen, M.S., Temereanca, S., Brown, E.N., Purdon, P.L., 2010. How dynamic models of brain connectivity can help solve the MEG inverse problem: analysis of source observability. Submitted.

Law, J.R., Flanery, M.A., Wirth, S., Yanike, M., Smith, A.C., Frank, L.M., et al., 2005. Functional magnetic resonance imaging activity during the gradual acquisition and expression of paired-associate memory. J. Neurosci. 25, 5720–5729.

Long, C.J., Desai, N.U., Hämäläinen, M., Temereanca, S., Purdon, P.L., Brown, E.N., 2006. A dynamic solution to the ill-conditioned magnetoencephalography (MEG) source localization problem. In: Proc. ISBI'06, 225–228.

McLachlan, G.J., Krishnan, T., 2008. The EM Algorithm and Extensions, second ed. Wiley, New York.

Mendel, J.M., 1995. Lessons in Estimation Theory for Signal Processing, Communication, and Control. Prentice Hall, Upper Saddle River, NJ.

Moran, D.W., Schwartz, A.B., 1999. Motor cortical representation of speed and direction during reaching. J. Neurophysiol. 82, 2676–2692.

Musallam, S., Corneil, B.D., Greger, B., Scherberger, H., Andersen, R.A., 2004. Cognitive control signals for neural prosthetics. Science 305, 258–262.

Okatan, M., Wilson, M.A., Brown, E.N., 2005. Analyzing functional connectivity using a network likelihood model of ensemble neural spiking activity. Neural Comput. 17, 1927–1961.

Ogawa, S., Tank, D.W., Menon, R., Ellermann, J.M., Kim, S.G., Merkle, H., et al., 1992. Intrinsic signal changes accompanying sensory stimulation: functional brain mapping with magnetic resonance imaging. Proc. Natl. Acad. Sci. U.S.A. 89, 5951–5955.

Paninski, L., 2009. Inferring synaptic inputs given a noisy voltage trace via sequential Monte Carlo methods. J. Comput. Neurosci. Submitted.

Paninski, L., Ahmadian, Y., Ferreira, D., Koyama, S., Rahnama, K., Vidne, M., et al., 2009. A new look at state-space models for neural data. J. Comput. Neurosci. (in press).

Pawitan, Y., 2001. In All Likelihood: Statistical Modelling and Inference Using Likelihood. Clarendon Press, Oxford, UK.

Penny, W., Ghahramani, Z., Friston, K., 2005. Bilinear dynamical systems. Philos. Trans. R. Soc. Lond. B 360, 983–993.

Prerau, M., Smith, A.C., Eden, U.T., Yanike, M., Suzuki, W.A., Brown, E.N., 2008. A mixed filter algorithm for cognitive state estimation from simultaneously recorded continuous and binary measures of performance. Biol Cybern 99, 1–14.

Prerau, M., Smith, A.C., Eden, U.T., Kubota, Y., Yanike, M., Suzuki, W., et al., 2009. Characterizing learning by simultaneous analysis of continuous and binary measures of performance. J. Neurophysiol. 102, 3060–3072.

Rieke, F., Warland, D., de Ruyter van Steveninck, R.R., Bialek, W., 1997. Spikes: Exploring the Neural Code. MIT Press, Cambridge, MA.

Robert, C.P., Casella, G., 2004. Monte Carlo Statistical Methods, second ed. Springer-Verlag, New York.

Roweis, S., Gharamani, Z., 1999. A unifying review of linear Gaussian models. Neural Comput. 11, 305–345.

Santhanam, G., Ryu, S.I., Yu, B.M., Afshar, A., Shenoy, K.V., 2006. A high-performance brain-computer interface. Nature 442, 195–198.

Serruya, M.D., Hatsopoulos, N.G., Paninski, L., Fellows, M.R., Donoghue, J.P., 2002. Instant neural control of a movement signal. Nature 416, 141–142.

Shephard, N., Pitt, M.K., 1997. Likelihood analysis of non-Gaussian measurement time series, Biometrika, 84, 653–667.

Shimazaki, H., Amari, S., Brown, E.N., Gruen, S., 2009. State-space analysis on time-varying correlations in parallel spike sequences. In: Proc. IEEE ICASSP (pp. 3501–3504), Taipei.

Shumway, R.H., Stoffer, D.S., 1982. An approach to time series smoothing and forecasting using the EM algorithm. J. Time Ser. Anal. 3, 253–264.

Shoham, S., Paninski, L.M., Fellows, M.R., Donoghue, J.P., Normann, R.A., 2005. Statistical encoding model for a primary motor cortical brain-machine interface. IEEE Trans. Biomed. Eng. 52, 1312–1322.

Smith, A.C., Brown, E.N., 2003. Estimating a state-space model from point process observations. Neural Comput. 15, 965–991.

Smith, A.C., Frank, L.M., Wirth, S., Yanike, M., Hu, D., Kubota, Y., et al., 2004. Dynamical analysis of learning in behavior experiments. J. Neurosci. 24, 447–461.

Smith, A.C., Stefani, M.R., Moghaddam, B., Brown, E.N., 2005. Analysis and design of behavioral experiments to characterize population learning. J. Neurophysiol. 93, 776–792.

Smith, A.C., Wirth, S., Suzuki, W., Brown, E.N., 2007. Bayesian analysis of interleaved learning and response bias in behavioral experiments. J. Neurophysiol. 97, 2516–2524.

Smith, A.C., Scalon, J., Wirth, S., Yanike, M., Suzuki, W., Brown, E.N., 2010. State-space algorithms for estimating spike rate functions. J. Comput. Intell. Neurosci. Online publication Article ID 426539. doi: 10.1155/2010/426539

Spall, J.C., 2003. Estimation via Markov chain Monte Carlo. IEEE Control Syst. Mag. 23 (2), 34–45.

Srinivasan, L., Eden, U.T., Willsky, A.S., Brown, E.N., 2006. A state-space analysis for reconstruction of goal-directed movements using neural signals. Neural Comput. 18, 2465–2494.

Srinivasan, L., Brown, E.N., 2007. A state-space framework for movement control to dynamic goals through brain-driven interfaces. IEEE Trans. Biomed. Eng. 54 (3), 526–535.

Srinivasan, L., Eden, U.T., Mitter, S.K., Brown, E.N., 2007. General-purpose filter design for neural prosthetic devices. J. Neurophysiol. 98, 2456–2475.

Stanley, G.B., Li, F.F., Dan, Y., 1999. Reconstruction of natural scenes from ensemble responses in the LGN. J. Neurosci. 19, 8036–8042.

Suzuki, W.A., Brown, E.N., 2005. Behavior and neurophysiological analyses of dynamic learning processes. Behav. Cogn. Neurosci. Rev. 4, 67–95.

Tanner, M.A., 1996. Tools for Statistical Inference, third ed. Springer, New York.

Taylor, D.M., Tillery, S.I.H., Schwartz, A.B., 2002. Direct cortical control of 3D neuroprosthetic devices. Science 296, 1829–1832.

Truccolo, W., Eden, U.T., Fellow, M., Donoghue, J.D., Brown, E.N., 2005. A point process framework for relating neural spiking activity to spiking history, neural ensemble and covariate effects. J. Neurophysiol. 93, 1074–1089.

Vogelstein, J., Watson, B., Packer, A., Yuste, R., Jedynak, B., Paninski, L., 2009. Spike inference from calcium imaging using sequential Monte Carlo methods. Biophys. J. 97, 636–655.

Wilson, M.A., McNaughton, B.L., 1993. Dynamics of the hippocampal ensemble code for space. Science 261, 1055–1058.

Wilson, M.A., McNaughton, B.L., 1994. Reactivation of hippocampal ensemble memories during sleep. Science 265, 676–679.

Wirth, S., Yanike, M., Frank, L.M., Smith, A.C., Brown, E.N., Suzuki, W.A., 2003. Single neurons in the monkey hippocampus and learning of new associations. Science 300, 1578–1584.

Wong, K.F.K., Ozaki, T., 2007. Akaike causality in state space: instantaneous causality between visual cortex in fMRI time series. Biol. Cybern. 97, 151–157.

Wu, W., Black, M.J., Mumford, D., Gao, Y., Bienenstock, E., Donoghue, J.P., 2004. Modeling and decoding motor cortical activity using a switching Kalman filter. IEEE Trans. Biomed. Eng. 51, 933–942.

Wu, W., Gao, Y., Bienenstock, E., Donoghue, J.P., Black, M.J., 2006. Bayesian population decoding of motor cortical activity using a Kalman filter. Neural Comput. 18, 80–118.

Wu, W., Kulkarni, J.E., Hatsopoulos, N.G., Paninski, L., 2009. Neural decoding of hand motion using a linear state-space model with hidden states. IEEE Trans. Neural Syst. Rehabil. Eng. 17, 370–378.

Zhang, K., Ginzburg, I., McNaughton, B.L., Sejnowski, T.J., 1998. Interpreting neuronal population activity by reconstruction: unified framework with application to hippocampal place cells. J. Neurophysiol. 79, 1017–1044.

Neural Decoding for Motor and Communication Prostheses

7

Byron M. Yu[*†**], **Gopal Santhanam**[*], **Maneesh Sahani**[**], **Krishna V. Shenoy**[*†***]

*Department of Electrical Engineering, †Neurosciences Program, Stanford University, Stanford, California, **Gatsby Computational Neuroscience Unit, UCL, London, ***Department of Bioengineering, Stanford University

7.1 INTRODUCTION

Neural prostheses, which are also termed brain–machine and brain–computer interfaces, offer the potential to substantially increase the quality of life for people suffering from motor disorders, including paralysis and amputation. Such devices translate electrical neural activity from the brain into control signals for guiding paralyzed upper limbs, prosthetic arms, and computer cursors. Several research groups have now demonstrated that monkeys (e.g., Serruya et al., 2002; Taylor et al., 2002; Carmena et al., 2003; Musallam et al., 2004; Santhanam et al., 2006; Mulliken et al., 2008; Velliste et al., 2008) and humans (e.g., Kennedy et al., 2000a,b; Leuthardt et al., 2004; Wolpaw and McFarland, 2004b; Hochberg et al., 2006; Kim et al., 2008; Schalk et al., 2008) can learn to move computer cursors and robotic arms to various target locations simply by activating neural populations that participate in natural arm movements.

Figure 7.1 illustrates the basic operating principle of neural prosthetic systems. Neural activity from various arm movement–related brain regions (e.g., Wise et al., 1997) is electronically processed to create control signals for enacting the desired movement. Non invasive sensors can collect neural signals representing the average activity of many neurons. When invasive permanently implanted arrays of electrodes are employed, as depicted, it is possible to use waveform shape differences to discriminate individual neurons (e.g., Lewicki, 1998; Santhanam et al., 2004). After determining how each neuron responds before and during a movement, which is typically accomplished by correlating arm movements made during a behavioral task with

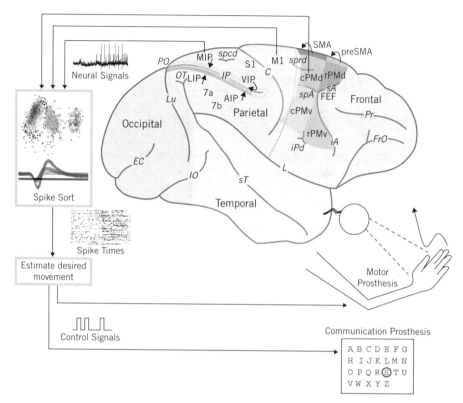

FIGURE 7.1

Concept sketch of cortically controlled motor and communication prostheses. Neural signals are recorded from cortical areas such as PMd, M1, and intraparietal area/parietal reach region (MIP/PRR) using intra-cortical electrode arrays. The times of action potentials (spikes) are identified and assigned to neural units. A decoder then translates the activity across a neural population into control signals for the prosthetic system. Please see this figure in color at the companion web site: www.elsevierdirect.com/companions/9780123750273

associated neural activity, estimation algorithms can be designed to decode the desired movement trajectory (*continuous decoder*) or movement target (*discrete decoder*) from the pattern of neural activity. The system can then generate control signals appropriate for continuously guiding a paralyzed or prosthetic arm through space (*motor prosthesis*) or positioning a computer cursor on the desired letter on a keyboard (*communication prosthesis*).

Motor prostheses aim to provide natural control of the paralyzed limb, via functional electrical stimulation of de-innervated muscle, or of a prosthetic replacement limb. In the case of upper-limb prostheses, natural control involves the precise movement of the arm along a desired path and with a desired speed profile. Such control is indeed a daunting ultimate goal, but presumably even intermediate steps along the way would lead to clinically viable systems. For example, simply being able to feed oneself would help thousands of tetraplegics (e.g., Velliste et al., 2008).

Communication prostheses do not aim to restore the ability to communicate in the form of natural voice or natural typing (i.e., moving the fingers). Instead, they aim to provide a fast and accurate communication mode, such as moving a computer cursor on an onscreen keyboard so as to type out words. Ideally performance could rival, or even surpass, the natural communication rate at which most people can speak or type. For example, "locked-in" ALS patients are altogether unable to converse with the outside world, and many other neurodegenerative diseases severely compromise the quality of speech. Being able to reliably type several words per minute on a computer will be a meaningful advance for these patients. In fact, many of the most severely disabled patients, who are the likely recipients of first-generation systems, will benefit from a prosthesis capable of typing even a few words per minute (e.g., Donoghue, 2008).

While motor and communication prostheses are quite similar conceptually, important differences critically affect their design. Motor prostheses must produce movement trajectories as accurately as possible to enact precisely the desired movement. The decoder translates, on a moment-by-moment basis, neural activity into a continuous-valued movement trajectory. In this case, the decoder solves a regression problem. In contrast, communication prostheses are concerned with information throughput from the subject to the world; this makes the speed and accuracy with which keys on a keyboard can be selected of primary importance. Although a continuously guided motor prosthesis can be used to convey information by moving to a key, only the key eventually struck actually contributes to information conveyance. The decoder, thus, acts as a classifier. This seemingly subtle distinction has implications that profoundly influence the type of neural activity and the class of decoder to be used.

In this chapter, we focus on the engineering goal of improving prosthetic system design, with particular emphasis on neural decoding algorithms. We make no claims here about these decoding algorithms furthering our basic scientific understanding of the relationship between neural activity and desired movements, although such findings may well inform next-generation algorithmic design. In the following sections, we will first describe the behavioral task and the different types of neural activity that are used to drive motor and communication prostheses. For both classes of prosthetic systems, we will then briefly review existing and detail new decoding algorithms that we recently published (Yu et al., 2007; Santhanam et al., 2009). Finally, we will consider future directions for further improving decoding performance and increasing the clinical viability of neural prosthetic systems.

7.2 PLAN AND MOVEMENT NEURAL ACTIVITY

Two types of neural activity are commonly used for driving prosthetic systems: plan and movement activity. Plan activity is present before arm movements begin, or even without ultimate arm movement, and is believed to reflect preparatory processing required for the fast and accurate generation of movement (e.g., Churchland et al., 2006a,b,c; Churchland and Shenoy, 2007a). This activity is readily observed before movement initiation in a "delayed reach task" (Figure 7.2). In this task, a trial begins when the subject touches and visually fixates a central target. After a randomized delay period (typically 0–1000 ms), a "go cue" indicates that a reach (and saccade) may begin. Figure 7.2 illustrates that plan activity in a premotor cortex (PMd) unit of a monkey performing this task is tuned for the direction of the upcoming movement, with plan activity arising soon after target onset and persisting throughout the delay period. Plan activity has also been shown to reflect movement extent (e.g., Messier and Kalaska, 2000; Churchland et al., 2006b), peak velocity (Churchland et al., 2006b), eye position (Batista et al., 2007), and attention (e.g., Boussaoud and Wise, 1993). Movement activity follows plan activity in a delayed reach task, being present 100–200 ms before and during the movement. It is tuned for both the direction (as shown in Figure 7.2) and speed of arm movement (e.g., Moran and Schwartz, 1999a). PMd neurons typically exhibit strong plan activity, as well as some movement activity; neurons in primary motor cortex (M1) typically exhibit the opposite pattern (e.g., Shen and Alexander, 1997a,b).

Neural activity was recorded using a 96-channel silicon electrode array (Blackrock Micosystems, Salt Lake City, UT) implanted in two rhesus macaque monkeys (monkeys G and H). Several different data sets are used in this chapter. A typical data set comprises 1000–2000 trials and 100–200 simultaneously recorded units. Of these units, 30–40% are typically single-neuron and 60–70% are multi-neuron. For more details about the data sets used, see Yu et al. (2007) and Santhanam et al. (2009).

Until recently, both motor and communication prostheses have focused primarily on movement activity. As depicted in Figure 7.3(a), movement activity can be elicited merely by "thinking" about moving the arm and, surprisingly, it has been found that no movement or electromyographic (EMG) activity

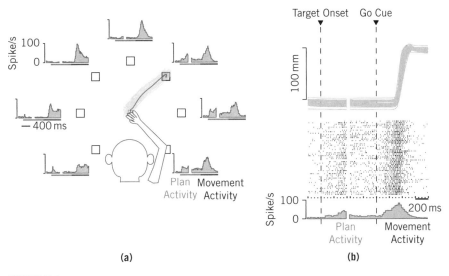

FIGURE 7.2

Plan and movement activity from a single PMd neuron in a delayed reach task. (a) Spike histograms showing average plan (gray bar) and movement (black bar) activity associated with center-out reaches to peripheral targets. Fifty representative reach trajectories to the upward-right target are shown in light gray (mean trajectory in black). (b) Top panel: same 50 representative reach trajectories as in (a) shown as a function of time (horizontal component only); middle panel: spike times associated with each of these 50 reaches—each row corresponds to a trial, black tick marks indicate spike times, light bar indicates movement onset; bottom panel: a histogram of the average response.

need result (e.g., Taylor et al., 2002). With animal models, such as monkeys, neural activity that can be generated in the absence of arm movements (or presumably even *with* arm movements when descending nerves are intact and without nerve block (Moritz et al., 2008)) is currently considered to be an adequate proxy for neural signals from paralyzed subjects (e.g., Taylor et al., 2002; Velliste et al., 2008). Movement activity is then decoded to generate instantaneous direction and speed signals, which are used to guide a computer cursor (e.g., Serruya et al., 2002; Taylor et al., 2002; Carmena et al., 2003) or a robotic arm (e.g., Velliste et al., 2008) to a target. Motor prostheses incorporate movement activity because the goal is to recreate the desired movement path and speed. Although largely reflecting movement target, plan activity can play an important role in full trajectory estimation by providing an estimate of where the movement will end, and thereby helping to constrain the instantaneous movement estimates (Yu et al., 2007). This will be detailed in Section 7.3.

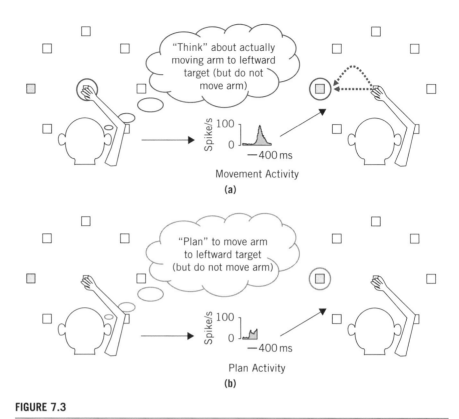

FIGURE 7.3

Differences between motor and communication prostheses. (a) For motor prostheses, movement activity that is generated by "thinking" about moving to the desired target is processed with a trajectory estimator, which guides a cursor (dark gray circle) along the desired path at the desired speed. (b) For communication prostheses, plan activity that is generated by "intending" to move to the desired target is then processed with a target estimator, which directly positions a cursor (medium gray circle) on the desired target.

In contrast, communication prostheses are not obliged to move along a continuous path in order to strike a target, such as a key on an onscreen keyboard (Figure 7.3(b)). Instead, if target location can be estimated directly from neural plan activity, the cursor can be positioned immediately on the desired key. Recent reports suggest that there may be a considerable performance benefit in using plan activity and direct-positional prosthesis control (e.g., Musallam et al., 2004; Santhanam et al., 2006). Figure 7.3(b) illustrates that plan activity can be elicited merely by "intending" to move the arm to a target/key location, and it is well established that it does not necessarily produce movements or EMG activity. Plan activity is then decoded to yield the desired key; the prosthetic cursor then immediately appears and selects it. In this way, communication prostheses can rely on plan activity alone.

7.3 CONTINUOUS DECODING FOR MOTOR PROSTHESES

In the late 1960 and early 1970, Fetz and colleagues discovered that non-human primates could learn to regulate the firing rate of individual cortical neurons (Fetz, 1969; Fetz and Finocchio, 1971; Fetz and Baker, 1972). These pioneering experiments relied on straightforward forms of real-time feedback, but clearly demonstrated that firing rates could be brought to requested levels without accompanying muscle contraction, even in M1. In the 1970 and early 1980, Schmidt, Humphrey, and colleagues proposed that neural activity could be used to directly control prostheses (Humphrey et al., 1970). By the late 1990, technological advances and a much better understanding of how cortical neurons contribute to limb movement (e.g., Georgopoulos et al., 1986; Schwartz, 1994) sparked renewed interest in developing clinically viable systems. This required an ongoing series of experiments with animal models and disabled human patients with the goal of learning fundamental design principles and quantifying performance.

Nicolelis and colleagues investigated one-dimensional (1D) control of a motor prosthesis by training rats to press a lever to receive a liquid reward (Chapin et al., 1999). The apparatus was then altered such that the lever was controlled by movement activity across a population of cortical neurons. In particular, the lever movement was estimated using a combination of principal components analysis and an artificial neural network. The researchers found that rats could still control the lever to receive their reward. Animals soon learned that actual forelimb movement was not needed and stopped movements altogether while continuing to move the lever with brain-derived activity. This proved the basic feasibility of motor prostheses.

Meanwhile, investigations of 2D and full 3D control essential for re-creating natural arm movements were underway. Schwartz and colleagues demonstrated that 2D and 3D hand location could be reconstructed fairly accurately from the movement activity of a population of simultaneously recorded M1 neurons in rhesus monkeys (Isaacs et al., 2000), and Nicolelis and colleagues reported similarly encouraging reconstructions using simultaneous recordings from parietal cortex, PMd, and M1 (Wessberg et al., 2000). Together with similar recording studies from Donoghue and colleagues (Maynard et al., 1999), the stage was set for the first "closed-loop" 2D and 3D motor prosthesis experiments.

Schwartz and colleagues pursued 3D cursor control with rhesus monkeys (Taylor et al., 2002). A few tens of M1 neurons were recorded with a chronically implanted electrode array, and the monkeys made 3D reaching movements to visual targets appearing in a 3D virtual reality environment. Neural responses were characterized in terms of the movement direction eliciting maximal response (also known as the *preferred direction*) and were combined to form a modified population vector. Let $\mathbf{v}^i \in \mathbb{R}^{p \times 1}$ be a vector pointing in the preferred direction of neural unit i, where p is the number of dimensions of the workspace (in this case, three). In its simplest form, the population vector $\hat{\mathbf{x}}_t \in \mathbb{R}^{p \times 1}$ at time t is defined as

$$\hat{\mathbf{x}}_t = \frac{\sum_{i=1}^{q} y_t^i \mathbf{v}^i}{\sum_{i=1}^{q} y_t^i}, \tag{7.1}$$

where $y_t^i \in \mathbb{R}$ is the activity of unit $i \in \{1, \ldots, q\}$ at time t. The population vector indicates the instantaneous movement of the hand or, in prosthesis mode, the prosthetic cursor.

Monkeys then entered "brain-control" mode, wherein the 3D prosthetic cursor was controlled by the neural population vector, as opposed to the location of the hand. This group has since gone on to demonstrate that monkeys can use these 3D control signals to feed themselves with an anthropomorphic robotic arm, and open and close a hand-like gripper (Velliste et al., 2008).

The fitting procedure for the preferred directions $\mathbf{v}^i$ and more sophisticated variants of the population vector are described elsewhere (Zhang et al., 1998; Moran and Schwartz, 1999b; Taylor et al., 2002). The population vector assumes a uniform distribution of preferred directions; when this assumption is not satisfied, the population vector is biased and an optimal linear estimator (OLE) may be preferred (Salinas and Abbott, 1994). The OLE takes the same form as Eq. (7.1); the only difference is how the parameters $\mathbf{v}^i$ are obtained. It was recently shown that, in a closed-loop setting, the subject could compensate for a bias in the population vector arising from a nonuniform distribution of preferred directions (Chase et al., 2009).

Donoghue and colleagues pursued 2D cursor control with rhesus monkeys (Serruya et al., 2002). A few tens of M1 neurons were recorded simultaneously with a chronically implanted electrode array while monkeys moved a manipulandum to guide the cursor. By recording spike activity while the monkeys tracked a continuously, pseudorandomly moving target, a linear filter could be learned to relate neural activity to cursor movement. A linear filter takes a linear combination of activity from different simultaneously recorded neurons to estimate physical state variables, such as cursor position or velocity.

The estimated state variable $\hat{x}_t \in \mathbb{R}$ at time t is

$$\hat{x}_t = b + \sum_{i=1}^{q} \sum_{\tau=0}^{N-1} y_{t-\tau}^i a_\tau^i \tag{7.2}$$

where $b \in \mathbb{R}$ is an additive constant, $y_t^i \in \mathbb{R}$ is the activity of neural unit $i \in \{1, \ldots, q\}$ at time t, and $\{a_0^i, \ldots a_{N-1}^i\}$ are the N filter weights for unit i. These filter weights are fit using least squares (Warland et al., 1997). For a multi-dimensional state variable, a separate set of filter weights and additive constants is fit for each dimension. Unlike the population vector and OLE, the linear filter can combine neural activity across multiple time steps and thus produce smoother trajectory estimates. This linear filter was used in a new task, where neural activity guided the prosthetic cursor to hit visual targets appearing at random locations. Together with collaborators and Cyberkinetics Neurotechnology Inc., this group has since gone on to show for the first time that a population of M1 neurons from tetraplegic patients can be used to control 2D cursors (Hochberg et al., 2006).

Nicolelis and colleagues pursued 2D cursor control, along with a form of prosthetic grasping, with rhesus monkeys (Carmena et al., 2003). Hundreds of M1, PMd, supplementary motor area (SMA), primary sensory area (S1), and posterior parietal neurons were recorded with chronically implanted electrode arrays while monkeys performed each of three behavioral tasks. The animals were trained to move a pole to control the position of an onscreen cursor and to grip the pole to control the size of the cursor, which indicated grip force. Linear filters (Eq. (7.2)) were trained to relate neural activity to a variety of motor parameters including hand position, velocity, and gripping force. These models were used in "brain control" mode to translate neural activity into cursor movement and cursor size. This study demonstrates (1) that grip force may be controlled (as does the more recent Velliste et al. (2008) report), which is essential once a prosthetic arm arrives as the desired location; and (2) that combined positioning and gripping are possible.

7.3.1 Recursive Bayesian Decoders

While linear filters and population vectors are effective, recursive Bayesian decoders have been shown to provide more accurate trajectory estimates in offline (i.e., open-loop) settings (Brown et al., 1998; Brockwell et al., 2004; Wu et al., 2004a, 2006) and online (i.e., closed-loop) settings (Wu et al., 2004b; Kim et al., 2008). In contrast to linear filters and population vectors, recursive Bayesian decoders are based on the specification of an explicit probabilistic model, comprising (1) a *trajectory model*, which describes how the arm state changes from one time step to the next, and (2) an *observation model*, which describes how the observed neural activity relates to the time-evolving arm state. The trajectory and observation models each include a probabilistic component, which attempts to capture trial-to-trial differences in the arm trajectory and neural spiking noise, respectively. These probabilistic relationships provide confidence regions for the arm state estimates, in contrast to linear filters and population vectors. If the modeling assumptions are satisfied, Bayesian estimation makes optimal use of the observed data. With an explicit probabilistic model, it is straighforward to relax the modeling assumptions and propose extensions, whereas such changes are not easily incorporated in a linear filter or a population vector. We will exploit this flexibility in the next subsection.

A widely used recursive Bayesian decoder is the Kalman filter, which is based on a linear Gaussian trajectory model:

$$\mathbf{x}_t \mid \mathbf{x}_{t-1} \sim \mathcal{N}\left(A\mathbf{x}_{t-1}, Q\right) \tag{7.3}$$

$$\mathbf{x}_1 \sim \mathcal{N}\left(\boldsymbol{\pi}, V\right) \tag{7.4}$$

and a linear Gaussian observation model:

$$\mathbf{y}_t \mid \mathbf{x}_t \sim \mathcal{N}\left(C\mathbf{x}_t, R\right), \tag{7.5}$$

where $\mathbf{x}_t \in \mathbb{R}^{p\times 1}$ is the arm state (which may include terms such as arm position, velocity, and acceleration) at time $t \in \{1,\dots,T\}$, and $\mathbf{y}_t \in \mathbb{R}_+^{q\times 1}$ is the movement-period activity across the q neural units at time t. The parameters $A \in \mathbb{R}^{p\times p}$, $Q \in \mathbb{R}^{p\times p}$, $\boldsymbol{\pi} \in \mathbb{R}^{p\times 1}$, $V \in \mathbb{R}^{p\times p}$, $C \in \mathbb{R}^{q\times p}$, and $R \in \mathbb{R}^{q\times q}$ do not depend on time and are fit to training data using maximum likelihood (Wu et al., 2006). Linear Gaussian models have been successfully applied to decoding the path of a foraging rat (Brown et al., 1998; Zhang et al., 1998), as well as arm trajectories in ellipse-tracing (Brockwell et al., 2004), pursuit-tracking (Wu et al., 2004a, 2006; Shoham et al., 2005), and "pinball" tasks (Wu et al., 2004a; Wu et al., 2006).

The task of decoding a continuous arm trajectory involves finding the likely sequences of arm states corresponding to the observed neural activity. At each time step t, we seek to compute the distribution of the arm state $\mathbf{x}_t$ given the movement neural activity $\mathbf{y}_1, \mathbf{y}_2, \dots, \mathbf{y}_t$ (denoted by $\{\mathbf{y}\}_1^t$) observed up to that time. This distribution is written $P\left(\mathbf{x}_t \mid \{\mathbf{y}\}_1^t\right)$ and termed the *state posterior*. State posteriors can be obtained by iterating the following two updates. First, the one-step prediction is found by applying Eq. (7.3) to the state posterior at the previous time step:

$$P\left(\mathbf{x}_t \mid \{\mathbf{y}\}_1^{t-1}\right) = \int P\left(\mathbf{x}_t \mid \mathbf{x}_{t-1}\right) P\left(\mathbf{x}_{t-1} \mid \{\mathbf{y}\}_1^{t-1}\right) d\mathbf{x}_{t-1}. \qquad (7.6)$$

Second, the state posterior at the current time step is computed using Bayes' rule

$$P\left(\mathbf{x}_t \mid \{\mathbf{y}\}_1^t\right) = \frac{P(\mathbf{y}_t \mid \mathbf{x}_t) P\left(\mathbf{x}_t \mid \{\mathbf{y}\}_1^{t-1}\right)}{P\left(\mathbf{y}_t \mid \{\mathbf{y}\}_1^{t-1}\right)}. \qquad (7.7)$$

Note that $P\left(\mathbf{y}_t \mid \mathbf{x}_t, \{\mathbf{y}\}_1^{t-1}\right)$ has been replaced by $P\left(\mathbf{y}_t \mid \mathbf{x}_t\right)$ to obtain Eq. (7.7) since, given the current arm state $\mathbf{x}_t$, the current observation $\mathbf{y}_t$ does not depend on the previous observations $\{\mathbf{y}\}_1^{t-1}$ (Eq. (7.5)). The terms in the numerator of Eq. (7.7) are the observation model from Eq. (7.5) and the one-step prediction from Eq. (7.6). The denominator of Eq. (7.7) can be obtained by integrating the numerator over $\mathbf{x}_t$.

The linear Gaussian trajectory and observation models (Eqs. (7.3) through (7.5)) can then be substituted into Eqs. (7.6) and (7.7) to obtain the key equations of the Kalman filter. Because all variables $\mathbf{x}_t$ and $\mathbf{y}_t$ defined by Eqs. (7.3) through (7.5) are jointly Gaussian, the one-step prediction $P\left(\mathbf{x}_t \mid \{\mathbf{y}\}_1^{t-1}\right)$ and state posterior $P\left(\mathbf{x}_t \mid \{\mathbf{y}\}_1^t\right)$ are also Gaussian in $\mathbf{x}_t$. Thus, the distributions are fully specified by their means and variances. Let $E\left[\mathbf{x}_t \mid \{\mathbf{y}\}_1^\tau\right]$ be denoted by $\mathbf{x}_t^\tau$ and let $\mathrm{Var}(\mathbf{x}_t \mid \{\mathbf{y}\}_1^\tau)$ be denoted by V_t^τ. The mean and variance of the one-step prediction are

$$\mathbf{x}_t^{t-1} = A\mathbf{x}_{t-1}^{t-1} \qquad (7.8)$$

$$V_t^{t-1} = AV_{t-1}^{t-1}A' + Q \qquad (7.9)$$

The mean and variance of the state posterior are

$$\mathbf{x}_t^t = \mathbf{x}_t^{t-1} + K_t(\mathbf{y}_t - C\mathbf{x}_t^{t-1}) \qquad (7.10)$$

$$V_t^t = (I - K_tC)V_t^{t-1}, \qquad (7.11)$$

where

$$K_t = V_t^{t-1}C'\left(CV_t^{t-1}C' + R\right)^{-1}$$

The recursions specified by Eqs. (7.8) through (7.11) start with $\mathbf{x}_1^0 = \pi$ and $V_1^0 = V$. Thus, the Kalman filter produces an arm trajectory estimate with mean $\mathbf{x}_1^1,\ldots,\mathbf{x}_T^T$ and corresponding variance $V_1^1,\ldots,V_T^T$. The variances define the confidence intervals corresponding to the estimated mean trajectory. As shown by Black and colleagues (Wu et al., 2006), the Kalman filter can be viewed as a linear filter in which (1) all past neural activity is taken into account, rather than just $N-1$ timesteps in the past (Eq. (7.2)), and (2) a confidence interval is produced to accompany the estimated trajectory.

7.3.2 Mixture of Trajectory Models

The function of the trajectory model is to build into the recursive Bayesian decoder prior knowledge about the form of the reaches. The model may reflect (1) the hard, physical constraints of the limb (for example, the elbow cannot bend backwards), (2) the soft, control constraints imposed by neural mechanisms (for example, the arm is more likely to move smoothly than in a jerky motion), and (3) the physical surroundings of the patient and his/her objectives in that environment. The degree to which the trajectory model reflects the kinematics of the actual reaches directly affects the accuracy with which trajectories can be decoded from neural data. When selecting a trajectory model, one is typically faced with a trade-off between how accurately a model captures the movement statistics and the computational demands of its corresponding decoder. For example, in real-time applications, we may decide to use a relatively simple trajectory model because of its low computational cost, even if it fails to capture some of the salient properties of the observed movements. While we may be able to identify a more complex trajectory model that could yield more accurate decoded trajectories, the computational demands of the corresponding decoder may be prohibitive in a real-time setting.

Here, we review a general approach to constructing trajectory models that can exhibit rather complex dynamical behaviors, whose decoder can be implemented to have the same running time as that of simpler trajectory models. The core idea was to combine simple trajectory models, each accurate within a limited regime of movement, in a probabilistic mixture of trajectory models (MTM) (Kemere et al., 2004; Yu et al., 2007). The development of the mixture framework was made possible by viewing the neural decoding problem as state estimation based on an explicit probabilistic model. We demonstrated the utility of this approach by developing a particular mixture of trajectory models suitable for goal-directed movements in settings with multiple goals. In the following, we will review (1) how to decode movements from neural activity under the general MTM framework, (2) the development of a particular MTM decoder for goal-directed movements, and (3) how this decoder can naturally incorporate prior goal information, when available, to improve the accuracy of the decoded trajectories.

7.3.2.1 *Mixture of Trajectory Models Framework*

Ideally, we wanted to construct a complete model of neural motor control that captures the hard physical constraints of the limb (Chan and Moran, 2006), the soft control constraints imposed by neural mechanisms, and the physical surroundings and context. One way to approximate such a complete model is to probabilistically combine trajectory models that are each accurate within a limited regime of movement. Examples of movement regimes include different parts of the workspace, different reach speeds, and different reach curvatures. At the onset of a new movement, the movement regime is unknown, or imperfectly known, so the full trajectory model is composed of a mixture of the individual, regime-specific trajectory models. Here, we review the development of a recursive Bayesian decoder based on a mixture of trajectory models (Yu et al., 2007).

As with the Kalman filter, we sought to compute the state posteriors $P\left(\mathbf{x}_t \mid \{\mathbf{y}\}_1^t\right)$, which define the decoded trajectory. If the actual movement regime $m^\star$ is perfectly known before the reach begins, then we can compute the state posterior based only on the individual trajectory model corresponding to that regime. This distribution is written $P\left(\mathbf{x}_t \mid \{\mathbf{y}\}_1^t, m^\star\right)$ and termed the *conditional state posterior*. However, in general the actual movement regime is unknown or imperfectly known, so we needed to compute $P\left(\mathbf{x}_t \mid \{\mathbf{y}\}_1^t, m\right)$ for each $m \in \{1,\ldots,M\}$, where M is the number of movement regimes (also referred to as mixture components).

To combine the M conditional state posteriors, we simply expanded $P\left(\mathbf{x}_t \mid \{\mathbf{y}\}_1^t\right)$ by conditioning on the movement regime m:

$$P\left(\mathbf{x}_t \mid \{\mathbf{y}\}_1^t\right) = \sum_{m=1}^{M} P\left(\mathbf{x}_t \mid \{\mathbf{y}\}_1^t, m\right) P\left(m \mid \{\mathbf{y}\}_1^t\right) \tag{7.12}$$

In other words, the state posterior is a weighted sum of the conditional state posteriors. The weights $P\left(m \mid \{\mathbf{y}\}_1^t\right)$ represent the probability that the actual movement regime is m, given the observed spike counts up to time t. Bayes' rule was then applied to these weights in Eq. (7.12), yielding the key equation for the MTM framework:

$$P\left(\mathbf{x}_t \mid \{\mathbf{y}\}_1^t\right) = \sum_{m=1}^{M} P\left(\mathbf{x}_t \mid \{\mathbf{y}\}_1^t, m\right) \frac{P\left(\{\mathbf{y}\}_1^t \mid m\right) P(m)}{P\left(\{\mathbf{y}\}_1^t\right)} \tag{7.13}$$

The conditional state posteriors $P\left(\mathbf{x}_t \mid \{\mathbf{y}\}_1^t, m\right)$ in Eq. (7.13) can be computed or approximated using any of a number of different recursive Bayesian decoding techniques, including the Kalman filter (Wu et al., 2006), Bayes' filter (Brown et al., 1998), and particle filters (Brockwell et al., 2004; Shoham et al., 2005). This is usually based on the recursions Eqs. (7.6) and (7.7). The likelihood terms $P\left(\{\mathbf{y}\}_1^t \mid m\right)$ in Eq. (7.13) can be expressed as

$$P\left(\{\mathbf{y}\}_1^t \mid m\right) = \prod_{\tau=1}^{t} P\left(\mathbf{y}_\tau \mid \{\mathbf{y}\}_1^{\tau-1}, m\right) \tag{7.14}$$

where

$$P\left(\mathbf{y}_t \mid \{\mathbf{y}\}_1^{t-1}, m\right) = \int P(\mathbf{y}_t \mid \mathbf{x}_t) P\left(\mathbf{x}_t \mid \{\mathbf{y}\}_1^{t-1}, m\right) d\mathbf{x}_t \tag{7.15}$$

The integral in Eq. (7.15) includes the observation model $P(\mathbf{y}_t \mid \mathbf{x}_t)$ and the one-step prediction $P\left(\mathbf{x}_t \mid \{\mathbf{y}\}_1^{t-1}, m\right)$ (Eq. (7.6)) using the parameters of mixture component m. If available, prior information about the identity of the movement regime can be incorporated naturally into the MTM framework via $P(m)$ in Eq. (7.13). This information must be available before the reach begins and may differ from trial to trial. If no such information is available, the same $P(m)$ (e.g., a uniform distribution) can be used across all trials.

The computational complexity of the MTM decoder is M times that of computing $P\left(\mathbf{x}_t \mid \{\mathbf{y}\}_1^t, m\right)$ and $P\left(\{\mathbf{y}\}_1^t \mid m\right)$ for a particular mixture component m. Because the computations for each mixture component can theoretically be carried out in parallel, it is possible to set up the MTM decoder so that its running time remains constant, regardless of the number of mixture components M. In other words, the MTM approach enables the use of more flexible, and potentially more accurate, trajectory models without a necessary penalty in decoder running time. Furthermore, the MTM decoder preserves the real-time properties of its constituent estimators and thus is suitable for real-time prosthetic applications.

7.3.2.2 *Mixture of Trajectory Models for Goal-Directed Movements*
For goal-directed movements, we proposed using the following mixture of linear Gaussian trajectory models

$$\mathbf{x}_t \mid \mathbf{x}_{t-1}, m \sim \mathcal{N}\left(A_m \mathbf{x}_{t-1} + \mathbf{b}_m, Q_m\right) \tag{7.16}$$

$$\mathbf{x}_1 \mid m \sim \mathcal{N}\left(\boldsymbol{\pi}_m, V_m\right) \tag{7.17}$$

where $m \in \{1, \ldots, M\}$ indexes reach target and M is the number of reach targets (Yu et al., 2007). In other words, each movement regime corresponds to movements heading toward a particular reach goal. As before, $\mathbf{x}_t \in \mathbb{R}^{p \times 1}$ is the arm state at time $t \in \{1, \ldots, T\}$ containing position, velocity, and acceleration terms. We used the observation model

$$y_t^i \mid \mathbf{x}_t \sim \text{Poisson}\left(e^{\mathbf{c}_i' \mathbf{x}_t + d_i} \Delta\right) \qquad (7.18)$$

where $y_t^i \in \{0, 1, 2, \ldots\}$ is the movement-period spike count for unit $i \in \{1, \ldots, q\}$ taken in a time bin of width Δ. For notational convenience, the spike counts across the q simultaneously recorded units were assembled into a $q \times 1$ vector $\mathbf{y}_t$, whose i^{th} element is y_t^i. The parameters $A_m \in \mathbb{R}^{p \times p}$, $\mathbf{b}_m \in \mathbb{R}^{p \times 1}$, $Q_m \in \mathbb{R}^{p \times p}$, $\boldsymbol{\pi}_m \in \mathbb{R}^{p \times 1}$, $V_m \in \mathbb{R}^{p \times p}$, $\mathbf{c}_i \in \mathbb{R}^{p \times 1}$, $d_i \in \mathbb{R}$ do not depend on time and are fit to training data (for details, see Yu et al. (2007)). Note that, whereas each mixture component indexed by m in the trajectory model (Eqs. (7.16) and (7.17)) can have different parameters leading to different arm state dynamics, the observation model (Eq. (7.18)) is the same for all m. It is also possible to probabilistically mix other related trajectory models, in which trajectories are directed toward a particular target location (Srinivasan et al., 2006; Kulkarni and Paninski, 2008).

For this model, the conditional state posteriors can be obtained using the recursions in Eqs. (7.6) and (7.7). Unlike for the Kalman filter, the observation model (Eq. (7.18)) is not linear Gaussian. This leads to distributions that are difficult to manipulate, and the integral in Eq. (7.6) cannot be computed analytically. We instead employed a modified Kalman filter that uses a Gaussian approximation during the measurement update step (Eq. (7.7)). We approximated the conditional state posterior as a Gaussian matched to the location and curvature of the mode of $P\left(\mathbf{x}_t \mid \{\mathbf{y}\}_1^t, m\right)$ (MacKay, 2003). This Gaussian approximation allows the integral in Eq. (7.6) to be computed analytically, since each mixture component of the full trajectory model (Eq. (7.16)) is linear Gaussian. This yields a Gaussian one-step prediction, which is fed back into Eq. (7.7).

7.3.2.3 *Incorporating Target Information from Plan Activity*
Up to this point, the neural activity discussed has been movement activity, which takes place around the time of movement and specifies the moment-by-moment details of the arm trajectory. In the delayed reach task, there is also neural activity present during an instructed delay period that directly precedes the "go cue" (termed *plan activity*). Neurons with plan activity are typically also active in the absence of an instructed delay during the reaction time

period (Crammond and Kalaska, 2000; Churchland et al., 2006c). Rather than specifying the moment-by-moment details of the trajectory, plan activity has been shown to reliably indicate the upcoming reach target (Shenoy et al., 2003; Hatsopoulos et al., 2004; Musallam et al., 2004; Yu et al., 2004; Santhanam et al., 2006). Both types of activity may be emitted by the same unit on a single trial, as can be seen in Figure 7.2.

The following describes how the reach target can be decoded from plan activity by applying Bayes' rule. Let $\mathbf{z}$ be a $q \times 1$ vector of spike counts across the q simultaneously recorded units in a prespecified time window during the delay period on a single trial. The distribution of spike counts (from training data) for each reach target m are fit to a Gaussian (Maynard et al., 1999; Yu et al., 2004):

$$\mathbf{z} \mid m \sim \mathcal{N}\left(\boldsymbol{\mu}_m, R_m\right) \tag{7.19}$$

where $\boldsymbol{\mu}_m \in \mathbb{R}^{q \times 1}$ is a vector of mean counts for reach target $m \in \{1, \ldots, M\}$, and the covariance matrix $R_m \in \mathbb{R}^{q \times q}$ is assumed to be diagonal. In other words, the units are assumed to be independent given the reach target m. In Section 7.4, we will describe how to relax this conditional independence assumption.

For any test trial, the probability that the upcoming reach target is m given the plan activity $\mathbf{z}$ can be computed by applying Bayes' rule:

$$P(m \mid \mathbf{z}) = \frac{P(\mathbf{z} \mid m)P(m)}{P(\mathbf{z})} = \frac{P(\mathbf{z} \mid m)}{\sum\limits_{m'} P(\mathbf{z} \mid m')} \tag{7.20}$$

where $P(m)$ in Eq. (7.20) is assumed to be uniform. The target information from the plan activity, $P(m \mid \mathbf{z})$, can be incorporated naturally in the MTM framework in the place of $P(m)$ in Eq. (7.13). The distribution $P(m)$ in Eq. (7.13) represents the prior knowledge (i.e., prior to movement onset) that the upcoming reach target is m. Because the plan activity entirely precedes movement onset and provides information about the upcoming reach target, it can be used to set $P(m)$ in Eq. (7.13) on a per-trial basis.

It is important to note that the most likely target from Eq. (7.20) is not simply assumed here to be the target of the upcoming reach. On a given trial, the plan activity may not definitively indicate the target of the upcoming reach (e.g., two different reach targets may have significant probability), or it may indicate an incorrect target for the upcoming reach. In this case, you want to allow the subsequent movement activity, to determine the target of the reach, or even correct the mistake, "in flight". Instead of making a hard target decode based on plan activity, the entire distribution $P(m \mid \mathbf{z})$ is retained and passed to the MTM framework.

7.3.3 Results

Here, we review the performance comparison of four decoders. The first is a state-of-the-art decoder presented by Kass and colleagues (Brockwell et al., 2004) based on a random-walk trajectory model (RWM) in acceleration. The second decoder is based on a single linear Gaussian trajectory model (STM) shared across reaches to all targets. It is defined by Eqs. (7.16) and (7.17) for special case of $M = 1$. The STM decoder uses the observation model shown in Eq. (7.18). The RWM and STM decoders provided points of comparison for the following two MTM decoders, both of which are based on Eqs. (7.16) and (7.18). Whereas the MTM_M decoder uses only movement activity, the MTM_DM decoder uses plan activity as well. In Eq. (7.13), the same $P(m)$ (in this case, a uniform distribution) is used across all trials for MTM_M. A different $P(m)$ is used on each trial for MTM_DM based on the prior target information extracted from plan activity.

For each of the four decoders, we first fit the model parameters to training data, as detailed in Yu et al. (2007). The test data for a single trial consisted of (1) the arm trajectory, taken from 50 ms before movement onset to 50 ms after movement end at $dt = 10$ ms time steps; (2) the movement period spike counts, taken in non-overlapping $\Delta = 10$ ms bins and temporally offset from the arm trajectory by an optimal lag found for each unit; and (3) the plan period spike counts, taken in a single 200 ms bin starting 150 ms after the appearance of the reach target. Arm trajectories in the test phase were used to evaluate offline the accuracy of the trajectories estimated from neural data. Note that natural arm movements accompanied the recorded movement activity (as is typical in offline studies), in contrast to most online closed-loop studies in which few or no arm movements are produced.

Figure 7.4 details, for a particular test trial, how the MTM-decoded trajectory was obtained and compares the trajectory estimates produced by the different decoders. From Eq. (7.12), the MTM-decoded trajectory $E\left[\mathbf{x}_t \mid \{\mathbf{y}\}_1^t\right]$ is a weighted sum of component trajectory estimates $E\left[\mathbf{x}_t \mid \{\mathbf{y}\}_1^t, m\right]$, one for each reach target indexed by $m \in \{1, \ldots, 8\}$. In Figure 7.4(b) and (c), the component trajectory estimates are plotted in the upper panels, while the middle panels show how the corresponding weights $P\left(m \mid \{\mathbf{y}\}_1^t\right)$ evolved during the course of the trial.

The values of the weights at time zero ($t = 0$) represent the probability that the upcoming reach target is m, before any movement neural activity is observed. The distribution of weights at $t = 0$ is precisely $P(m)$ in Eq. (7.13).

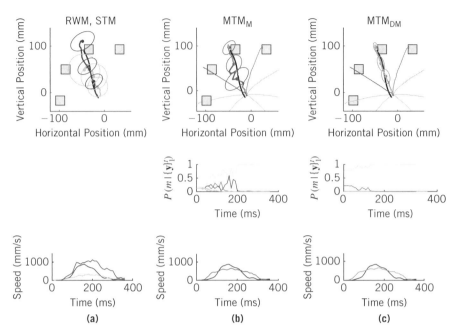

FIGURE 7.4

A representative test trial in which the use of plan activity improved the MTM-decoded trajectory. The upper panels compare the actual trajectory (all panels, black) with the decoded trajectories for (a) RWM (dark green) and STM (light green), (b) MTM$_M$ (red), and (c) MTM$_{DM}$ (orange). Ellipses denote 95% confidence intervals at three different time steps. Yellow squares represent the visual reach targets presented to the monkey in actual dimensions. The upper panels in (b) and (c) also show the eight component trajectory estimates for the MTM (cyan, blue, magenta for the three components with the largest weights; gray for the other five components). Their corresponding weights, as they evolve during the trial, are plotted in the middle panels. The lower panels compare the actual and estimated single-trial speed profiles using the same color conventions as in the upper panels. Time zero corresponds to 60 ms before movement onset (i.e., one time step before we begin to decode movement). Note that the red and orange traces in the upper panels are overlaid with the cyan trace. For this trial, E_{rms} was 11.8, 26.9, 10.9, and 10.5 mm for the RWM, STM, MTM$_M$, and MTM$_{DM}$, respectively. Monkey G, 98 units. (Experiment G20040508, trial ID 474). Please see this figure in color at the companion web site: www.elsevierdirect.com/companions/9780123750273
Source: Reproduced with permission from Yu et al., 2007.

In Figure 7.4(b), we assumed that there was no information available about the identity of the upcoming reach target before the reach began (i.e., no plan activity), so all eight targets were equiprobable (i.e., $P(m) = 1/8$ for $m \in \{1,\ldots,8\}$). As time proceeded, these weights were updated as more and more movement activity was observed. Recall that $P\left(m \mid \{\mathbf{y}\}_1^t\right)$ represents the probability that the actual reach target is m, given the observed neural activity up to time t.

During the first 200 ms, the actual reach target (cyan) was more likely than the other seven reach targets at nearly every time step; however, there was some competition with the neighboring reach targets (blue and magenta). It was only after approximately 200 ms that the decoder became certain of the actual reach target (i.e., $P\left(m \mid \{\mathbf{y}\}_1^t\right)$ approached unity) and remained certain for the rest of the trial. A weighted sum of the eight component trajectory estimates (upper panel) using these weights (middle panel) yields the MTM-decoded trajectory (upper panel, red; E_{rms}: 10.9 mm). To measure decoding performance, we computed the root mean square position error (E_{rms}) between the actual and decoded trajectories.

If plan activity is available, it can be used to set a nonuniform $P(m)$ in Eq. (7.13) on a per-trial basis, as previously discussed. The only difference between Figure 7.4(b) and (c) is that the MTM decoder used plan activity in the latter but not the former. In Figure 7.4(c) (middle panel), the weights at $t = 0$ represent the probabilities of each reach target based only on plan activity, before any movement activity was observed. In this case, the plan activity indicated that the actual reach target (cyan) was more probable than the other targets. This prior knowledge of the identity of the upcoming reach target was then taken into account when updating the weights $P\left(m \mid \{\mathbf{y}\}_1^t\right)$ during the course of the trial as more and more movement activity was observed. Note that using plan activity only affected $P(m)$ in Eq. (7.13); the conditional state posteriors $P\left(\mathbf{x}_t \mid \{\mathbf{y}\}_1^t, m\right)$ and the likelihood terms $P\left(\{\mathbf{y}\}_1^t \mid m\right)$ remained unchanged. As the means of the conditional state posteriors, the component trajectory estimates therefore also remained unchanged, as can be verified by comparing Figure 7.4(b) and (c) (upper panels). For the trial shown in Figure 7.4, the use of plan activity reduced the competition between the actual reach target (cyan) and the neighboring targets (blue and magenta). Compared to Figure 7.4(b) (middle panel), the weight for the actual reach target (cyan) in Figure 7.4(c) (middle panel) was higher at every time point, the clearest effect seen during the first 200 ms. In other words, by using plan activity, the decoder was more certain of the actual reach target throughout the trial. In Figure 7.4(c), a weighted sum of the eight component trajectory estimates (upper panel) using these weights (middle panel) yields the MTM-decoded trajectory (upper panel, orange; E_{rms}: 10.5 mm).

By comparing the MTM-decoded trajectories with the actual trajectory in Figure 7.4(b) and (c) (upper panels), we see that the use of plan activity decreased the decoding error and tightened the confidence ellipses for this trial. Both MTM-decoded trajectories had lower decoding error than the RWM (E_{rms}: 11.8 mm) and STM (E_{rms}: 26.9 mm), whose decoded trajectories are plotted in Figure 7.4(a) (upper panel). On this trial, the RWM decoder produced a reasonably accurate decoded trajectory, while the STM-decoded

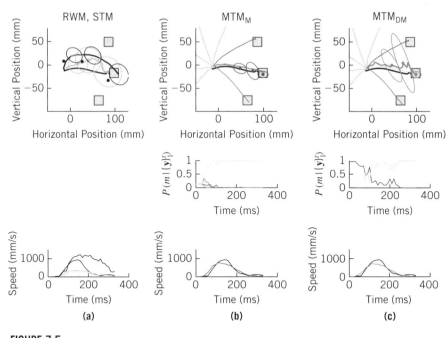

FIGURE 7.5

A representative test trial in which the movement activity corrected an incorrect target identification from the plan activity. Figure conventions are identical to those in Figure 7.4. For this trial, E_{rms} was 27.7, 24.2, 11.2, and 13.0 mm for RWM, STM, MTM$_M$, and MTM$_{DM}$, respectively. Monkey G, 98 units. (Experiment G20040508, trial ID 676). Please see this figure in color at the companion web site: www.elsevierdirect.com/companions/9780123750273

Source: Reproduced with permission from Yu et al., 2007.

trajectory proceeded slowly outwards with wide confidence intervals. These decoders can also be used to estimate the bell-shaped speed profile of the actual reach (Figure 7.4, lower panels).

In contrast to Figure 7.4, Figure 7.5 shows a trial where movement activity alone was able to quickly determine the actual reach target without much competition from neighboring targets. This can be seen in Figure 7.5(b) (middle panel), where the weight corresponding to the actual reach target (cyan) rose to unity after approximately 100 ms and stayed there for the remainder of the trial. As a result, the resulting MTM decoded trajectory (upper panel, red; E_{rms}: 11.2 mm) was quite accurate. As in Figure 7.4, we can incorporate plan activity if available; however, in this case, the dominant weight at $t = 0$ (blue) did *not* correspond to the actual reach target (cyan), as seen in Figure 7.5(c) (middle panel). In other words, the plan activity incorrectly indicated the identity of the upcoming reach target. However, as these weights were updated by the observation of movement activity, this "error" was soon corrected (within approximately 100 ms). From that point on, the weight corresponding to the actual reach target dominated. Despite this error at the beginning of the trial, the MTM decoded trajectory in Figure 7.5(c) (upper panel, orange; E_{rms}: 13.0 mm) still headed to the correct target and provided a reasonably accurate estimate of the arm trajectory. The larger confidence ellipses for MTM_{DM} compared to MTM_M reflect the competition between the actual (cyan) and neighboring (blue) reach targets. The decoded trajectories for the RWM (dark green) and STM (light green) are shown in Figure 7.5(a) for comparison. As in Figure 7.4, both MTM-decoded trajectories yielded lower decoding error than the RWM (E_{rms}: 27.7 mm) and STM (E_{rms}: 24.2 mm). Furthermore, as shown in the lower panels, the speed profiles estimated by the MTM decoders (red and orange) tracked the actual bell-shaped speed profile (black) more closely than those estimated by the RWM (dark green) and STM (light green) decoders.

Figures 7.4 and 7.5 together illustrate the benefits of the joint use of plan and movement activity. When one type of activity is unable to definitively identify (or incorrectly identifies) the actual reach target, the MTM framework allows the other type of activity to strengthen (or overturn) the target identification in a probabilistic manner. In Figure 7.4, the movement activity alone was unable to definitively identify the actual reach target during the first 200 ms, as there was competition with a neighboring target. When prior target information from plan activity was incorporated, the decoder was more certain of the actual reach target throughout the trial. In Figure 7.5, the plan activity incorrectly indicated the identity of the upcoming reach target. However, the movement activity overturned this incorrect target identification early on and rescued the decoder from incurring a large E_{rms} on this trial.

Having demonstrated how the MTM framework produces trajectory estimates in individual trials, we now review the quantification and comparison of the average performance of the four decoders (RWM, STM, MTM_M, MTM_{DM}) across entire data sets. Figure 7.6 illustrates the following two main results, which hold true across both monkeys. First, a mixture of linear Gaussian trajectory models (MTM_M) provides lower decoding error than either of the nonmixture trajectory models (RWM and STM) (Wilcoxon paired-sample test, $p < 0.01$). Compared to the STM decoder, the MTM_M decoder reduced E_{rms} from 22.5 to 13.9 mm (22.8 to 11.8 mm) in monkey G (H). Second, the use of prior target information $P(m)$ in the MTM framework (MTM_{DM}) can further decrease decoding error (Wilcoxon paired-sample test, $p < 0.01$). Compared to the MTM_M decoder, the MTM_{DM} decoder reduced E_{rms} from 13.9 to 11.1 mm (11.8 to 10.5 mm) in monkey G (H). Because the MTM decoder is inherently parallelizable (as described in Section 7.3.2.1),

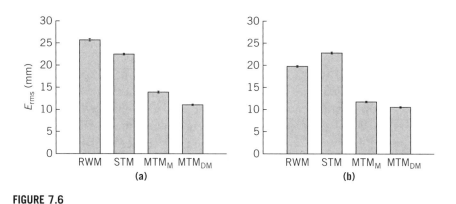

FIGURE 7.6

E_{rms} (mean $\pm$ SE) comparison for the RWM, STM, MTM$_M$, and MTM$_{DM}$ decoders.
(a) Monkey G (98 units). (b) Monkey H (99 units).
Source: Reproduced with permission from Yu et al., 2007.

these performance gains can be obtained *without* an associated increase in decoder running time. The superior performance of the MTM$_M$ compared to the RWM and STM can be attributed to the fact that the MTM better captures the kinematics of goal-directed reaches (see Figure A2 in Yu et al. (2007)). If plan activity is available, this additional source of information can be naturally incorporated in the MTM framework to further improve decoding performance (MTM$_{DM}$).

7.4 DISCRETE DECODING FOR COMMUNICATION PROSTHESES

There has been considerable research on cortically controlled communications prostheses since at least the 1980s. This interest has been especially prevalent among the noninvasive community, and the intra-cortical community has recently begun to explore this domain as well. Using noninvasive EEG recordings, researchers have explored several approaches for engineering prostheses that allow discrete target selection. Some commonly studied categories of EEG signals include slow cortical potentials (SCPs) (Hinterberger et al., 2004), sensorimotor (μ and β) rhythms (McFarland et al., 2003; Wolpaw and McFarland, 2004a), and evoked potentials (P300) (Serby et al., 2005). More recently, minimally invasive communication prostheses relying on subdural electrocorticographic (ECoG) surface arrays

(Leuthardt et al., 2004; Schalk et al., 2008) and extra-cortical local field potentials (Kennedy et al., 2004) have been developed.

In the invasive intra-cortical electrode domain, Kennedy and colleagues demonstrated that just one or two neurons from the motor cortex of locked-in ALS patients could be used to move a cursor across a virtual keyboard to type out messages (Kennedy and Bakay, 1998; Kennedy et al., 2000a,b). Patients reported simply "imagining" or "thinking" about moving various parts of their bodies, and eventually the computer cursor itself, to guide the cursor.

In this section, we focus on decoding algorithms for communication prostheses based on large populations of neurons, typically recorded using intra-cortical electrode arrays. One possibility is to adapt a motor prosthesis from Section 7.3 for use as a communication prosthesis. Schwartz and colleagues demonstrated that the continuous trajectory output from a prosthetic system designed to reach to discrete targets could be processed by an algorithm that chooses the most likely target location (Taylor et al., 2003). If only a very early portion of the trajectory is needed to make the discrete classification, the system can simply cut the trial short and prepare to decode a new target.

More recent studies have shown that directly classifying the neural activity (rather than taking a two-stage approach of first estimating cursor position, then classifying) can lead to greater speed and accuracy of target selection (Shenoy et al., 2003; Musallam et al., 2004; Santhanam et al., 2006). These studies were performed using plan activity in medial intraparietal area (MIP), area 5 and PMd in rhesus monkeys. In these studies, the authors computed the most likely target $\widehat{m}$ given the observed neural activity $\mathbf{z}$:

$$\widehat{m} = \underset{m}{\mathrm{argmax}} \; P(m \mid \mathbf{z}) \tag{7.21}$$

where $P(m \mid \mathbf{z})$ is given in Eq. (7.20). The performance of the decoder (measured in terms of speed and accuracy of target selection) depends on many factors, including the duration and placement of the time window in which the spikes $\mathbf{z}$ are counted, the arrangement of targets in the monkey's workspace, and the spike count model $P(\mathbf{z} \mid m)$ (Eq. (7.20)) (Santhanam et al., 2006). The choice of $P(\mathbf{z} \mid m)$ is the topic of this section.

7.4.1 Independent Gaussian and Poisson Models

The most commonly used probabilistic models for spike counts are Gaussian (Eq. (7.19)) (Maynard et al., 1999; Hatsopoulos et al., 2004; Santhanam et al., 2006; Yu et al., 2007) and Poisson (Shenoy et al., 2003; Hatsopoulos et al.,

2004; Santhanam et al., 2006)

$$P(\mathbf{z} \mid m) = \prod_{i=1}^{q} P(z_i \mid m) \tag{7.22}$$

$$z_i \mid m \sim \text{Poisson}(\lambda_{i,m})$$

where $z_i \in \{0, 1, 2, \ldots\}$ is the spike count for neural unit $i \in \{1, \ldots, q\}$, and $\lambda_{i,m} \in \mathbb{R}_+$ is the mean spike count for unit i and reach target $m \in \{1, \ldots, M\}$. By construction, the units in Eq. (7.22) are assumed to be independent given the reach target. For the Gaussian model, it is common to make the same conditional independence assumption by constraining R_m in Eq. (7.19) to be diagonal.

The assumption of conditional independence in both the Gaussian (Eq. (7.19)) and Poisson (Eq. (7.22)) cases is largely for practical reasons. In the Gaussian case, there are typically too few training trials to fit a full covariance matrix, especially with a large number of units (many tens to hundreds). In other words, a full covariance matrix is underconstrained; this leads to overfitting of the training data, thereby reducing decoding performance for the test data. Furthermore, the full covariance that is fit is often not invertible, which is problematic when used for decoding in Eq. (7.21) (Maynard et al., 1999). In the Poisson case, there is no standard multivariate generalization.

Historically, Poisson-based algorithms have been found to perform better than Gaussian-based algorithms in decoding applications (Hatsopoulos et al., 2004; Santhanam et al., 2006). The observed spike counts are, by definition, non-negative integers and can be relatively small in some cases (e.g., <5 spikes). In this regime, a Poisson distribution is typically a better fit to the data than a Gaussian distribution. One might suspect that the data are not well suited for a Gaussian distribution, especially since a Gaussian distribution has nonzero probability density for noninteger values, as well as negative values. It has been shown that taking the square root of the counts can improve the Gaussian fit (Thacker and Bromiley, 2001). We employ this square root transform when using Gaussian-based decoding algorithms here.

7.4.2 Factor Analysis Methods

While the assumption of conditional independence provides a simple and effective decoder, we found that its performance began to deteriorate in a closed-loop setting where monkeys planned to different targets in rapid succession (Santhanam et al., 2006). Upon inspection of the neural activity recorded across multiple trials while the monkey planned to the same target, we observed that the firing rates across the neural population fluctuated together in a systematic way that was dependent on chain position (i.e., the number of targets that had already been planned in rapid succession) (Gilja et al., 2005; Kalmar et al., 2005). In other words, it appeared that the activity of the units was *not* conditionally independent, which suggested that the conditional independence assumption may have been limiting the accuracy of the decoder. This provided the motivation for introducing conditional dependencies into the spike count model. However, as described above, using a fully dependent Gaussian distribution (i.e., a Gaussian distribution with a full covariance matrix) is problematic due to practical considerations. In this section, we explore a compromise by parameterizing the covariance matrix so as to introduce conditional dependencies without having to fit a full covariance matrix.

In addition to chain position, there is evidence that other factors may modulate PMd plan activity from trial to trial (for the same reach target) in a systematic way across the neural population. These factors include reach curvature (Hocherman and Wise, 1991), reach speed (Churchland et al., 2006b), force (Riehle and Requin, 1994), the required accuracy (Gomez et al., 2000), and the type of grasp (Godschalk et al., 1985). There may also be unobserved influences, including attentional states (Musallam et al., 2004; Chestek et al., 2007) that further modulate cortical activity, even when observable behavior is held fixed. These various influences may inadvertently mask the signal of interest (i.e., reach target) and thereby reduce decoding accuracy.

Figure 7.7(a) provides an example of the trial-to-trial variability present in the activity of two recorded neural units. For each unit and reach target, the distribution of plan activity across many trials is shown. Note the large variability in spike counts across trials for each target, which results in overlap between the ranges of spike counts for the reach target. Part of this variability can be attributed to the multitude of potential factors described above. These factors presumably affect the activity of each unit in a different, yet reliable, way (e.g., as the intended reach speed increases, the firing rate for unit 1 increases while that for unit 2 decreases) and are thus *shared* across the neural population. The other part of this variability, which will simply be termed "spiking noise," can arise from various sources, including channel noise in the spiking process, variability in active dendritic integration, synaptic failure, general quantal (in the vesicle sense) release variation, and axonal failure in incoming spikes (Faisal et al., 2008). These sources of variability presumably yield nonsystematic, and therefore *independent*, perturbations across the neural population. The challenge for the decoder is to avoid mistaking these variations (both shared and independent) as being the signature for an entirely different reach target.

Figure 7.7(b) shows simulated data for two neurons that exhibit independent variability and no shared variability for a particular reach target. The spike counts for each neuron are different in each trial, but there is no inherent correlation between the observations with regard to the magnitude or polarity of these perturbations. In contrast, Figure 7.7(c) shows simulated data from two neurons that exhibit shared variability *and* independent variability. The inset shows a hypothetical factor, such as intended reach speed, undergoing modulation across trials for the same reach target (dark gray points correspond to fast reaches; medium gray dots, to slow reaches). In this construction, when

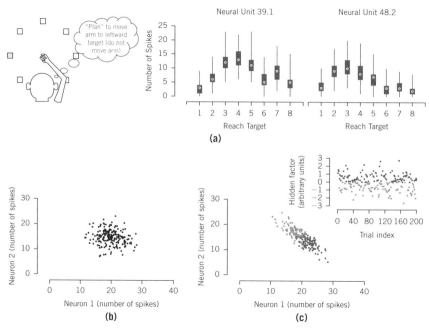

FIGURE 7.7

(a) Schematic of delayed reach task where subject reaches to one of eight reach targets. Data from two real neural units are shown where spikes were counted in a 200-ms window after target presentation for each trial. Trials were segregated by reach target, and a box plot was generated denoting the median spike count (white circles), the 25th to 75th percentile of the spike counts (dark gray rectangles), and the ranges of the outliers (thin "whiskers") for each target. (b) Simulation of two neurons whose trial-to-trial spike counts were perturbed independently of one another. (c) Simulation of two neurons whose spike counts were primarily perturbed by a shared latent factor (see inset). The diagonal orientation relative to the main axes is a tell-tale sign of a correlated relationship between these two neurons. Please see this figure in color at the companion web site: www.elsevierdirect.com/companions/9780123750273

Source: Reproduced with permission from Santhanam et al., 2009.

the speed is greater than average, neuron 1 emits slightly more spikes than average and neuron 2 emits slightly fewer than average. When the factor is below average, the reverse is true. By assuming conditional independence (Eqs. (7.19) and (7.22)), both the shared and independent sources of variability are treated as independent across the neural population. However, if we explicitly account for this shared component of variability, we can potentially increase decoding accuracy, as discussed below.

In reality, we do not have direct access to the inset in Figure 7.7(c), since many different uncontrollable factors may be involved and many of them are simply unobservable (e.g., cognitive attentiveness to the task). Although we cannot explicitly identify these factors, we attempted to infer a set of shared underlying factors for each trial, along with the mapping between these factors and the observed data. By "mapping," we mean the precise polarity and magnitude by which a factor perturbs the response of a particular neuron. Such mapping defines the systematic relationship between the latent factors and the observed neural activity, providing a model of how the spike counts of a population of neurons are expected to covary.

7.4.2.1 *Modeling Shared Variability Using Factor Analysis*

For a particular reach target, the independent Gaussian (Eq. (7.19)) and Poisson (Eq. (7.22)) models treat all trial-to-trial variability as independent between neural units. A model that could successfully capture spike count correlations across the population might well yield a more accurate decoding algorithm. As discussed above, fitting a Gaussian model with a full covariance matrix to a large number of units would require an impractically large number of trials. Instead, we recently proposed to use a parametrized form of covariance matrix (Santhanam et al., 2009), which models correlations in a limited number of principal directions. This is the central topic of this section. The form we used derives from a standard statistical technique called factor analysis (FA) (Everitt, 1984), where the observed quantities are modeled as a linear function of unobserved, shared factors.

For our prosthetic system, the observed variables are the neural spiking data that we recorded from the electrode array. The underlying factors represent the physical, cognitive, or cortical network states of the subject that are uncontrolled or effectively uncontrollable. We used the larger number of observed variables to help triangulate the smaller number of unobserved factors in the system. FA is a well-established technique of multivariate statistics, and we review the method here in the context of our prosthetic decoding application.

For each experimental trial, as before, we collected the (square root of the) number of spikes observed from each neural unit within the neural integration window into the q-dimensional vector $\mathbf{z}$. We also considered an abstract set of latent factors, which were assumed to be Gaussian-distributed and assembled into a vector $\mathbf{x} \in \mathbb{R}^{p \times 1}$. Without loss of generality, we took these p factors to be independent and to have unit variance around $\mathbf{0}$.[1] The observation $\mathbf{z}$ was also taken to be Gaussian-distributed, but with a mean that depended on the latent factors $\mathbf{x}$. The complete model is as follows:

$$\mathbf{x} \sim \mathcal{N}(\mathbf{0}, I) \tag{7.23}$$

$$\mathbf{z} \mid \mathbf{x}, m \sim \mathcal{N}(C_m \mathbf{x} + \boldsymbol{\mu}_m, R_m) \tag{7.24}$$

As in Eq. (7.19), $\boldsymbol{\mu}_m \in \mathbb{R}^{q \times 1}$ contains the number of spikes that each neural unit would produce for target $m \in \{1, \ldots, M\}$ if there were only target-related variability in the system, and $R_m \in \mathbb{R}^{q \times q}$ is a diagonal matrix that captures the independent variability present in $\mathbf{z}$ from trial to trial. However, there is now an additional mechanism by which trial-to-trial variability can emerge in the neural observations: specifically, the neural data vector $\mathbf{z}$ is perturbed by a linear function of the underlying factors. The matrix $C_m \in \mathbb{R}^{q \times p}$ defines the mapping between the factors and the observations. Given that the underlying factors are shared between the observed neural units, there is a predefined correlation structure between the neural units built into the model. Integrating over all possible $\mathbf{x}$, we obtained the likelihood of the data $\mathbf{z}$ for a reach target m:

$$\mathbf{z} \mid m \sim \mathcal{N}\left(\boldsymbol{\mu}_m, C_m C_m' + R_m\right) \tag{7.25}$$

Equation (7.25) shows that FA effectively partitions the data covariance matrix into two components. The first term in the covariance, $C_m C_m'$, captures the *shared* variability across the neural population. The second term, R_m, captures the remaining unexplained variability, which is assumed to be *independent* across units. It may include biophysical spiking noise and other nonshared sources of variability. Note that FA can be viewed as a generalization of principal component analysis (PCA). As the diagonal elements of R_m approach zero, the space spanned by the columns of C_m will be equivalent to the space found by PCA (Roweis and Ghahramani, 1999).

---

[1] The following is why there is no loss of generality when assuming $\mathbf{x} \sim \mathcal{N}(\mathbf{0}, I)$. Suppose that $\mathbf{x} \sim \mathcal{N}(\mathbf{0}, V)$ for some $V \neq I$. In other words, suppose that the latent factors are dependent and have nonunit variance. This model is formally equivalent to one in which $\mathbf{x} \sim \mathcal{N}(\mathbf{0}, I)$ and $\mathbf{z} \mid \mathbf{x}, m \sim \mathcal{N}\left(C_m V^{\frac{1}{2}} \mathbf{x} + \boldsymbol{\mu}_m, R_m\right)$, in the sense that both models would give the same decoded target $\hat{m}$ and data likelihood $P(\mathbf{z})$.

7.4.2.2 *Decoding Reach Targets Using Factor Analysis*

The model parameters (μ_m, C_m, R_m for each target m) are fit by maximum likelihood using the neural data $\mathbf{z}$ and the reach target m across all trials in the training data set. Although $P(\mathbf{z} \mid m)$ is Gaussian (Eq. (7.25)), the parametric form of the covariance makes direct calculation of the parameters difficult. Instead, we applied the expectation-maximization (EM) algorithm (Dempster et al., 1977), which iteratively updates the model parameters and latent factors until convergence is reached. The update equations for fitting FA using EM can be found elsewhere (Roweis and Ghahramani, 1999; Santhanam et al., 2009).

When decoding test trials, we used the maximum likelihood estimator previously described (cf. Eqs. (7.20) and (7.21)), where $P(\mathbf{z} \mid m)$ is given in Eq. (7.25). We refer to this approach as FA$_{sep}$ since there is a separate C_m and R_m per reach target m.[2] To reiterate, the FA model provides a method by which we can approximate the covariance structure of our observed variables without fitting a full covariance matrix. Once we have fit the model, the decoding procedure does not rely on finding the underlying factor vector $\mathbf{x}$, but rather operates solely with the distribution $P(\mathbf{z} \mid m)$.

An important question is how to select p, the number of underlying factors. With too many factors in the model, the model parameters will be fit to any idiosyncratic correlation structure in the training data and these features may not appear in the test data. If p is too small (i.e., too few factors), the model will be unable to capture all underlying correlation present in the training data. One must ensure that the model is not overfit to the training data and that it generalizes well for test data. Selecting the optimal p, or p^*, is part of the process of model selection. It is important to keep in mind that the underlying factors identified for each model are abstract. They are a vehicle by which we are able to describe shared variability—the exact number of factors selected does not necessarily have physical significance. The model selection procedure is a simple, necessary step to ensure that a model fit with the training trials will perform well when decoding the test data.

We used the standard approach of partitioning data into training and validation sets to find the value of p beyond which overfitting became a problem (Hastie et al., 2003). Each data set was split approximately into two equal halves, one to serve as a training set and the other as a test set. The trials in the training set were further divided in order to perform a 5-fold or 10-fold cross-validation. For example, in 5-fold cross-validation, roughly four-fifths

---

[2]Note that the overall model for $\mathbf{z}$ is very similar to the "mixture of factor analyzers" proposed by Ghahramani and Hinton (1997), except that FA$_{sep}$ allows for a separate covariance R_m per mixture component.

of the training trials would be used to train the FA model and one-fifth to validate the model's prediction error. This would be repeated five times until all of the original training set was used for validation. A series of models, each with a different number of underlying factors p, were tested using this cross-validation method. The p^* was selected by identifying the model with the lowest decoding error. The performance of this one model was then assessed on the test data to compare the classification performance of the optimal FA model against the independent Gaussian and Poisson models.

7.4.2.3 *A Unified Latent Space*

The FA$_{\text{sep}}$ approach in Eq. (7.25) is a straightforward extension of the independent Gaussian model of (Eq. (7.19)). It defines a separate latent space for each reach target indexed by m (Eq. (7.24)). To visualize the data points across different reach targets in a unified latent space, we developed a second, novel approach to FA that defines a single output mapping and incorporates the separation of reach targets through the shared latent space (Santhanam et al., 2009). The model is

$$\mathbf{x} \mid m \sim \mathcal{N}\left(\boldsymbol{\mu}_m, I\right) \tag{7.26}$$

$$\mathbf{z} \mid \mathbf{x} \sim \mathcal{N}\left(C\mathbf{x}, R\right) \tag{7.27}$$

where $\boldsymbol{\mu}_m$ is now a p-dimensional vector that describes the reach target influence on the neural data within the lower-dimensional space. The mapping between the low- and high-dimensional spaces (defined by $C \in \mathbb{R}^{q \times p}$) is shared across reach targets, thereby unifying the effect of target-related and non-target-related factors on neural data. For each trial, the shared target-related effect on the neural observations is $C\boldsymbol{\mu}_m$, while the shared non-target-related factors perturb the observations by $C(\mathbf{x} - \boldsymbol{\mu}_m)$. As before, R is diagonal and describes the independent variability. We refer to this model as FA$_{\text{cmb}}$, since the output mapping is "combined" across reach targets.

We fit the model parameters ($\boldsymbol{\mu}_m$ for each target m, C, R), again using the EM algorithm. The EM update equations for FA$_{\text{cmb}}$ can be found in Santhanam et al. (2009). To decode the reach target for test trials, we rewrote Eqs. (7.26) and (7.27) as

$$\mathbf{z} \mid m \sim \mathcal{N}\left(C\boldsymbol{\mu}_m, CC' + R\right) \tag{7.28}$$

and applied the same maximum likelihood decoder shown in Eqs. (7.20) and (7.21).

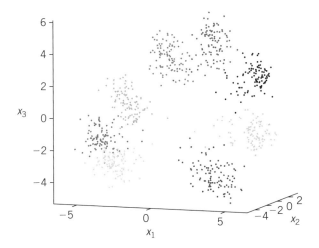

FIGURE 7.8

Visualization of FA$_{cmb}$ with a three-dimensional latent space. Each point corresponds to the inferred **x** for a given trial. Axis scaling is determined by the fitted model. The various shading of the data points denotes the upcoming reach target. Clusters correspond to different reach targets. Analysis performed on data set G20040428. Please see this figure in color at the companion web site: www.elsevierdirect.com/companions/9780123750273

Source: Reproduced with permission from Santhanam et al., 2009.

Figure 7.8 is an example of how these unified factors might manifest themselves over many trials.[3] There are three latent factors that capture both target and non-target-related influences on the neural data **z** (which is not shown and is much higher-dimensional). Each differently shaded cloud of points corresponds to a different reach target, with trials to a given target clustered next to one another. Note that even in this three-dimensional instantiation of FA$_{cmb}$, there is visible separation between the clouds, suggesting that it is possible to discriminate between targets.

By comparing Eqs. (7.25) and (7.28), we can gain further insight into the relationship between FA$_{sep}$ and FA$_{cmb}$. Both models define a mixture of M Gaussians in the high-dimensional space in which **z** lies. In both cases, the Gaussian covariances take the same parameterized form of low-rank plus diagonal. Whereas FA$_{sep}$ can have a different covariance for each reach target indexed by m (e.g., each Gaussian cloud can be oriented in a different direction), they are constrained to be the same in FA$_{cmb}$. In FA$_{sep}$,

---

[3]We fix the number of factors in this example to be three to allow for convenient plotting of the data; ordinarily, the number of factors (p) would be optimized using the model selection procedure described earlier.

the Gaussians can be centered anywhere in the high-dimensional space. In FA$_{cmb}$, however, they must lie within the lower-dimensional space spanned by the columns of C, which is also used to describe the shared variability. In other words, the single, low-dimensional space in FA$_{cmb}$ must be large enough (i.e., have a high enough dimensionality) to capture both between-class and within-class spread (where the latter is synonymous with shared variability). In FA$_{sep}$, each of the M low-dimensional spaces needs only to be large enough to capture the within-class spread for that reach target. As will be reviewed in the next section, we found that the optimal latent dimensionality of FA$_{cmb}$ was typically larger than that of FA$_{sep}$.

7.4.3 Results

We first chose the number of latent factors (p) that maximized cross-validated decoding performance. Conceptually, we had to ensure that we had enough factors to sufficiently describe the shared variability in the data but not so many that we overfit the training data. This was a necessary step before we could compare FA models to the independent Gaussian and Poisson models. We examined two data sets, G20040401 and H20040916, which comprised 185 and 200 total trials per reach target, respectively. This yielded ~80 training and ~20 validation trials per reach target for each cross-validation fold. A spike count window of 250 ms was used for this particular analysis, although we saw similar results for smaller spike count windows as well.

Figure 7.9(a) shows the validation performance for these two data sets as a function of p when fitting the FA$_{sep}$ algorithm. The performance for $p = 0$ corresponds to an independent Gaussian model where there is no C_m matrix and only a diagonal covariance matrix, R_m, per reach target. The best validation performance for FA$_{sep}$ was obtained for low p: $p^* = 1$ for G20040401 and $p^* = 3$ for H20040916. The results show that as p was increased beyond p^*, performance steadily declined. Overfitting was an issue for even relatively small values of p. Why might this be? There are surely many factors that influence the upcoming reach—direction, distance, curvature, speed, force, and so forth. However, for our data set, reach trajectories to the same target were highly similar. Moreover, the shared variability (captured by trial-to-trial modulation of the factors) may be small relative to the independent variability. Given a small number of training trials for each reach target (~80) relative to the number of units (~100), it is likely that we were unable to identify more latent factors without poorly estimating the model parameters and overfitting. The penalty for overfitting is evident: by $p = 30$, all of the benefits from using FA$_{sep}$ over the independent Gaussian approach ($p = 0$) are lost.

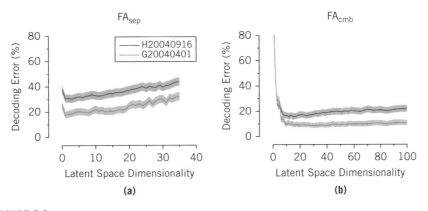

FIGURE 7.9

Model selection for FA_{sep} and FA_{cmb}. The shaded area denotes the 95% confidence interval (Bernoulli process) around the mean performance (embedded line). (a) Optimal cross-validated performance for FA_{sep} occurs for a low number of factors. (b) Optimal cross-validated performance for FA_{cmb} occurs between 10 and 20 factors. Please see this figure in color at the companion web site: www.elsevierdirect.com/companions/ 9780123750273
Source: Reproduced with permission from Santhanam et al., 2009.

We performed the same sweep of p for FA_{cmb}, and the validation performance is plotted in Figure 7.9(b). The best validation performance for the combined FA model was obtained for higher p than for FA_{sep}: $p^* = 31$ for G20040401 and $p^* = 12$ for H20040916. This makes sense because the latent space for FA_{cmb} must capture both within-class and between-class spread, whereas each of the M latent spaces for FA_{sep} need only capture within-class spread, as previously described. While at first it seems that the optimal number of factors for G20040401 is substantially higher than that for H20040916, one can see that the validation performance reached a floor around $p^* = 15$ for the G20040401 data set. There was negligible improvement when the dimensionality was increased past that point. This, combined with the fact that performance degradation due to overfitting in the H20040916 data set was relatively minor, demonstrated that FA_{cmb} is less sensitive to the exact choice of p than is FA_{sep}.

Next, we compared the performance of four different decoders grouped into two pairs. One pair consisted of the independent Gaussian and Poisson models. The second pair was the FA-based models, FA_{sep} and FA_{cmb}. We computed decoding accuracy based on a spike count window of 250 ms. For the independent algorithms, half of the trials in each data set were used to train the model parameters, and the model was used to predict the reach target for the other half. For the FA-based algorithms, the training set was first used

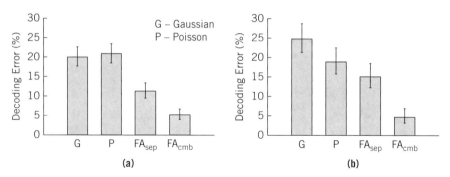

FIGURE 7.10

Comparison of decoding algorithms on two example data sets. Error bars denote the 95% confidence interval (Bernoulli process) for each measurement. (a) Data set G20040429. (b) Data set H20040928.
Source: Reproduced with permission from Santhanam et al., 2009.

to select the optimal dimensionality p^* using 5-fold cross-validation, and the training set was then used in its entirety to fit a model with p^* dimensionality. As with the independent algorithms, the fitted model was finally applied to the test data to compute the average decoding performance.

Figure 7.10 shows the relative performance of these four algorithms. There were a grand total of 1024 and 528 test trials for the G20040429 and H20040928 data sets, respectively. For data set G20040429, the independent Gaussian and Poisson algorithms had roughly the same performance, while for data set H20040928 the independent Gaussian algorithm had the worst performance, with ~5% worse accuracy than that of the independent Poisson algorithm. The FA_{sep} algorithm reduced the decoding error for monkey G's data set by nearly 10%, but showed only a modest improvement for monkey H's data set. The FA_{cmb} algorithm showed by far the best performance. Our novel decoder was able to drastically reduce the decoding error to only ~5%.

7.5 DISCUSSION

We described two classes of neural prosthetic systems—motor and communication—that aim to assist disabled patients by translating their thoughts into actions in the outside world. While many challenges remain (Andersen et al., 2004; Schwartz et al., 2006; Donoghue, 2008; Ryu and Shenoy, 2009), one of the major challenges is interpreting, or decoding, the neural signals. In this chapter, we reviewed current decoding approaches and detailed advances that we recently published (Yu et al., 2007; Santhanam et al., 2009). For motor prostheses, we showed that arm trajectories can be more accurately decoded by probabilistically mixing trajectory models, each of which is accurate within a limited regime of movement. The MTM is a general approach to constructing trajectory models that can exhibit rather complex dynamical behaviors, whose decoder can be implemented to have the same running time as as those of simpler trajectory models. Furthermore, prior information about the identity of the upcoming movement regime can be incorporated in a principled way. For communication prostheses, we showed that the reach target can be more accurately decoded by taking into account the correlation structure across the neural population. These correlations arise because of unobserved or uncontrolled aspects of the behavioral task, possibly due to the subject being more attentive or more fatigued. Using factor analysis methods, we exploited this statistical structure in a maximum likelihood classifier.

For goal-directed reaches, the observed neural activity provides two categorically different types of information about the arm trajectory to be estimated. One type is informative of the moment-by-moment details of the arm trajectory (dynamic); the other is informative of the identity of the upcoming reach target (static). If the moment-by-moment details of the arm trajectory can be decoded perfectly using only the neural activity present during movement, there is no need for target information in motor prostheses. However, the moment-by-moment details of the arm trajectory and the target identity are each decoded with varying levels of uncertainty. When both types of information are available, it is desirable to combine them in a way that takes into account their relative uncertainty and yields a coherent arm trajectory estimate. The MTM framework provides a principled way to combine these two types of information.

In this work, we extracted both types of information from the same cortical areas: PMd and M1. The type of information being decoded depends on *when* the neural activity occurs relative to the reach, which we assumed to be known. In settings where the subject is free to decide when to reach, it will be necessary to implement a state machine (Shenoy et al., 2003; Achtman et al., 2007; Kemere et al., 2008) that determines the type of information being conveyed by the neural activity at each time point.

The MTM can be viewed as a discrete approximation of a single, unified trajectory model with nonlinear dynamics. One may consider directly decoding using a nonlinear trajectory model without discretization. Although this makes the decoding problem more difficult, there are effective approximate methods that can be applied (Yu et al., 2006b, 2008). In addition to the MTM and decoding using a nonlinear model, other important extensions to the basic Kalman filter include a switching state model (Srinivasan et al., 2007), a switching observation model (Wu et al., 2004a; Srinivasan et al.,

2007), and an adaptive Kalman filter whose parameters change over time (Wu and Hatsopoulos, 2008).

The advances presented in Section 7.4 can be applied in a straightforward way to the MTM. In Section 7.3, we used the independent Gaussian model (Eq. (7.19), assuming conditional independence across different units) to obtain the prior target information $P(m \mid \mathbf{z})$ (Eq. (7.20)). The results from Section 7.4 suggest that a more accurate target prior can be obtained by using either a Poisson model (Eq. (7.22)) or a Gaussian model that captures the correlated activity across the neural population (Eq. (7.25) or (7.28)). This could provide performance improvements beyond that shown in Section 7.3.

In Section 7.4, we introduced latent variables to capture unobserved or uncontrolled aspects of the behavioral task (e.g., the subject's level of attention or fatigue) in the context of discrete decoding for communication prostheses. The same idea can be applied to continuous decoding for motor prostheses. Paninski and colleagues showed that a Kalman filter with latent variables can outperform the standard Kalman filter (Wu et al., 2009).

For discrete decoding, we found that the independent Poisson model yielded similar or better decoding accuracy compared to the independent Gaussian model (see Figure 7.10). It is natural to ask whether the accuracy of the independent Poisson model can be improved by if one takes into account correlated activity across the neural population by replacing Eq. (7.24) or (7.27) with a Poisson observation model. In Santhanam et al. (2009), we developed a family of models called "factor analysis with Poisson output" (FAPO). We found that, while these models often outperformed the independent Poisson model, they tended to have higher levels of decoding error than their Gaussian counterparts (FA_{sep} and FA_{cmb}). There are several possible explanations for this result, including the fixed mean–variance relationship for a Poisson distribution and the approximations required for fitting FAPO models (Santhanam et al., 2009).

Although we sought to maximize classification performance in Section 7.4, we trained a generative model by maximizing the likelihood of the observed data. One would hope for a monotonic relationship between classification performance and data likelihood, but there is no guarantee that this will be the case. On the other hand, discriminative classifiers, such as support vector machines (Vapnik, 1995), directly maximize classification performance. One advantage of using a generative model is that we can recover and attempt to interpret the latent variables (see Figure 7.8), whereas discriminative methods tend to yield a "black box" that is difficult to interrogate. Jordan and colleagues compared these two types of classifiers and found that generative classifiers tended to outperform discriminative classifiers in the regime of small training set sizes, while the reverse was true for larger training set sizes

(Ng and Jordan, 2002). We leave the exploration of discriminative classifiers for communication prostheses, and their comparison to generative classifiers, as a topic of future work.

7.6 **FUTURE DIRECTIONS**

There has been considerable progress in the development of neural prosthetic systems in recent years, including FDA-approved clinical trials (Hochberg et al., 2006; Kim et al., 2008). However, many critical research challenges remain before invasive, electrode-based systems are ready for "clinical prime time" (Donoghue et al., 2007; Ryu and Shenoy, 2009). Among them are increasing decoding performance, minimizing surgical invasiveness, increasing electrode lifetime and recording stability (Polikov et al., 2005), and developing fully implantable, low-power electronics (Sodagar et al., 2007; Chestek et al., 2009; Harrison et al., 2009). In this chapter, we considered only the first of these challenges, neural decoding.

Within the scope of neural decoding, there remain many challenges, including (1) furthering our understanding of the time course of plan and movement activity and their relationship to arm movement, and (2) designing decoders that are effective in a closed-loop setting. For the first challenge, in both Sections 7.3 and 7.4 we collapsed the timecourse of plan activity by taking spike counts within a large time window. However, from Figure 7.2, it appears that the time-varying nature of plan activity can, in principle, be leveraged to improve the prior target information (Section 7.3) and target classification (Section 7.4). For movement activity, most decoders (including population vectors, linear filters, and all recursive Bayesian decoders mentioned in this work) assume a simple relationship (e.g., linear cosine tuning or some variant thereof) between arm state and neural activity. However, there is mounting evidence that this relationship is more complex than previously appreciated (Churchland and Shenoy, 2007b; Scott, 2008), and leads to model mismatch (see Figure A2 in Yu et al., 2007) in the training phase and subsequent decoding error in the testing phase. In brief, we need methods for studying the time course of plan and movement activity across a neural population that does not collapse neural activity across a large window, or assume an overly restrictive relationship with arm movement. One possible approach is to extract a low-dimensional *neural trajectory* that summarizes the activity recorded simultaneously from many units (Yu et al., 2006a, 2009; Churchland et al., 2007, 2010). An advantage with this approach is that neural trajectories are extracted using neural activity alone, and no assumption needs to be made about the relationship between the neural activity and arm movement. Figure 7.11 shows neural trajectories extracted from 61 units in PMd and M1 during a delayed reach task. Each trace corresponds to a single trial of planning and executing a reach to the upper-right target (medium gray traces) or lower-left target (light gray traces). For different reach targets, the trajectories head off toward different parts of the low-dimensional space in a reliable fashion. After the neural trajectories are extracted, they can then be related to the subject's behavior, such as arm kinematics and muscle activity (Afshar et al., 2008). Such analyses should provide insight into how the time course of plan activity should be incorporated and how to better capture the relationship between movement activity and desired arm movement in prosthetic decoders.

For the second challenge, Kass and colleagues have shown that the open-loop performance of decoding algorithms does not necessarily reflect their closed-loop performance (Chase et al., 2009). The reason is that the subject receives visual (and potentially other sources of) feedback in a closed-loop

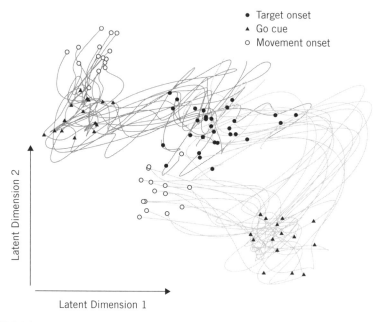

FIGURE 7.11

Neural trajectories extracted by Gaussian-process factor analysis (GPFA). Each trace corresponds to a single trial of planning and executing a reach to the upper-right target (dark gray traces) or the lower-left target (light gray traces). Indicated are time of reach target onset (black dots), go cue (triangles), and movement onset (open circles). Please see this figure in color at the companion web site: www.elsevierdirect.com/companions/9780123750273

setting, which enables the subject to compensate for systematic errors in the decoded cursor movements. Furthermore, there is increasing evidence that neurons can change their firing properties between open-loop and closed-loop settings (Taylor et al., 2002; Carmena et al., 2003), and that the brain can learn the mapping between neural activity and cursor movement (Jarosiewicz et al., 2008; Ganguly and Carmena, 2009). This suggests that we not only should carefully compare existing decoders in a closed-loop setting (Kim et al., 2008; Chase et al., 2009), but should also design decoding algorithms that are specifically tailored for closed-loop use. Currently missing from decoding algorithm development are control theoretic concepts (Scott, 2004), such as feedback, signal delays (in the nervous system and in the signal processing machinery of the prosthetic system), and system stability. To design a decoder that accounts for sensory feedback, we need to develop a deeper understanding of how the subject alters his neural activity based on feedback received to achieve task success. Because the cursor motion is different from one experimental trial to the next, simply averaging the neural activity across trials will likely obscure neural signatures of the brain attempting to correct for errors in cursor motion. Instead, we need statistical methods that can characterize the time course of neural population activity on a *single-trial* basis (Yu et al., 2006a, 2009; Churchland et al., 2007, 2010). Such methods will allow us to study (1) how corrections in cursor motion are brought about by changes in the pattern of neural population activity, and (2) how the pattern of neural activity shifts when transitioning from an open-loop to a closed-loop setting.

Acknowledgments

We thank Stephen Ryu, Afsheen Afshar, Caleb Kemere, Vikash Gilja, and Teresa Meng for their contributions to the original research articles that are reviewed in this chapter, John Cunningham for his contributions to the development of the GPFA algorithm (Figure 7.11), Sandra Eisensee for administrative assistance, Melissa Howard for surgical assistance and veterinary care, and Drew Haven for computing support.

This work was supported by NIH-NINDS-CRCNS 5-R01-NS054283, NDSEG Fellowships, NSF Graduate Research Fellowships, Gatsby Charitable Foundation, Christopher and Dana Reeve Foundation, Burroughs Wellcome Fund Career Award in the Biomedical Sciences, Stanford Center for Integrated Systems, NSF Center for Neuromorphic Systems Engineering at Caltech, Office of Naval Research, Sloan Foundation and Whitaker Foundation.

References

Achtman, N., Afshar, A., Santhanam, G., Yu, B.M., Ryu, S.I., Shenoy, K.V., 2007. Free paced high-performance brain-computer interfaces. J. Neural. Eng. 4, 336–347.

Afshar, A., Yu, B.M., Santhanam, G., Ryu, S.I., Shenoy, K.V., Sahani, M., 2008. Evidence for smooth and inertial dynamics in the evolution of neural state. Soc. Neurosci. Abstr. (319.8).

Andersen, R.A., Burdick, R.W., Musallam, S., Pesaran, B., Cham, J.G., 2004. Cognitive neural prosthetics. Trends Cogn. Sci. 8, 486–493.

Batista, A.P., Santhanam, G., Yu, B.M., Ryu, S.I., Afshar, A., Shenoy, K.V., 2007. Reference frames for reach planning in macaque dorsal premotor cortex. J. Neurophysiol. 98, 966–983.

Boussaoud, D., Wise, S.P., 1993. Primate frontal cortex: neuronal activity following attentional versus intentional cues. Exp. Brain. Res. 95 (1), 15–27.

Brockwell, A.E., Rojas, A.L., Kass, R.E., 2004. Recursive Bayesian decoding of motor cortical signals by particle filtering. J. Neurophysiol. 91 (4), 1899–1907.

Brown, E.N., Frank, L.M., Tang, D., Quirk, M.C., Wilson, M.A., 1998. A statistical paradigm for neural spike train decoding applied to position prediction from the ensemble firing patterns of rat hippocampal place cells. J. Neurosci. 18 (18), 7411–7425.

Carmena, J.M., Lebedev, M.A., Crist, R.E., O'Doherty, J.E., Santucci, D.M., Dimitrov, D.F., et al., 2003. Learning to control a brain-machine interface for reaching and grasping by primates. PLoS Biol. 1 (2), 193–208.

Chan, S.S., Moran, D.W., 2006. Computational model of a primate arm: from hand position to joint angles, joint torques and muscle forces. J. Neural. Eng. 3, 327–337.

Chapin, J.K., Moxon, K.A., Markowitz, R.S., Nicolelis, M.A.L., 1999. Real-time control of a robot arm using simultaneously recorded neurons in the motor cortex. Nat. Neurosci. 2, 664–670.

Chase, S.M., Schwartz, A.B., Kass, R.E., 2009. Bias, optimal linear estimation, and the differences between open-loop simulation and closed-loop performance of spiking-based brain-computer interface algorithms. Neural Netw. 22 (9), 1203–1213.

Chestek, C.A., Batista, A.P., Santhanam, G., Yu, B.M., Afshar, A., Cunningham, J.P., et al., 2007. Single-neuron stability during repeated reaching in macaque premotor cortex. J. Neurosci. 27, 10742–10750.

Chestek, C. A., Gilja, V., Nuyujukian, P., Kier, R. J., Solzbacher, F., Ryu, S. I., Harrison, R. R., and Shenoy, K. V. (2009). HermesC: Low-power wireless neural recording system for freely moving primates. IEEE Trans Neural Syst Rehabil Eng. 17 (4), 330–338.

Churchland, M.M., Afshar, A., Shenoy, K.V., 2006a. A central source of movement variability. Neuron 52, 1085–1096.

Churchland, M.M., Santhanam, G., Shenoy, K.V., 2006b. Preparatory activity in premotor and motor cortex reflects the speed of the upcoming reach. J. Neurophysiol. 96 (6), 3130–3146.

Churchland, M.M., Shenoy, K.V., 2007a. Delay of movement caused by disruption of cortical preparatory activity. J. Neurophysiol. 97, 348–359.

Churchland, M.M., Shenoy, K.V., 2007b. Temporal complexity and heterogeneity of single-neuron activity in premotor and motor cortex. J. Neurophysiol. 97, 4235–4257.

Churchland, M.M., Yu, B.M., Ryu, S.I., Santhanam, G., Shenoy, K.V., 2006c. Neural variability in premotor cortex provides a signature of motor preparation. J. Neurosci. 26 (14), 3697–3712.

Churchland, M.M., Yu, B.M., Sahani, M., Shenoy, K.V., 2007. Techniques for extracting single-trial activity patterns from large-scale neural recordings. Curr. Opin. Neurobiol. 17 (5), 609–618.

Churchland, M.M., Yu, B.M., Cunningham, J.P., Sugrue, L.P., Cohen, M.R., Corrado, G.S., Newsome, W.T., Clark, A.M., Hosseini, P., Scott, B.B., Bradley, D.C., Smith, M.A., Kohn, A., Movshon, J.A., Armstrong, K.M., Moore, T., Chang, S.W., Snyder, L.H., Lisberger, S.G., Priebe, N.J., Finn, I.M., Ferster, D., Ryu, S.I., Santhanam, G., Sahani, M., and Shenoy, K.V., 2010. Stimulus onset quenches neural variability: a widespread cortical phenomenon. Nat Neurosci. 13 (3), 369–378.

Crammond, D.J., Kalaska, J.F., 2000. Prior information in motor and premotor cortex: activity during the delay period and effect on pre-movement activity. J. Neurophysiol. 84, 986–1005.

Dempster, A.P., Laird, N.M., Rubin, D.B., 1977. Maximum likelihood from incomplete data via the EM algorithm (with discussion). J. R. Stat. Soc. Ser. B 39, 1–38.

Donoghue, J.P., 2008. Bridging the brain to the world: a perspective on neural interface systems. Neuron 60, 511–521.

Donoghue, J.P., Nurmikko, A., Black, M., Hochberg, L.R., 2007. Assistive technology and robotic control using motor cortex ensemble-based neural interface systems in humans with tetraplegia. J. Physiol. 579 (3), 603–611.

Everitt, B.S., 1984. An Introduction to Latent Variable Models. Chapman and Hill, London.

Faisal, A.A., Selen, L.P.J., Wolpert, D.M., 2008. Noise in the nervous system. Nat. Rev. Neurosci. 9, 292–303.

Fetz, E.E., 1969. Operant conditioning of cortical unit activity. Science 163, 955–957.

Fetz, E.E., Baker, M.A., 1972. Operantly conditioned patterns of precentral unit activity and correlated responses in adjacent cells and contralateral muscles. J. Neurophysiol. 36, 179–204.

Fetz, E.E., Finocchio, D.V., 1971. Operant conditioning of specific patterns of neural and muscular activity. Science 174, 431–435.

Ganguly, K., Carmena, J.M., 2009. Emergence of a stable cortical map for neuroprosthetic control. PLoS Biol. 7, e1000153.

Georgopoulos, A.P., Schwartz, A.B., Kettner, R.E., 1986. Neuronal population coding of movement direction. Science 233, 1416–1419.

Ghahramani, Z., Hinton, G., 1997. The EM algorithm for mixtures of factor analyzers. Technical Report CRG-TR-96-1, Department of Computer Science, University of Toronto.

Gilja, V., Kalmar, R.S., Santhanam, G., Ryu, S.I., Yu, B.M., Afshar, A., et al., 2005. Trial-by-trial mean normalization improves plan period reach target decoding. Soc. Neurosci. Abstr. (519.18).

Godschalk, M., Lemon, R.N., Kuypers, H.G., van der Steen, J., 1985. The involvement of monkey premotor cortex neurones in preparation of visually cued arm movements. Behav. Brain Res. 18, 143–157.

Gomez, J.E., Fu, Q., Flament, D., Ebner, T.J., 2000. Representation of accuracy in the dorsal premotor cortex. Eur. J. Neurosci. 12 (10), 3748–3760.

Harrison, R.R., Kier, R.J., Chestek, C.A., Gilja, V., Nuyujukian, P., Ryu, S.I., et al., 2009. Wireless neural recording with single low-power integrated circuit. IEEE Trans. Neural. Syst. Rehabil. Eng. 17 (4), 322–329.

Hastie, T., Tibshirani, R., Friedman, J.H., 2003. The Elements of Statistical Leaning, second ed. Springer, New York.

Hatsopoulos, N., Joshi, J., O'Leary, J.G., 2004. Decoding continuous and discrete motor behaviors using motor and premotor cortical ensembles. J. Neurophysiol. 92, 1165–1174.

Hinterberger, T., Schmidt, S., Neumann, N., Mellinger, J., Blankertz, B., Curio, G., et al., 2004. Brain-computer communication and slow cortical potentials. IEEE Trans. Biomed. Eng. 51(6), 1011–1018.

Hochberg, L.R., Serruya, M.D., Friehs, G.M., Mukand, J.A., Saleh, M., Caplan, A.H., et al., 2006. Neuronal ensemble control of prosthetic devices by a human with tetraplegia. Nature 442, 164–171.

Hocherman, S., Wise, S.P., 1991. Effects of hand movement path on motor cortical activity in awake, behaving rhesus monkeys. Exp. Brain Res. 83, 285–302.

Humphrey, D.R., Schmidt, E.M., Thompson, W.D., 1970. Predicting measures of motor performance from multiple cortical spike trains. Science 170, 758–762.

Isaacs, R.E., Weber, D.J., Schwartz, A.B., 2000. Work toward real-time control of a cortical neural prosthesis. IEEE Trans. Rehabil. Eng. 8, 196–198.

Jarosiewicz, B., Chase, S.M., Fraser, G.W., Velliste, M., Kass, R. E., Schwartz, A.B., 2008. Functional network reorganization during learning in a brain-computer interface paradigm. Proc. Natl. Acad. Sci. USA 105 (49), 19486–19491.

Kalmar, R.S., Gilja, V., Santhanam, G., Ryu, S.I., Yu, B.M., Afshar, A., et al., 2005. PMd delay activity during rapid sequential-movement plans. Soc. Neurosci. Abstr. (519.17).

Kemere, C., Santhanam, G., Yu, B.M., Afshar, A., Ryu, S.I., Meng, T.H., et al., 2008. Detecting neural state transitions using hidden Markov models for motor cortical prostheses. J. Neurophysiol. 100, 2441–2452.

Kemere, C., Shenoy, K.V., Meng, T.H., 2004. Model-based neural decoding of reaching movements: a maximum likelihood approach. IEEE Trans. Biomed. Eng. 51 (6), 925–932.

Kennedy, P., Andreasen, D., Ehirim, P., King, B., Kirby, T., Mao, H., et al., 2004. Using human extra-cortical local field potentials to control a switch. J. Neural. Eng. 1 (2), 72–77.

Kennedy, P.R., Bakay, R.A.E., 1998. Restoration of neural output from a paralyzed patient by a direct brain connection. NeuroReport 9, 1707–1711.

Kennedy, P.R., Bakay, R.A.E., Moore, M.M., Adams, K., Goldwaithe, J., 2000a,b. Direct control of a computer from the human central nervous system. IEEE Trans. Rehabil. Eng. 8, 198–202.

Kim, S.P., Simeral, J.D., Hochberg, L.R., Donoghue, J.P., Black, M.J., 2008. Neural control of computer cursor velocity by decoding motor cortical spiking activity in humans with tetraplegia. J. Neural. Eng. 5, 455–476.

Kulkarni, J.E., Paninski, L., 2008. State-space decoding of goal-directed movements. IEEE Signal. Process. Mag. 25 (1), 78–86.

Leuthardt, E.C., Schalk, G., Wolpaw, J.R., Ojemann, J.G., Moran, D.W., 2004. A brain-computer interface using electrocorticographic signals in humans. J. Neural. Eng. 1, 63–71.

Lewicki, M.S., 1998. A review of methods for spike sorting: the detection and classification of neural action potentials. Network: Comput. Neural Syst. 9 (4), R53–R78.

MacKay, D.J.C., 2003. Information Theory, Inference, and Learning Algorithms. Cambridge University Press, Cambridge, UK.

Maynard, E.M., Hatsopoulos, N.G., Ojakangas, C.L., Acuna, B.D., Sanes, J.N., Normann, R.A., et al., 1999. Neuronal interactions improve cortical population coding of movement direction. J. Neurosci. 19, 8083–8093.

McFarland, D.J., Sarnacki, W.A., Wolpaw, J.R., 2003. Brain-computer interface (BCI) operation: optimizing information transfer rates. Biol. Psychol. 63 (3), 237–251.

Messier, J., Kalaska, J.F., 2000. Covariation of primate dorsal premotor cell activity with direction and amplitude during a memorized-delay reaching task. J. Neurophysiol. 84, 152–165.

Moran, D.W., Schwartz, A.B., 1999a. Motor cortical activity during drawing movements: population representation during spiral tracing. J. Neurophysiol. 82, 2693–2704.

Moran, D.W., Schwartz, A.B., 1999b. Motor cortical representation of speed and direction during reaching. J. Neurophysiol. 82, 2676–2692.

Moritz, C.T., Perlmutter, S.I., Fetz, E.E., 2008. Direct control of paralysed muscles by cortical neurons. Nature 456, 639–642.

Mulliken, G.H., Musallam, S., Andersen, R.A., 2008. Decoding trajectories from posterior parietal cortex ensembles. J. Neurosci. 28 (48), 12913–12926.

Musallam, S., Corneil, B.D., Greger, B., Scherberger, H., Andersen, R.A., 2004. Cognitive control signals for neural prosthetics. Science 305, 258–262.

Ng, A.Y., Jordan, M.I., 2002. On Discriminative Vs. Generative Classifiers: A Comparison of Logistic Regression and Naive Bayes. In: Dietterich, T.G., Becker, S., Ghahramani, Z. (Eds.), Advances in Neural Information Processing Systems 14. MIT Press, Cambridge, MA.

Polikov, V.S., Tresco, P.A., Reichert, W.M., 2005. Response of brain tissue to chronically implanted neural electrodes. J. Neurosci. Methods 148, 1–18.

Riehle, A., Requin, J., 1994. Are extent and force independent movement parameters? preparation- and movement-related neuronal activity in the monkey cortex. Exp. Brain Res. 99, 56–74.

Roweis, S., Ghahramani, Z., 1999. A unifying review of linear Gaussian models. Neural Comput. 11 (2), 305–345.

Ryu, S.I., Shenoy, K.V., 2009. Human cortical prostheses: lost in translation? Neurosurg. Focus 27 (1), E5.

Salinas, E., Abbott, L.F., 1994. Vector reconstruction from firing rates. J. Comput. Neurosci. 1, 89–107.

Santhanam, G., Ryu, S.I., Yu, B.M., Afshar, A., Shenoy, K.V., 2006. A high-performance brain-computer interface. Nature 442, 195–198.

Santhanam, G., Sahani, M., Ryu, S.I., Shenoy, K.V., 2004. An extensible infrastructure for fully automated spike sorting during online experiments. In: Proceedings of the 26th Annual Conference of the IEEE EMBS, 4380–4384.

Santhanam, G., Yu, B.M., Gilja, V., Ryu, S.I., Afshar, A., Sahani, M., et al., 2009. Factor-analysis methods for higher-performance neural prostheses. J. Neurophysiol. 102, 1315–1330.

Schalk, G., Miller, K.J., Anderson, N.R., Wilson, J.A., Smyth, M.D., Ojemann, J.G., et al., 2008. Two-dimensional movement control using electrocorticographic signals in humans. J. Neural Eng. 5 (1), 75–84.

Schwartz, A.B., 1994. Direct cortical representation of drawing. Science 265, 540–542.

Schwartz, A.B., Cui, X.T., Weber, D.J., Moran, D.W., 2006. Brain-controlled interfaces: movement restoration with neural prosthetics. Neuron 52 (1), 205–220.

Scott, S.H., 2004. Optimal feedback control and the neural basis of volitional motor control. Nat. Rev. Neurosci. 5, 532–546.

Scott, S.H., 2008. Inconvenient truths about neural processing in primary motor cortex. J. Physiol. 586, 1217–1224.

Serby, H., Yom-Tov, E., Inbar, G.F., 2005. An improved p300-based brain-computer interface. IEEE Trans. Neural Syst. Rehabil. Eng. 13 (1), 89–98.

Serruya, M.D., Hatsopoulos, N.G., Paninski, L., Fellows, M.R., Donoghue, J.P., 2002. Instant neural control of a movement signal. Nature 416, 141–142.

Shen, L., Alexander, G.E., 1997a. Neural correlates of a spatial sensory-to-motor transformation in primary motor cortex. J. Neurophysiol. 77, 1171–1194.

Shen, L., Alexander, G.E., 1997b. Preferential representation of instructed target location versus limb trajectory in dorsal premotor area. J. Neurophysiol. 77, 1195–1212.

Shenoy, K.V., Meeker, D., Cao, S., Kureshi, S.A., Pesaran, B., Mitra, P., et al., 2003. Neural prosthetic control signals from plan activity. NeuroReport 14, 591–596.

Shoham, S., Paninski, L.M., Fellows, M.R., Hatsopoulos, N.G., Donoghue, J.P., Normann, R.A., 2005. Statistical encoding model for a primary motor cortical brain-machine interface. IEEE Trans. Biomed. Eng. 52 (7), 1313–1322.

Sodagar, A.M., Wise, K.D., Najafi, K., 2007. A fully-integrated mixed-signal neural processor for implantable multi-channel cortical recording. IEEE Trans. Biomed. Eng. 54, 1075–1088.

Srinivasan, L., Eden, U.T., Mitter, S.K., Brown, E.N., 2007. General-purpose filter design for neural prosthetic devices. J. Neurophysiol. 98, 2456–2475.

Srinivasan, L., Eden, U.T., Willsky, A.S., Brown, E.N., 2006. A state-space analysis for reconstruction of goal-directed movements using neural signals. Neural Comput. 18 (10), 2465–2494.

Taylor, D.M., Tillery, S.I.H., Schwartz, A.B., 2002. Direct cortical control of 3D neuroprosthetic devices. Science 296, 1829–1832.

Taylor, D.M., Tillery, S.I.H., Schwartz, A.B., 2003. Information conveyed through brain-control: cursor vs. robot. IEEE Trans. Neural Syst. Rehabil. Eng. 11, 195–199.

Thacker, N.A., Bromiley, P.A., 2001. The Effects of a Square Root Transform on a Poisson Distributed Quantity. Technical Report 2001-010, Imaging Science and Biomedical Engineering Divison, University of Manchester.

Vapnik, V.N., 1995. The Nature of Statistical Learning Theory. Springer Verlag, New York.

Velliste, M., Perel, S., Spalding, M.C., Whitford, A.S., Schwartz, A.B., 2008. Cortical control of a prosthetic arm for self-feeding. Nature 453 (7198), 1098–1101.

Warland, D.K., Reinagel, P., Meister, M., 1997. Decoding visual information from a population of retinal ganglion cells. J. Neurophysiol. 78, 2336–2350.

Wessberg, J., Stambaugh, C.R., Kralik, J.D., Beck, P.D., Laubach, M., Chapin, J.K., et al., 2000. Real-time prediction of hand trajectory by ensembles of cortical neurons in primates. Nature 408, 361–365.

Wise, S.P., Boussaoud, D., Johnson, P.B., Caminiti, R., 1997. Premotor and parietal cortex: corticocortical connectivity and combinatorial computations. Annu. Rev. Neurosci. 20, 25–42.

Wolpaw, J., McFarland, D., 2004a. Control of a two-dimensional movement signal by a noninvasive brain-computer interface in humans. Proc. Natl. Acad. Sci. USA 101 (51), 17849–17854.

Wolpaw, J.R., McFarland, D.J., 2004b. Control of a two-dimensional movement signal by a noninvasive brain-computer interface in humans. Proc. Natl. Acad. Sci. USA 101 (51), 17849–17854.

Wu, W., Black, M.J., Mumford, D., Gao, Y., Bienenstock, E., Donoghue, J.P., 2004a. Modeling and decoding motor cortical activity using a switching Kalman filter. IEEE Trans. Biomed. Eng. 51 (6), 933–942.

Wu, W., Gao, Y., Bienenstock, E., Donoghue, J.P., Black, M.J., 2006. Bayesian population decoding of motor cortical activity using a Kalman filter. Neural Comput. 18 (1), 80–118.

Wu, W., Hatsopoulos, N.G., 2008. Real-time decoding of nonstationary neural activity in motor cortex. IEEE Trans. Neural Syst. Rehabil. Eng. 16 (3), 213–222.

Wu, W., Kulkarni, J.E., Hatsopoulos, N.G., Paninski, L., 2009. Neural decoding of hand motion using a linear state-space model with hidden states. IEEE Trans. Biomed. Eng. 17 (4), 370–378.

Wu, W., Shaikhouni, A., Dongohue, J.P., Black, M.J., 2004b. Closed-loop neural control of cursor motion using a Kalman filter. In: Proceedings of the 26th Annual Conference of the IEEE EMBS, 4126–4129.

Yu, B.M., Afshar, A., Santhanam, G., Ryu, S.I., Shenoy, K.V., Sahani, M., 2006a. Extracting dynamical structure embedded in neural activity. In: Weiss, Y., Schölkopf, B., Platt, J. (Eds.), Advances in Neural Information Processing Systems 18 MIT Press, Cambridge, MA, pp. 1545–1552.

Yu, B.M., Cunningham, J.P., Santhanam, G., Ryu, S.I., Shenoy, K.V., Sahani, M., 2009. Gaussian-process factor analysis for low-dimensional single-trial analysis of neural population activity. J. Neurophysiol. 102, 614–635.

Yu, B.M., Cunningham, J.P., Shenoy, K.V., Sahani, M., 2008. Neural Decoding of Movements: From Linear to Nonlinear Trajectory Models. Volume 4984 of Lecture Notes in Computer Science. Springer, Berlin/Heidelberg, pp. 586–595.

Yu, B.M., Kemere, C., Santhanam, G., Afshar, A., Ryu, S.I., Meng, T.H., et al., 2007. Mixture of trajectory models for neural decoding of goal-directed movements. J. Neurophysiol. 97, 3763–3780.

Yu, B.M., Ryu, S.I., Santhanam, G., Churchland, M.M., Shenoy, K.V., 2004. Improving neural prosthetic system performance by combining plan and peri-movement activity. In: Proceedings of the 26th Annual Conference of the IEEE EMBS, 4516–4519.

Yu, B.M., Shenoy, K.V., Sahani, M., 2006b. Expectation propagation for inference in non-linear dynamical models with Poisson observations. In: Proceedings IEEE Nonlinear Statistical Signal Processing Workshop.

Zhang, K., Ginzburg, I., McNaughton, B.L., Sejnowski, T.J., 1998. Interpreting neuronal population activity by reconstruction: unified framework with application to hippocampal place cells. J. Neurophysiol. 79, 1017–1044.

Inner Products for Representation and Learning in the Spike Train Domain

António R. C. Paiva[*], **Il Park**[*†], **and José C. Príncipe**[†]

[*]*Scientific Computing and Imaging Institute, University of Utah, Salt Lake City, UT 84112*

[*†]*J. Crayton Pruitt Family Department of Biomedical Engineering and Computational NeuroEngineering Laboratory, University of Florida, Gainesville, FL 32611*

[†]*Computational NeuroEngineering Laboratory, University of Florida, Gainesville, FL 32611*

8.1 INTRODUCTION

Spike trains are a representation of neural activity. In neurophysiological studies, spike trains are obtained by detecting intra- or extra-cellularly the action potentials (i.e., "spikes"), but preserving only the time instant at which they occur. By neglecting the stereotypical shape of an action potential, spike trains contain an abstraction of the neurophysiological recordings, preserving only the spike times (i.e., the time at which action potentials occur) (Dayan and Abbott, 2001). Then, a spike train s is simply a sequence of ordered spike times $s = \{t_n \in \mathcal{T} : n = 1, \ldots, N\}$, with spike times in the interval $\mathcal{T} = [0, T]$ corresponding to the duration of the recording period. However, this abstraction poses a major problem in the way spike trains are manipulated. Indeed, put in this way, spike trains can be more readily modeled as realizations of point processes (Snyder, 1975). This means that any information, if present, is contained only in the location of the spike times. This is a very different perspective from the usual (functional) random processes, where the information is coded in the amplitude of the signal. For this reason, the usual operations of filtering, classification, and clustering are not directly applicable to spike trains.

A commonly used methodology for spike train analysis and processing starts by converting the spike train to its *binned* counterpart and proceeding as with functional random processes. The binned spike train is obtained by sliding a window in constant intervals over the spike train and counting the number of spikes within the window (Dayan and Abbott, 2001). The fundamental property of this transformation is that it maps randomness in time (expressed in the spike times) to randomness in amplitude of a discrete-time

Statistical Signal Processing for Neuroscience and Neurotechnology. DOI: 10.1016/B978-0-12-375027-3.00008-9

random process. Hence, after binning the spike trains, statistical analysis and machine learning methods can be applied as usual, making this methodology very straightforward and thus highly attractive. On the other hand, the impact of the *lossy* binning transformation must be taken into account. Note that, by binning, information about the exact spike times is lost because an implicit time quantization takes place (with the quantization step equal to the width of the sliding window). Therefore, any subsequent analysis will have limited accuracy, as demonstrated, for example, by Park et al. (2008) while comparing the usual cross-correlogram to an approach that avoids this quantization. This loss of information is of special concern in studies focusing on temporal dynamics because differences in spike timing smaller than the binning step will likely not be detected.

An alternative methodology is to utilize machine learning tools using a statistical characterization of the point process model. In concept, this methodology is very similar to the use of Bayesian methods in machine learning (Jordan, 1998; Jensen, 2001). This approach is supported by the extensive characterizations of point processes in the statistical literature (Snyder, 1975; Karr, 1986; Daley and Vere-Jones, 1988; Reiss, 1993), but their analysis from realizations (such as spike trains) typically requires knowledge or the assumption of an underlying model (Barbieri et al., 2001; Brown et al., 2001; Eden et al., 2004). Thus, the result is inaccurate if the model chosen is incomplete (i.e., if it is not capable of fully describing time-dependencies in the spike train). To characterize spike trains from more general models the standard approach has been to utilize generalized linear models (GLM) (Kass et al., 2005; Truccolo et al., 2005), which have many parameters and require large amounts of data for parameter estimation. Hence, GLMs are not appropriate for analysis of spike train data from a single trial. More fundamentally, statistical approaches do not provide an effective way to handle multiple spike trains. To avoid handling high-dimensional joint distribution, statistical approaches systematically assume the spike trains to be independent, which neglects potentially synergistic information (Schneidman et al., 2003; Narayanan et al., 2005). Finally, using a statistical characterization directly, general machine learning methods are not easily developed and are very computationally expensive unless Gaussian approximations are used (Wang et al., 2009).

A number of other spike train methods are available, such as frequency methods, Volterra series modeling, and spike train distance measures. However, these approaches are limited in scope and in the class of problems they are capable of tackling, whereas we are interested in developing a framework for general machine learning. Frequency-based methods (Baccalá and Sameshima, 1999; Pesaran et al., 2002; Hurtado et al., 2004) can be useful for analysis of the frequency contents and to search for repetitive patterns, and phase synchrony, but they do not directly allow machine learning. Moreover, stationarity must be assumed, at least piecewise, which limits the dynamics that the frequency decomposition can expose. Volterra series (Marmarelis, 2004; Song et al., 2007) have been utilized to model neural systems by expressing the system in a series of "kernels" that combine the input spike train(s) at multiple time instants. Intuitively, the Volterra series is the generalization of the Taylor series to systems with memory. The Volterra series decomposition can be utilized to reproduce a neural process or to study a complex system by analyzing the (simpler) "kernels" (Marmarelis, 2004). However, it cannot be utilized for general machine learning since, for example, its intrinsic supervised formulation prevents its use for any unsupervised learning method. Another recent approach is based on spike train measures. Some of the most widely utilized distance measures include Victor-Purpura's distance (Victor and Purpura, 1996, 1997) and van Rossum's distance (van Rossum, 2001). A key advantage of using spike train distances is that they avoid problems with binning by operating with the spike times directly. Moreover, several problems can be posed and solved resorting only to distances. On the other hand, distances do not provide the necessary "mathematical structure" for general machine learning. Most methods require other operations, such as the ability to compute means and projections, which are not possible with distance alone.

In this chapter we present a general framework to develop analysis and machine learning methods for spike trains. The core concept of the framework is the definition of spike train inner product operators. By defining these operators, we build on the mathematical theory of reproducing kernel Hilbert spaces (RKHS) and the ideas from kernel methods, thus allowing a multitude of analysis and learning algorithms to be easily developed. The importance of RKHS theory and the corresponding induced inner products have shown in a number of problems, such as statistical signal processing and detection (Parzen, 1959; Kailath, 1971; Kailath and Duttweiler, 1972). Moreover, inner products in RKHS spaces are the fundamental construct behind kernel machine learning (Wahba, 1990; Vapnik, 1995; Schölkopf et al., 1999). However, rather that tacitly choosing kernels, as is typically done in kernel methods (Schölkopf et al., 1999), inner products will be defined here from functional representations of spike trains. We motivate these representations from two perspectives: as a biological modeling problem and as a statistical description. The biological modeling approach highlights the potential biological mechanisms that take place at the neuron level and that are quantified by the inner product. On the other hand, by interpreting the representation from a statistical perspective, one relates to other work in the literature. Hence, for spike trains, the two perspectives are closely related. One of the most interesting consequences of this observation is that it exposes the advantages and limitations of some neural models.

It must be noted that applications of this framework are not restricted to neural spike trains. Recently, there has been great interest in using spike trains directly for biologically inspired computation paradigms, rather than the more traditional firing rate network models (Dayan and Abbott, 2001). Examples of such paradigms are liquid-state machine computational models (Maass and Bishop, 1998; Maass et al., 2002), spiking neural networks (Maass and Bishop, 1998; Bohte et al., 2002), and neuromorphic hardware design (Jabri et al., 1996). In all of these applications, and regardless of the nature of the process generating the spike trains, there is a need to analyze and decode the information that the spike trains express.

8.2 FUNCTIONAL REPRESENTATIONS OF SPIKE TRAINS

Two complementary forms of functional spike train representation are presented in this section. By functional representation we mean a function of time that characterizes the spike train. These functional representations will be utilized later in defining inner products for spike trains. By first introducing them we aim (1) to show the connection between neuron modeling and statistical approaches, and (2) to make clear how the inner product definitions in Section 8.3 reflect synapse processing dynamics and statistical knowledge.

8.2.1 Synaptic Models

In spite of the "digital" characteristics embodied in spike trains, neurons are analog entities. Hence, some form of "spike-to-analog" conversion must take place in the synapses, which converts spike trains into a functional form. Indeed, a spike propagated and arriving at the axon terminal will trigger the release of neurotransmitters through the synaptic cleft onto to receptors in the dendrite of the efferent neuron (Dowling, 1992). This induces an increase in the dendrite's conductance, which allows the transfer of ions between the surrounding and the dendrite, resulting in the post-synaptic potential. Immediately after, the neurotransmitters unbind and are metabolized, and the difference in potential between the dendrite and the surrounding makes ions flow toward the base potential, terminating the post-synaptic potential.

If the diffusion of neurotransmitters and the induced increase in conductance is assumed instantaneous, this process can be described through a very simple model. A spike arriving at the axon terminal results in the increase in conductance g:

$$g \rightarrow g + \Delta g \tag{8.1a}$$

where the constant Δg denotes the increase in conductance. Then the unbinding of neurotransmitters leads to a decrease in conductance:

$$\tau_s \frac{dg}{dt} = -g \tag{8.1b}$$

considering a decrease controlled by a constant membrane time constant τ_s. Similarly, the model can be written in terms of the membrane potential.

For a spike train, the dynamics expressed by Eqs. (8.1a) and (8.1b) can be written quite simply. If the input spike train is written as

$$s = \sum_{i=1}^{N} \delta(t - t_i) \tag{8.2}$$

with $\delta(\cdot)$, the Dirac delta impulse function, then the post-synaptic potential can be written as

$$v(t) = \sum_{i=1}^{N} h(t - t_i) \tag{8.3}$$

where h is the stereotypical evoked post-synaptic potential. With the parameters above, $h(t) = (\Delta g) \exp[-t/\tau_s] u(t)$, where $u(t)$ denotes the Heaviside step function.

It is interesting to verify that Eq. (8.3) is the same as the filtered spike train definition in van Rossum (van Rossum, 2001, Eq. (2.2)). Van Rossum defines the distance between two filtered spike trains $v_i(t)$ and $v_j(t)$ as

$$D(v_i, v_j) = \sqrt{\frac{1}{\tau_s} \int_0^\infty [v_i(t) - v_j(t)]^2 dt} \tag{8.4}$$

That is, van Rossum's distance is merely the Euclidean distance between the evoked post-synaptic potentials according to this simple model.

One of the fundamental limitations of the previous model is that it neglects the limited availability of receptors in the efferent dendrite. When two spikes arrive at a synapse within a short time interval (i.e., such that the conductance has not returned to zero), the second spike will induce a smaller change in the conductance because many of the receptors are already bound to neurotransmitters because of the previous spike. This observation can be inserted in the model by modifying Eq. (8.1a) to

$$g \rightarrow g + \frac{\Delta g}{g_{max}}(g_{max} - g) \tag{8.5}$$

where g_{max} is the maximum value the conductance can attain. The equation expressing the unbinding, Eq. (8.1b), remains the same. Eq. (8.5) means that, for small conductance values, the conductance is increased by $\approx \Delta g$, but the increase saturates as g approaches g_{max}. Put differently, this equation behaves much like Eq. (8.1a) followed by a nonlinear function, such as the hyperbolic tangent. Hence, Eq. (8.5) can be approximated as

$$g \rightarrow f(g + \Delta g) \tag{8.6}$$

where f denotes a nonlinear function, appropriately scaled by g_{max}. For example, both $f(x) = g_{max} \tanh(x/g_{max})$ or $f(x) = g_{max}(1 - \exp(-x^2/(2g_{max}^2)))$ are valid possibilities. In fact, any monotonic increasing function in $\mathbb{R}^+$ that is zero at zero and saturates as the argument goes toward infinity, can potentially be utilized under the appropriate scaling. Actually, nonlinear "sigmoid" functions that verify these requirements have been measured in neurophysiological recordings (Freeman, 1975). Correspondingly, the post-synaptic potential can be written as

$$v^\dagger(t) = f\left(\sum_{i=1}^{N} h(t - t_i)\right) \tag{8.7}$$

affected by the same nonlinear function.[1] Note that the argument of the nonlinearity f equals the evoked membrane potential given by the previous model (Eq. (8.3)). An example of the evoked membrane potentials by the different models is given in Figure 8.1.

Obviously, more complex synaptic models can be formulated (see, for example, Gerstner and Kistler (2002) for an extensive review). A common

---

[1] Actually, Eq. (8.7) is only an approximation of the model characterized by Eq. (8.6) because the nonlinearity also shapes the dynamics of the unbinding (Eq. (8.1b)), and the actual dynamics will depend on the exact choise of f.

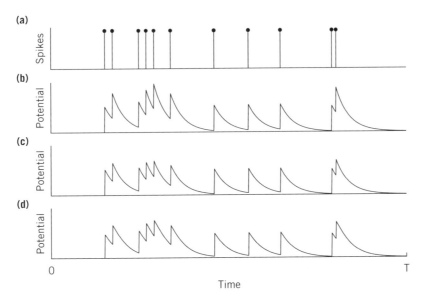

FIGURE 8.1

Comparison of evoked membrane potentials with different synapse models. The input spike train to the synapse model is shown in (a). The potentials in (b) through (d) correspond, respectively, to the linear model (Eq. (8.3)), the nonlinear model (potential corresponding to the conductance given by Eq. (8.5)), and the nonlinear model through nonlinear transformation (Eq. (8.7)). The same scale was used for the y-axis in (b) through (d).

modification is to account for the diffusion time of the neurotransmitters across the synapse and their binding with the dendrite receptors. Similarly, the model can be augmented to account for the depletion of neurotransmitters in the axon terminal (Tsodyks and Markram, 1997).

Overall, these synaptic models aim to emulate the filtering process that occurs at the synapses. In its simpler form, such as shown by Eq. (8.3), and even if the diffusion time is taken into account, the filter is linear. In modeling the depletion of neurotransmitters and/or receptors, however, the filter becomes nonlinear. A very important observation is that nonlinear mechanisms incorporate a dependency on previous spikes, whereas a linear filter expresses only their superposition. In other words, the output of a nonlinear filter can reflect interactions in time between spikes. This observation can be made explicit, for example, if one substitutes f in Eq. (8.7) by its Taylor series expansion and develops the expression, which reveals that the postsynaptic potential depends on interactions between the smoothed spikes (see Appendix A for details).

As in van Rossum (2001), functional Euclidean distances can also be defined between the post-synaptic potentials evoked according to any of these models. In spike trains from zebra finch recordings, Houghton (2009) showed an improvement in classification performance when more complex models were used. The improvement was particularly significant when accounting for the depletion of receptors, which suggests the importance of multiple interacting spike trains.

8.2.2 Intensity Estimation

Although spike trains are sets of discrete events in time, the underlying point process has an intrinsic functional representation: the intensity function, or more generally the *conditional* intensity function. The conditional intensity function is a powerful functional descriptor because, rather than only explaining the spike train per se, it aims to describe the statistics of the underlying process, which is more closely tied to the information the spike train encodes.

The conditional intensity function is defined for $t \in \mathcal{T} = (0, T]$ as

$$\lambda_s(t|H_t) = \lim_{\epsilon \to 0} \frac{1}{\epsilon} \Pr[N(t + \epsilon) - N(t) = 1|H_t] \qquad (8.8)$$

where $N(t)$ is the counting process (i.e., the number of spikes in the interval $(0, t]$), and H_t is the history, up to time t, of the point process realization and any other random variables or processes the point process depends on. The conditional intensity function fully characterizes a point process because it describes the probability of an event (i.e., a spike) given its past.

Estimation of a general conditional intensity function is very difficult, if not impossible, because the definition in Eq. (8.8) can accommodate an infinite number of dependencies. Even if this is number is small, one has to estimate how each parameter affects the probability of an event for any time in the future, and how these influences may change because of correlations in the parameters.

Nevertheless, one can attempt to estimate the conditional intensity function by constraining its functional form. An example is the statistical estimation framework recently proposed by Truccolo et al. (2005). Briefly, the spike train is first represented as a realization of a Bernoulli random process (i.e., a discrete-time random process with 0/1 samples), and then a generalized linear model (GLM) is used to fit a conditional intensity function to the spike train. The GLM considers that the logarithm of the conditional intensity function has the form (Truccolo et al., 2005),

$$\log \lambda_{s_i}(\hat{t}_n | H_n^i) = \sum_{m=1}^{q} \theta_m g_m(v_m(\hat{t}_n)) \tag{8.9}$$

where $\hat{t}_n$ is the nth discrete-time instant, $\{g_m\}$ are general linear transformations of independent functions $\{v_m(\cdot)\}$, $\{\theta_m\}$ are the parameters of the GLM and q is the number of parameters. The terms $\{g_m(v_m(\hat{t}_n))\}$ are called the predictor variables in the GLM framework and, if one considers the conditional intensity to depend linearly on the spiking history, then the $\{g_m\}$ can be simply delays. In general the intensity can depend nonlinearly on the history or external factor such as stimuli. Other functional forms of the log conditional intensity function can also be utilized. Examples are the inhomogeneous Markov interval (IMI) process model proposed by Kass and Ventura (2001), and the point process filter proposed by Brown et al. (2001).

These approaches, however, have a main drawback: For practical use, the functional form of the conditional intensity function must be heavily constrained to only a few parameters, or long spike trains must be available for estimation of the parameters. The first approach has been utilized by Brown et al. (2001). Conversely, in the framework proposed by Truccolo et al. (2005), a large number of parameters need to be estimated since q must be larger that the average inter-spike interval.

The estimation of the conditional intensity function is greatly simplified if a (inhomogeneous) Poisson process model is assumed. This is because there is no history dependence. Hence, in this case the conditional intensity function is only a function of time $\lambda_s(t)$ and is called the intensity function. Several methods exist in the point process literature for intensity function estimation. The obvious approach is to divide the duration of the spike train $\mathcal{S}(\mathcal{T}) = [0, T]$ in small intervals and average the number of spikes in each interval over realizations. In essence, binning aims at exactly this, without the average over realizations. A common alternative is kernel smoothing (Richmond et al., 1990; Reiss, 1993; Dayan and Abbott, 2001), which is much more data efficient, as demonstrated by Kass et al. (2003). Given a spike train s comprising spike times $\{t_m \in \mathcal{T} : m = 1, \ldots, N\}$, the estimated intensity function is

$$\hat{\lambda}_{s_i}(t) = \sum_{m=1}^{N_i} h(t - t_m^i) \tag{8.10}$$

where h is the smoothing function. This function must be non-negative and integrate to one over the real line (just like a probability density function (pdf)).

Commonly used smoothing functions are the Gaussian, Laplacian and α-function, among others. Note that binning is nothing but a special case of this procedure in which h is a rectangular window and the spike times are first quantized according to the width of the rectangular window (Dayan and Abbott, 2001).

Comparing Eqs. (8.10) and (8.3) the resemblance between them is clear. However, based on the context in which they were introduced, those equations provide two seemingly very different perspectives. Equation (8.3) aims to describe the evoked membrane potential due to a spike. On the other hand, Eq. (8.10) suggests that this mechanism is equivalent to intensity estimation. Conversely, it is possible to think statistically of the former synapse model as an approximation of the expectation of the dendrite response to a spike train. Perhaps more important, the connection between these equations proves that the synapse model described by Eq. (8.3) is limited to Poisson spike trains. In other words, this model is memoryless. Thus, potential information contained in interactions between spikes in a spike train will be neglected. In contrast, the model characterized by Eq. (8.7) has memory due to the nonlinearity. An mentioned in the previous section and shown in Appendix A, the nonlinear function introduces higher-order correlations in the smoothed spike train obtained from the linear model.

In summary, although spike train's content could be decoded using a statistical characterization, there is a severe trade-off between complexity of the model and inference, and its estimation ability. For these reasons, this is not the goal of this chapter. Instead, we build a general framework for machine learning using inner products. A fundamental reason to define inner products as spike train operators is that we do not need an *explicit* statistical characterization. Instead, we need only to ensure that the inner product detects the *implicit* statistical similarities between two spike trains.

8.3 INNER PRODUCTS FOR SPIKE TRAINS

As mentioned in the introduction, because spike trains are realizations of point processes, it is not straightforward to devise machine learning algorithms to operate with them. In this section we show that inner products are fundamental operators for signal processing, and that their definition for spike trains provides the necessary structure for machine learning.

Using a few signal processing and machine learning applications as examples, let us demonstrate why inner products are so important (Figure 8.2). In filtering, the output is the convolution of the input with an impulse response; for principal component analysis (PCA), one needs to be able to project the

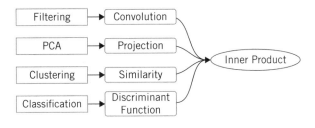

FIGURE 8.2

Inner products are fundamental operators for signal processing and machine learning.

data; for clustering, the concept of similarity (or dissimilarity) between points is needed; and for classification, it is necessary to define discriminant functions that separate the classes. In all of these applications, the operators needed are either implemented directly by an inner product or can be constructed with one. Convolution implements an inner product at each time instant between a mirrored and shifted version of the input function and the system's impulse response. Projection is by definition an inner product between two objects. An inner product is also a similarity measure, and dissimilarity measures (such as distances) can be defined given an inner product. Discriminant functions cannot be obtained directly with an inner product, but they can be nonlinearly combined to approximate it to the desired precision, as happens for example in neural networks where the linear projections in the processing elements are implemented by inner products.

Many more examples can be shown to rely entirely on inner products, or can be formulated and solved using only inner products. In fact, using inner products to solve general machine learning problems is the enabling idea behind kernel methods (Schölkopf et al., 1999). There are many studies in the literature that show how this idea can be utilized in statistical signal processing (Parzen, 1959; Kailath, 1971; Kailath and Duttweiler, 1972; Kailath and Weinert, 1975), smoothing and spline methods (Wahba, 1990), classification (Vapnik, 1995; Schölkopf et al., 1999), and so forth. Similarly, what we propose is to establish such a general framework for spike train signal processing and pattern recognition by defining inner product operators.

On the other hand, there has been much work recently on spike train distance measures (van Rossum, 2001; Victor, 2002; Houghton and Sen, 2008). Hence, a natural question is, Are spike train metrics sufficient for general machine learning with spike trains? The answer is no. The reason for this is that most machine learning methods require the mathematical structure of a vector space, which is not provided by a metric but implicitly attained with the definition of an inner product. For example, it is not possible to cluster spike trains using a k-means-like algorithm with only a distance, because there is no way to compute the center points (the "means" in k-means). However, computing the mean is trivial in a inner product (vector) space. Moreover, the definition of spike train inner products naturally induces at least two spike train metrics, as shown later. Clearly, it is possible to define a distance without having an inner product associated with it (see, for example, Victor and Purpura (1996); Christen et al. (2006); Kreuz et al. (2007)). Yet, in addition to the limitations just discussed, these approaches tend to obscure the assumed point process model.

8.3.1 Defining Inner Products for Spike Trains

In this section we discuss how to define inner products for spike trains and the advantages and limitations of each approach. There are two main approaches. One is to start by defining an inner product for spike times and then utilize the linearity in the inner product space, and the fact that spike trains are sets of spike times, to extend the inner product to spike trains (Carnell and Richardson, 2005; Schrauwen and Campenhout, 2007; Paiva et al., 2009). An alternative methodology is to define inner products on functional representations of spike trains (Paiva et al., 2009). Both of these methodologies are discussed next.

8.3.1.1 *Inner Products for Spike Times*
In the first approach there are two key design elements: the definition or choice of the inner product between spike times and how these are combined to obtain the final spike train inner product. Often, the inner product for spike times is simply set by choosing a symmetric and positive definite kernel to operate on them (Carnell and Richardson, 2005; Schrauwen and Campenhout, 2007; Paiva et al., 2009). Using RKHS theory (see Appendix C for a brief introduction), it immediately follows that a kernel obeying these conditions will induce a reproducing kernel Hilbert space, for which the kernel represents the evaluation of the inner product (Aronszajn, 1950). Conceptually, these kernels operate in much the same way as the kernels operating on data samples in machine learning (Schölkopf et al., 1999). Then, depending on the application, the spike time kernels can be combined differently to capture information that matches one's understanding or a priori knowledge of the problem. For example, defining the inner product between two spikes as

$$\langle \delta(t - t_m), \delta(t - t_n) \rangle = \exp\left(-\frac{|t_m - t_n|}{\tau} \right) = \kappa(t_m, t_n) \qquad (8.11)$$

and using the fact that a spike train can be written as a sum of spikes, Eq. (8.2), one obtains the spike train inner product (Carnell and Richardson, 2005),

$$
\begin{aligned}
K(s_i, s_j) &= \left\langle s_i, s_j \right\rangle \\
&= \left\langle \sum_{m=1}^{N_i} \delta(t - t_m^i), \sum_{n=1}^{N_j} \delta(t - t_n^j) \right\rangle \\
&= \sum_{m=1}^{N_i} \sum_{n=1}^{N_j} \left\langle \delta(t - t_m^i), \delta(t - t_n^j) \right\rangle \\
&= \sum_{m=1}^{N_i} \sum_{n=1}^{N_j} \kappa(t_m^i, t_n^j)
\end{aligned}
\tag{8.12}
$$

Another example is the kernel

$$
K^*(s_i, s_j) = \begin{cases}
\max_{l=0,1,\dots,(N_i-N_j)} \sum_{n=1}^{N_j} \kappa(t_{n+l}^i - t_n^j), & N_i \geq N_j \\
\max_{l=0,1,\dots,(N_j-N_i)} \sum_{n=1}^{N_i} \kappa(t_n^i - t_{n+l}^j), & N_i < N_j
\end{cases}
\tag{8.13}
$$

which quantifies the presence of related spike patterns between spike trains s_i, s_j (Chi and Margoliash, 2001; Chi et al., 2007). Since this kernel measures if spike trains have a one-to-one correspondence of a given sequence of spike times, it can be utilized, for example, to study temporal precision and reliability in neural spike trains in response to stimulus, or to detect/classify these stimuli.

The prime advantage of this approach is that it allows for spike train inner products to be easily designed for a specific application by utilizing available a priori knowledge about the problem, as the second example illustrates. On the one hand, this is also a disadvantage. If the inner product is designed without this information, as in the first example, it is not easy to determine what the inner product is quantifying. More fundamentally, it is unclear what the implicitly assumed point process model is or if the inner product has any biological interpretation. From this point of view, the methodology presented next is clearly more advantageous.

8.3.1.2 *Functional Representations of Spike Trains*
Rather than defining the spike train inner products directly, in the following we define inner products on functional representations of spike trains. This

bottom-up construction from fundamental data descriptors is unlike the previously mentioned approach and is rarely taken in machine learning, but it explicitly shows the implied synaptic model and/or the assumed point process model. Hence, this explicit construction is very informative about the inner product and derived constructs because it clearly exposes the limitations of a given definition. This knowledge can be of paramount importance in both theoretical and neurophysiological studies because the outcome of these studies depends on the implied model and the expressive power of the measures utilized to quantify it.

Consider two spike trains, $s_i, s_j \in \mathcal{S}(\mathcal{T})$, and denote their corresponding evoked membrane potentials by $v_{s_i}(t)$ and $v_{s_j}(t)$, with $t \in \mathcal{T}$. For spike trains with finite duration (i.e., $t \in \mathcal{T}$, with $\mathcal{T}$ any finite interval), the membrane potentials are bounded, and therefore

$$\int_{\mathcal{T}} v_{s_i}^2(t)dt < \infty \tag{8.14}$$

That is, membrane potentials are square integral functions in $\mathcal{T}$. Hence, $v_{s_i}(t), v_{s_j}(t) \in L_2(\mathcal{T})$. Consequently, an inner product for spike trains can be taken as the usual inner product of membrane potentials in $L_2(\mathcal{T})$ space:

$$V(s_i, s_j) = \left\langle v_{s_i}(t), v_{s_j}(t) \right\rangle_{L_2(\mathcal{T})} = \int_{\mathcal{T}} v_{s_i}(t) v_{s_j}(t) dt \tag{8.15}$$

Clearly, the inner product can quantify only spike train characteristics expressed in the membrane potential. That is, the limitations of the inner product are set by the chosen synapse model.

Similarly, inner products for spike trains can be easily defined from their statistical descriptors. Considering spike trains s_i and s_j and the conditional intensity functions of the underlying point processes, denoted by $\lambda_{s_i}(t|H_t^i)$ and $\lambda_{s_j}(t|H_t^j)$, the inner product can be defined as

$$I(s_i, s_j) = \left\langle \lambda_{s_i}(t|H_t^i), \lambda_{s_j}(t|H_t^j) \right\rangle_{L_2(\mathcal{T})} = \int_{\mathcal{T}} \lambda_{s_i}(t|H_t^i) \lambda_{s_j}(t|H_t^j) dt \tag{8.16}$$

where, as before, the usual inner product in $L_2(\mathcal{T})$ was utilized. Note that, just like membrane potentials, conditional intensity functions are square integrable in a finite interval. This is the *cross-intensity* (CI) kernel defined in

Paiva et al. (2009). The advantage in defining the inner product from the conditional intensity functions is that the resulting *inner product incorporates the statistics of the point processes directly*. Moreover, in the general form given, this inner product can be utilized for *any* point process model since the conditional intensity function is a complete characterization of the point process (Cox and Isham, 1980; Daley and Vere-Jones, 1988).

Both of these definitions are very general but, in practice, some particular form must be considered. Some of these particular cases are presented and discussed next.

For the simplest synaptic model presented, with membrane potential given by Eq. (8.3), it is possible to evaluate the inner product without explicit computation of the integral. Substituting Eq. (8.3) in the inner product definition Eq. (8.15), and using the linearity of the integral, yields

$$V(s_i, s_j) = \sum_{m=1}^{N_i} \sum_{n=1}^{N_j} \int_T h(t - t_m^i) h(t - t_n^j) dt = \sum_{m=1}^{N_i} \sum_{n=1}^{N_j} k(t_m^i, t_n^j) dt \qquad (8.17)$$

where k is the auto-correlation of h. As remarked at the end of Section 8.2.2, it is possible to reason about the membrane potential in Eq. (8.3) as an estimation of the intensity function through kernel smoothing Eq. (8.10). Indeed, if we consider the cross-intensity inner product Eq. (8.16) constrained to inhomogeneous Poisson processes, and substitute the kernel smoothing intensity estimator in Eq. (8.10), we obtain the result in Eq. (8.17). This shows that the inner product in Eq. (8.17) is limited to Poisson point processes, and will fail to distinguish between spike trains with the same intensity function even if they were generated by point processes with different statistics. For this reason, this inner product was called the *memoryless cross-intensity* (mCI) kernel in Paiva et al. (2009). It is noteworthy that in this case the inner product also matches the definition obtained directly from the spike times, shown in Eq. (8.12). However, the meaning of the inner product, what it quantifies, and its limitations are now easily understood in light of the latter derivation.

The limitations of the previous inner product are easily verified, both from its implied synapse model and its assumed point process. A more expressive inner product can be obtained if the synapse model with saturation is utilized instead. The inner product takes the same general form of Eq. (8.15) but with the membrane potential now given by equation (8.7), and can be written as,

$$V^\dagger(s_i, s_j) = \int_T v_{s_i}^\dagger(t) v_{s_j}^\dagger(t) dt, = \int_T f\left(v_{s_i}(t)\right) f\left(v_{s_j}(t)\right) dt \qquad (8.18)$$

where $v_{s_i}(t), v_{s_j}(t)$ denotes the potentials one obtains from the linear synapse model (Eq. (8.3)). Equivalently, the dual interpretation of the potentials $v_{s_i}(t), v_{s_j}(t)$ as intensity functions of inhomogeneous Poisson processes can be utilized to write the above inner product as

$$V^{\dagger}(s_i, s_j) = \int_{\mathcal{T}} f\left(\lambda_{s_i}(t)\right) f\left(\lambda_{s_j}(t)\right) dt \tag{8.19}$$

This inner product is conceptually similar to the *nonlinear cross-intensity* (nCI) kernel defined by Paiva et al. (2009),

$$I_{\sigma}^{\dagger}(s_i, s_j) = \int_{\mathcal{T}} \mathcal{K}_{\sigma}\left(\lambda_{s_i}(t), \lambda_{s_j}(t)\right) dt \tag{8.20}$$

where $\mathcal{K}_{\sigma}$ is a symmetric positive definite kernel with kernel size parameter σ. The difference is that in Eq. (8.19) the nonlinear coupling of the intensity functions is shown explicitly, whereas in the nCI kernel it is merely implied by $\mathcal{K}$. Hence, an advantage of the derivation given here is that it explicitly shows the implied synapse model, which supports the plausibility and significance of the inner product known from spike train signal processing. Moreover, from RKHS theory it is known that $\mathcal{K}$ denotes an inner product of the transformed linear membrane potentials, but the transformation is usually unknown. Taking the definition in Eq. (8.19), we can define $\mathcal{K}(x, y) = f(x)f(y)$, which connects the two definitions and gives explicitly the nonlinear transformation into the inner product space.

As remarked in Sections 8.2.1 and 8.2.2, an important characteristic of the synapse model with binding saturation (Eq. (8.7)) is the dependence on higher-order interactions between multiple spikes in time. That is, the nonlinearity introduces memory. Statistically, this means that, if kernel smoothing is used as the intensity function estimator in Eq. (8.19), this inner product is sensitive to point processes with memory. This statement has been verified empirically for the nCI (Paiva et al., 2009). In theory, depending on the nonlinearity utilized, the memory depth of the inner product can be infinity. However, in practice the decaying dynamics of the conductance after a spike and the saturation in the nonlinearity forces the influence of past spikes to die out. This also means that, because of the decreasing weighting of the higher-order spike interactions, the inner product will lose specificity as the memory requirements to distinguish spike trains from different point processes increase.

As mentioned earlier, in its general form, $I(s_i, s_j)$ (Eq. (8.16)) is an optimal inner product because the conditional intensity functions completely characterize the underlying point process. On the other hand, as discussed in Section 8.2.2, estimation of general conditional intensity functions is not feasible in practice, and even their estimation under constrained scenarios is troublesome. Although both $V^\dagger$ and $I^\dagger$ also have constrained memory dependencies, their estimation is very simple and offers a shortcut around the current difficulties in the estimation of conditional intensity functions, providing a trade-off between memory depth and specificity.

8.3.2 Properties of the Defined Inner Products

In this section some properties of the inner products defined in the previous section are presented. The main purpose of the properties shown is to prove the defined inner products are well defined and have an associated reproducing kernel Hilbert space (RKHS), which provides the necessary mathematical structure for machine learning.

Property 8.3.1 *In Eqs. (8.15) and (8.16)*

1. *The inner products are symmetric.*
2. *The inner products are linear operators in the space of the evoked membrane potentials or intensity functions, respectively.*
3. *The inner product of a spike train with itself is non-negative and zero if and only if the spike train is empty (i.e., has no spikes).*

This property asserts that the proposed definitions verify the axioms of an inner product. Actually, because the defined inner products operate on elements of $L_2(\mathcal{T})$ and correspond to the usual dot product in that space, this property is a direct consequence of the properties inherited from L_2.

Property 8.3.2 *The inner products in Eqs. (8.15) and (8.16) correspond to symmetric positive definite kernels in spike train space.*

The proof is given in Appendix B. Through the work of Moore (1916) and from the Moore-Aronszajn theorem (Aronszajn, 1950), the following two properties result as corollaries.

Property 8.3.3 *For any set of n spike trains (n ≥ 1), the inner product matrix, with the ijth element given by inner product of spike trains s_i and s_j is symmetric and non-negative definite.*

Property 8.3.4 *There exists a Hilbert space for which the defined spike train's inner products is a reproducing kernel.*

Properties 8.3.2 through 8.3.4 are equivalent in the sense that any of them imply the other two. To verify this, note that property 8.3.3, by definition, is merely a restatement of property 8.3.2,[2] and a kernel being symmetric and positive definite is a necessary and sufficient condition to property 8.3.4 (Aronszajn, 1950).

An important consequence of these properties, explicitly stated in property 8.3.4, is the existence of a unique RKHS associated with each of the inner products defined earlier. This is of importance because it ensures the existence of a complete metric space,[3] and it abstracts the design of machine learning methods from the exact inner product definition.

Property 8.3.5 *The inner products in Eqs. (8.15) and (8.16) verify the Cauchy-Schwarz inequality:*

$$V^2(s_i, s_j) \leq V(s_i, s_i)V(s_j, s_j) \, and$$
$$I^2(s_i, s_j) \leq I(s_i, s_i)I(s_j, s_j) \tag{8.21}$$

for any two spike trains $s_i, s_j \in \mathcal{S}(\mathcal{T})$.

As before, the proof is given in Appendix B. The Cauchy-Schwarz inequality will be shown to lead to a spike train metric in the next section.

Property 8.3.6 *The inner products defined on functions in $L_2(\mathcal{T})$, using either membrane potentials or conditional intensity functions, are congruent[4] to the corresponding inner product in the RKHS.*

This property is a direct consequence of the approach we used to define the inner product. In other words, in our case we defined the inner product between spike trains as the inner product of their functional representations in $L_2(\mathcal{T})$. Hence, by definition, that inner product establishes the existence of the RKHS and its inner product, making them equal. See Paiva et al. (2009, section 5.1) for further details.

---

[2] Note that we are considering a "positive definite" kernel in the sense of Aronszajn (1950) and Parzen (1959). That is, a kernel (i.e., a function of two arguments), $\kappa : X \times X \to \mathbb{R}$, is positive definite if and only if

$$\sum_{i=1}^{n}\sum_{j=1}^{n} a_i a_j \kappa(x_i, x_j) \geq 0$$

for any $n > 0$, $x_1, \ldots, x_n \in X$ and $a_1, \ldots, a_n \in \mathbb{R}$.

[3] A complete metric space is one in which every Cauchy sequence is convergent. The metric space is naturally induced with the inner product definition, basically ensuring that any sequence of points and their limit are contained in the space.

[4] See Appendix C for the definition of congruent inner products.

8.3.3 Distances Induced by Inner Products

Metrics, and distances in particular, can be very useful in classification and analysis of data. Spike trains are no exception. The importance of spike train metrics can be observed from the attention they have received in the literature (Victor and Purpura, 1997; van Rossum, 2001; Schreiber et al., 2003). However, when used without further mathematical structure, their usefulness is limited. In this section we demonstrate that by first defining inner products we naturally induce two distances in the inner product space. More than presenting spike train distances, we aim to further support the argument that defining inner products is a principled way to build, from the bottom up, the required structure for general machine learning.

Given an inner product space, two distances, at least, can be immediately defined. The first utilizes the fact that an inner product naturally induces a norm,[5] which in turn provides a distance between differences of elements. The second distance derives from the Cauchy-Schwarz inequality.

Consider the inner product between spike trains defined in Eq. (8.15). (The same idea applies directly for *any* spike train inner product.) The natural norm induced by this inner product is given by

$$\left\| v_{s_i}(t) \right\| = \sqrt{\left\langle v_{s_i}(t), v_{s_i}(t) \right\rangle_{L_2(T)}} = \sqrt{V(s_i, s_j)} \tag{8.22}$$

Hence, the *norm distance* between spike trains $s_i, s_j \in \mathcal{S}(T)$ is

$$d_{ND}(s_i, s_j) = \left\| v_{s_i} - v_{s_j} \right\| = \sqrt{\left\langle v_{s_i} - v_{s_j}, v_{s_i} - v_{s_j} \right\rangle} \tag{8.23a}$$

$$= \sqrt{\left\langle v_{s_i}, v_{s_i} \right\rangle - 2\left\langle v_{s_i}, v_{s_j} \right\rangle + \left\langle v_{s_j}, v_{s_j} \right\rangle} \tag{8.23b}$$

$$= \sqrt{V(s_i, s_i) - 2V(s_i, s_j) + V(s_j, s_j)} \tag{8.23c}$$

where the linearity of the inner product is utilized from Eqs. (8.23a) to (8.23b). Note that, even if we do not have an explicit functional decomposition of the inner product, as occurs for the inner products (or kernels) defined in Section 8.3.1.1, the same idea applies as long as symmetry and positive definiteness holds. In this case, the inner products in Eqs. (8.23a) through (8.23c) are simply the inner products in the induced RKHS.

---

[5]One can also reason in terms of the norm in the RKHS, because an RKHS is a Hilbert space and therefore, has a norm. In fact, the two norms are equivalent because of the congruence between the spaces (Paiva et al., 2009).

An alternative distance can be obtained based on the Cauchy-Schwarz inequality (property 8.3.5). We can define the *Cauchy-Schwarz (CS) distance* between two spike trains as

$$d_{CS}(s_i, s_j) = \arccos \frac{V(s_i, s_j)^2}{V(s_i, s_i)V(s_j, s_j)} \tag{8.24}$$

From the properties of norm and the Cauchy-Schwarz inequality, it is easy to verify that either distance satisfies the three distance axioms: symmetry, positiveness, and the triangle inequality. Indeed, the norm distance is basically the generalization of the idea behind the Euclidean distance to a space of functions. On the other hand, the CS distance measures the "angular distance" between spike trains, as suggested by the usual dot product between two vectors $\vec{x}, \vec{y}$ written as $\vec{x} \cdot \vec{y} = \|\vec{x}\| \|\vec{y}\| \cos(\theta)$, with θ the angle between the two vectors. Hence, a major difference between the norm distance and the CS distance is that the latter is not an Euclidean measure but a Riemannian metric, which measures the geodesic distance between points in a curved space.

As observed by Houghton (2009), it is possible to define a van Rossum-like distance (van Rossum, 2001) for each synaptic model. Clearly, taking the inner product in the definition of the norm distance to be the inner product between membrane potentials given by each synapse model, Eq. (8.15), one obtains the distances suggested by Houghton. In other words, van Rossum's distance as originaly defined (van Rossum, 2001) is a particular case of the norm distance. More specifically, van Rossum's distance considers the inner product in Eq. (8.15), with the potential given by Eq. (8.3), using a causal decaying exponential function as the smoothing function h. Also, it was noted before that other smoothing functions could be utilized (Schrauwen and Campenhout, 2007). However, the approach taken here requires only an inner product and thus is far more general. Moreover, the dual perspective in terms of conditional intensity functions clearly identifies that, regardless of the choice of smoothing function h, van Rossum's distance is sensitive at most to Poisson process statistics. On the other hand, using nonlinear synapse models one introduces memory into the inner product, explaining Houghton's results both neurophysiologically and statistically.

A spike train metric closely related to the CS distance has also been suggested in the literature. Indeed, the Cauchy-Schwarz distance can be compared with the "correlation measure" between spike trains proposed by Schreiber et al. (2003). We can observe that the latter corresponds to the argument of the arc cosine and thus denotes the cosine of an angle between spike trains. In the case of the metrics proposed by Schreiber et al., the norm and inner product were given by Eq. (8.15), with the potentials given by Eq. (8.3) using the Gaussian smoothing function. However, the metric suffers from the same limitation as van Rossum's distance because it implicitly assumes Poisson processes. With our methodology the metric can be easily extended. Moreover, notice that Schreiber and colleagues' "correlation measure" is only a premetric since it does not verify the triangle inequality. In d_{CS} this is ensured by the arc cosine function.

8.4 APPLICATIONS

To exemplify the importance of the developments shown here, in the following we derive two fundamental machine learning algorithms for spike trains: principal component analysis (PCA) and Fisher linear discriminant (FLD). These two algorithms serve to demonstrate that the framework presented here allows the development of both supervised and unsupervised learning methods. The primary aim is to demonstrate the ease with which machine learning algorithms for spike trains can be developed once the mathematical structure created by an inner product and the induced RKHS is established.

8.4.1 Unsupervised Learning: Principal Component Analysis

Principal component analysis (PCA) is widely used for data analysis and dimensionality reduction (Diamantaras and Kung, 1996). It is derived for spike trains directly in the RKHS induced by any of the inner products defined. This approach highlights that optimization with spike trains is possible by the definition of an inner product and, more specifically, through the mathematical structure provided by the induced RKHS. This is also the traditional approach in functional analysis literature (Ramsay and Silverman, 1997) and has the advantage of being completely general, regardless of the spike train inner product.

8.4.1.1 *Derivation*

In the following, let P denote a spike train inner product inducing an RKHS $\mathcal{H}$, such as any of the inner products defined earlier. Consider a set of spike trains, $\{s_i \in \mathcal{S}(\mathcal{T}), i = 1,\ldots,N\}$, for which we wish to determine the principal components. Computing the principal components of the spike trains directly is not feasible because we do not know how to define a principal component (PC); however, this is a trivial task in the RKHS induced by a spike train inner product.

Let $\{\Lambda_{s_i} \in \mathcal{H}, i = 1, \ldots, N\}$ denote the set of elements in the RKHS $\mathcal{H}$ (associated with P) corresponding to the given spike trains—that is, the transformation or mapping of the spike trains into $\mathcal{H}$. The mean of the transformed spike trains is

$$\bar{\Lambda} = \frac{1}{N} \sum_{i=1}^{N} \Lambda_{s_i} \tag{8.25}$$

Thus, the centered transformed spike trains (i.e., with the mean removed) can be obtained:

$$\tilde{\Lambda}_{s_i} = \Lambda_{s_i} - \bar{\Lambda} \tag{8.26}$$

PCA finds an orthonormal transformation providing a compact description of the data. Determining the principal components of spike trains in the RKHS can be formulated as the problem of finding the set of orthonormal vectors in the RKHS such that the projection of the centered transformed spike trains $\{\tilde{\Lambda}_{s_i}\}$ has *maximum variance*. This means that the principal components can be obtained by solving an optimization problem in the RKHS. A function $\xi \in \mathcal{H}$ (i.e., $\xi : \mathcal{S}(\mathcal{T}) \longrightarrow \mathbb{R}$) is a principal component if it maximizes the cost function

$$J(\xi) = \sum_{i=1}^{N} \left[\mathrm{Proj}_{\xi} (\tilde{\Lambda}_{s_i}) \right]^2 - \rho \left(\|\xi\|^2 - 1 \right) \tag{8.27}$$

where $\mathrm{Proj}_{\xi} (\tilde{\Lambda}_{s_i})$ denotes the projection of the ith centered transformed spike train onto ξ, and ρ is the Lagrange multiplier for the constraint $\left(\|\xi\|^2 - 1 \right)$ that the principal components have unit norm. To evaluate this cost function one needs to be able to compute the projection and the norm of the principal components. In an RKHS, an inner product is the projection operator and the norm is naturally defined (see Eq. (8.22)). Thus, the above cost function can be expressed as

$$J(\xi) = \sum_{i=1}^{N} \left\langle \tilde{\Lambda}_{s_i}, \xi \right\rangle^2 - \rho (\langle \xi, \xi \rangle - 1) \tag{8.28}$$

In practice, we always have a finite number of spike trains. Hence, by the representer theorem (Kimeldorf and Wahba, 1971; Schölkopf et al., 2001), ξ is restricted to the subspace spanned by the centered transformed spike trains

$\{\tilde{\Lambda}_{s_i}\}$. Consequently, there exist coefficients $b_1, \ldots, b_N \in \mathbb{R}$ such that

$$\xi = \sum_{j=1}^{N} b_j \tilde{\Lambda}_{s_j} = \mathbf{b}^T \tilde{\boldsymbol{\Lambda}} \tag{8.29}$$

where $\mathbf{b}^T = [b_1, \ldots, b_N]$ and $\tilde{\boldsymbol{\Lambda}} = \left[\tilde{\Lambda}_{s_1}, \ldots, \tilde{\Lambda}_{s_N} \right]^T$. Substituting in Eq. (8.28) yields

$$\begin{aligned} J(\xi) &= \sum_{i=1}^{N} \left(\sum_{j=1}^{N} b_j \left\langle \tilde{\Lambda}_{s_i}, \tilde{\Lambda}_{s_j} \right\rangle \right) \left(\sum_{k=1}^{N} b_k \left\langle \tilde{\Lambda}_{s_i}, \tilde{\Lambda}_{s_k} \right\rangle \right) \\ &\quad + \rho \left(1 - \sum_{j=1}^{N} \sum_{k=1}^{N} b_j b_k \left\langle \tilde{\Lambda}_{s_i}, \tilde{\Lambda}_{s_k} \right\rangle \right) \\ &= \mathbf{b}^T \tilde{\mathbf{P}}^2 \mathbf{b} + \rho \left(1 - \mathbf{b}^T \tilde{\mathbf{P}} \mathbf{b} \right) \end{aligned} \tag{8.30}$$

where $\tilde{\mathbf{P}}$ is the inner product matrix of the centered spike trains—that is, the $N \times N$ matrix with elements

$$\begin{aligned} \tilde{\mathbf{P}}_{ij} &= \left\langle \tilde{\Lambda}_{s_i}, \tilde{\Lambda}_{s_j} \right\rangle = \left\langle \Lambda_{s_i} - \bar{\Lambda}, \Lambda_{s_j} - \bar{\Lambda} \right\rangle \\ &= \left\langle \Lambda_{s_i}, \Lambda_{s_j} \right\rangle - \frac{1}{N} \sum_{l=1}^{N} \left\langle \Lambda_{s_i}, \Lambda_{s_l} \right\rangle - \frac{1}{N} \sum_{l=1}^{N} \left\langle \Lambda_{s_l}, \Lambda_{s_j} \right\rangle + \frac{1}{N^2} \sum_{l=1}^{N} \sum_{n=1}^{N} \left\langle \Lambda_{s_l}, \Lambda_{s_n} \right\rangle \\ &= P(s_i, s_j) - \frac{1}{N} \sum_{l=1}^{N} P(s_i, s_l) - \frac{1}{N} \sum_{l=1}^{N} P(s_l, s_j) + \frac{1}{N^2} \sum_{l=1}^{N} \sum_{n=1}^{N} P(s_l, s_n) \end{aligned} \tag{8.31}$$

In matrix notation,

$$\tilde{\mathbf{P}} = \mathbf{P} - \frac{1}{N}(\mathbf{1}_N \mathbf{1}_N^T \mathbf{P} + \mathbf{P} \mathbf{1}_N \mathbf{1}_N^T) + \frac{1}{N^2} \mathbf{1}_N \mathbf{1}_N^T \mathbf{P} \mathbf{1}_N \mathbf{1}_N^T \tag{8.32}$$

where $\mathbf{P}$ is the spike train inner product matrix, $\mathbf{P}_{ij} = \left\langle \Lambda_{s_i}, \Lambda_{s_j} \right\rangle = P(s_i, s_j)$, and $\mathbf{1}_N$ is the $N \times 1$ vector with all ones (such that $\mathbf{1}_N \mathbf{1}_N^T$ is an $N \times N$ matrix of ones). This means that $\tilde{\mathbf{P}}$ can be computed directly from $\mathbf{P}$ without the need to explicitly remove the mean of the transformed spike trains.

From Eq. (8.30), finding the principal components makes the problem simply one of estimating the coefficients $\{b_i\}$ that maximize $J(\xi)$. Since $J(\xi)$ is a quadratic function, its extrema can be found by equating the gradient to zero. Taking the derivative with regard to $\mathbf{b}$ (which characterizes ξ) and setting it to zero results in

$$\frac{\partial J(\xi)}{\partial \mathbf{b}} = 2\tilde{\mathbf{P}}^2 \mathbf{b} - 2\rho \tilde{\mathbf{P}} \mathbf{b} = 0 \tag{8.33}$$

and thus corresponds to the usual eigendecomposition problem[6]:

$$\tilde{\mathbf{P}} \mathbf{b} = \rho \mathbf{b} \tag{8.34}$$

This means that any eigenvector of the centered inner product matrix is a solution to Eq. (8.33). Thus, the eigenvectors determine the coefficients of Eq. (8.29) and characterize the principal components. It is easy to verify that, as expected, the variance of the projections onto each principal component equals the corresponding eigenvalue squared. So the ordering of ρ specifies the relevance of the principal components.

To compute the projection of a given input spike train s onto the kth principal component (corresponding to the eigenvector with the kth largest eigenvalue), we need only compute the inner product of Λ_s with ξ_k in the RKHS. That is,

$$\begin{aligned}
\text{Proj}_{\xi_k}(\Lambda_s) &= \langle \Lambda_s, \xi_k \rangle_{\mathcal{H}} \\
&= \sum_{i=1}^{N} b_{ki} \langle \Lambda_s, \tilde{\Lambda}_{s_i} \rangle \\
&= \sum_{i=1}^{N} b_{ki} \left(P(s, s_i) - \frac{1}{N} \sum_{j=1}^{N} P(s, s_j) \right)
\end{aligned} \tag{8.35}$$

In summary, the algorithm proceeds as follows:

1. Compute $\tilde{\mathbf{P}}$, the centered inner product matrix, according to Eq. (8.32).
2. Compute the k leading eigenvectors of $\tilde{\mathbf{P}}$ (i.e., the eigenvectors corresponding to the largest eigenvalues).

---

[6]Note that the simplification in the eigendecomposition problem is valid whether or not if the inner product matrix is invertible, since $\tilde{\mathbf{P}}^2$ and $\tilde{\mathbf{P}}$ have the same eigenvectors and the eigenvalues of $\tilde{\mathbf{P}}^2$ are the eigenvalues of $\tilde{\mathbf{P}}$ squared.

3. Compute the projections onto the k principal components, according to Eq. (8.35).

One can easily see that the above result is very similar to kernel PCA (Schölkopf et al., 1998). Indeed, kernel PCA is a well-known example of functional PCA applied in an RKHS of discrete data (Ramsay and Silverman, 1997), which is in essence what we do here.

8.4.1.2 *Results*

To demonstrate the algorithm and empirically verify the differences between the various inner products, we show some results with synthetic spike trains generated from two homogeneous renewal point processes. The goal of the simulation is, first, to show that spike train PCA can be utilized to statistically analyze a set of spike trains in an unsupervised manner, and, second, to verify which inner products yield a projection consistent with our expectation. That is, an inner product should be able to differentiate the different statistical characteristics in the spike trains because of the different renewal processes, from intrinsic spike train variability, and therefore yield two separate clusters.[7]

The spike trains used in the simulation were one second long with mean spike rate of 20 spikes/s. The inter-spike interval was gamma-distributed, with shape parameters $\theta = 0.5$ and $\theta = 3$, for the two renewal processes. The algorithm is applied on two data sets, one for computation of the principal components (a "training set"), and the other for testing. The evaluation and testing data sets comprised a total of 50 and 200 spike trains, respectively, half from each renewal process. The simulated spike trains are shown in Figure 8.3.

The PCA algorithm was applied using three different spike train inner products, defined in Eqs. (8.17), (8.18) and (8.20). In the first case, k was chosen to be the Laplacian function, $k(x,y) = \exp(-|x-y|/\tau)/2\tau$ with $\tau = 50\,\text{ms}$, which corresponds to a causal exponential smoothing function h with the same width (Paiva et al., 2009). In the second case, the same smoothing function, also with $\tau = 50\,\text{ms}$, and the tanh nonlinearity function were used. For the nCI kernel (Equation (8.20)), the fast implementation described in Paiva (2008) was used, with smoothing width $\tau = 50\,\text{ms}$ and Gaussian kernel $\mathcal{K}_\sigma$ with $\sigma = 1$. These parameters, and the choice of smoothing function and k, allow the algorithms to be directly compared, with differences only due to the actual form of the inner product. From an implementation perspective, the only difference between different inner products is in the computation of

---

[7] Although in general PCA does not preserve separability, this can be expected in this case because the clusters due to each renewal process should be compact. That is, the inner product should yield higher values between spike trains from the same renewal process (low in-cluster variability) and lower values between spike trains from different renewal processes (high between-cluster variability).

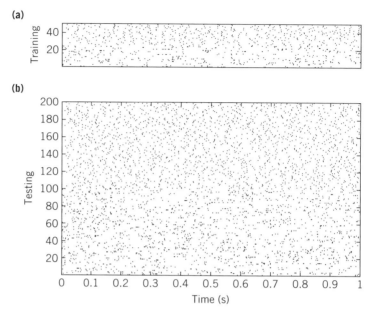

FIGURE 8.3

Spike trains used for computing the principal components (a) and for testing (b). In either case, the bottom half corresponds to spike trains drawn from a renewal process with shape parameter $\theta = 0.5$; the top half, from a renewal process with $\theta = 3$. Notice the difference in variability of the inter-spike intervals, with spike trains in the top half ($\theta = 3$) being much more regular because of the more compact distribution of the inter-spike intervals.

the centered inner product matrix (Eq. (8.31). The remainder of the algorithm remains unaltered.

For the inner product defined in Eq. (8.17), the result of the eigendecomposition and projection of the spike trains onto the first two principal components is shown in Figure 8.4. The first observation is that there is no clear distinction in spike train variability in the first two principal components, but rather the variability decreases gradually. Moreover, both the eigenvectors and the projections shown in Figure 8.4(b)–(d) reveal an overlap of the spike trains from the two renewal processes, which is noticeable only in the higher dispersion of spike trains from the first renewal model ($\theta = 0.5$) due to the more irregular firing. This means that the inter-process variability is smaller than the within-process variability. In other words, the inner product does not distinguish between spike trains with different inter-spike interval statistics, but senses only the difference in variability. This result is not surprising. As pointed out in Section 8.3.1.2, this inner product is memoryless and can discriminate only differences in intensity, as in this case.

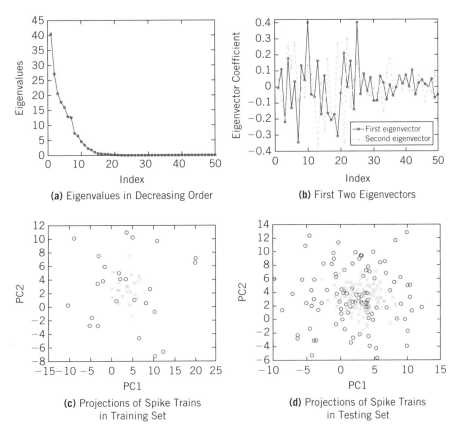

(a) Eigenvalues in Decreasing Order

(b) First Two Eigenvectors

(c) Projections of Spike Trains
in Training Set

(d) Projections of Spike Trains
in Testing Set

FIGURE 8.4

Eigendecomposition (top) and projected spike trains on the first two principal components (bottom) for PCA computed using the inner product in Eq. (8.17). (Different marks in the projection plots denote spike trains from different renewal processes.)

Similarly, we computed the principal components of the same set of spike trains using the inner product defined in Eq. (8.18). However, in this case the choice of the g_{max} scaling parameter was very important. This parameter controls the transition point between the linear and saturation regions of the nonlinearity. More fundamentally, g_{max} controls the weighting of the higher-order spike interaction in time (see Eq. (8.46)) and, therefore, the ability to sense different inter-spike interval statistics. The results are shown in Figure 8.5. It is easy to verify that lower values of g_{max} yield more meaningful (i.e., better separated) projections of the spike trains from the two renewal processes. This is noticeable, for example, in the top left plot, which show a better separation between the first principal components. However,

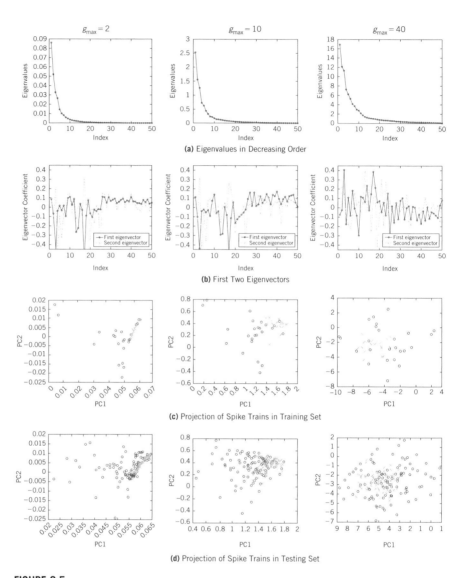

FIGURE 8.5

Eigendecomposition (top) and projected spike trains on the first two principal components (bottom) for PCA computed using the inner product in Eq. (8.18). Each column corresponds to the results for a given value of g_{max} shown at the top. (Different marks in the projection plots denote spike trains from different renewal processes.)

if g_{max} is decreased further the results start to worsen (not shown). This is understandable since the nonlinearity will almost always be in the saturated region and thus the functional representations will be almost always equal. Conversely, as g_{max} is increased, the results converge toward those shown in

Figure 8.4. This occurs because as g_{max} is increased the nonlinearity operates in its linear region, having almost no effect.

Finally, we computed the principal components using the inner product in Eq. (8.20)—that is, the nCI kernel proposed in Paiva et al. (2009). The results are shown in Figure 8.6. Although this inner product has a kernel size parameter (of $\mathcal{K}_\sigma$), the results were very similar with any value within the range 0.1 to 10. In the distribution of the eigenvalues, the significantly higher variability captured in the first principal component versus the remaining is obvious. In comparison with the previous inner products, only the results in Figure 8.5 for $g_{max} = 2$ approximate this behavior. The importance of the first principal component can be verified from the first eigenvector

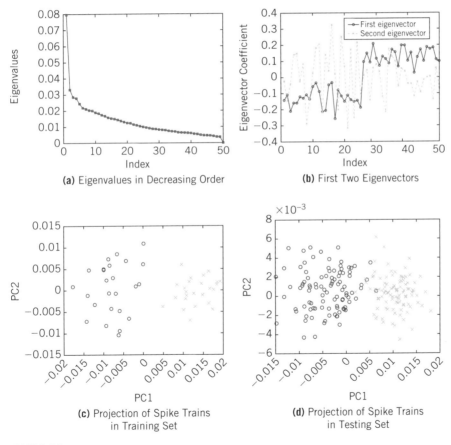

(a) Eigenvalues in Decreasing Order

(b) First Two Eigenvectors

(c) Projection of Spike Trains
in Training Set

(d) Projection of Spike Trains
in Testing Set

FIGURE 8.6

Eigendecomposition (top) and projected spike trains on the first two principal components (bottom) for PCA, computed using the inner product in Eq. (8.20). (Different marks in the projection plots denote spike trains from different renewal processes.)

of the eigendecomposition (Figure 8.6(b)), which shows a clear distinction between spike trains generated from different renewal processes. This observation is noticeable in the projections of spike trains from the two processes (Figure 8.6(c)–(d)). It is obvious that in this case the within-process variability is smaller than between-process variability. This seems a more faithful representation of the data since the within-process variability is only "noise" (due to stochasticity in the realizations). The separation between projections of the spike trains from different renewal processes was also noticeable in the results in Figure 8.5 (for $g_{max} = 2$), although not as strikingly. Moreover, differences in variability between processes were clearly represented, because in this latter case the nonlinear kernel $\mathcal{K}$ measures differences rather than the products in the smoothed spike train, which factors out the variability between realizations as long as they are similar.

8.4.2 Supervised Learning: Fisher's Linear Discriminant

Classification of spike trains is very important. It can be used to decode the applied stimulus from a spike train. In neurophysiological studies, this has been utilized to study spike coding properties (Victor and Purpura, 1996; Machens et al., 2001, 2003). Indeed, it was the need to classify spike trains that first motivated several spike train distances (Victor and Purpura, 1996).

In the following, we derive Fisher's linear discriminant for spike trains. Note that in spite of the linear decision boundary, the discrimination occurs in the RKHS. Hence, depending on the inner product, the decision boundary can easily be nonlinear in the spike train (or its functional representation) space. From a different perspective, one can think that, given an appropriate inner product, the classes can be made linearly separable, as occurs in kernel methods. The approach shown turns out to be very similar to that of kernel Fisher discriminant proposed by Mika et al. (1999).

8.4.2.1 *Derivation*

As before, we will use P to denote a generic spike train inner product inducing an RKHS $\mathcal{H}$ with mapping Λ to it. Consider two sets of spike trains, $C_1 = \{s_i^1 \in \mathcal{S}(\mathcal{T}), i = 1, \dots, N_1\}$ and $C_2 = \{s_j^2 \in \mathcal{S}(\mathcal{T}), j = 1, \dots, N_2\}$, one for each class. With some abuse of notation, the set of all spike trains will be denoted $C = C_1 \cup C_2 = \{s_n \in \mathcal{S}(\mathcal{T}), n = 1, \dots, N\}$, with $N = N_1 + N_2$. Without loss of generality, only the case for two classes is considered here. For the general case, one can repeatedly apply the procedure discussed next between each class versus all others, or extend the derivation to find a subspace-preserving discrimination.

The Fisher linear discriminant in the RKHS associated with the inner product is given by the element $\psi \in \mathcal{H}$, such that the means μ_k and variances σ_k^2

of the data classes projected onto ψ maximize the criterion:

$$J_F(\psi) = \frac{(\mu_1 - \mu_2)^2}{\sigma_1^2 + \sigma_2^2} \tag{8.36}$$

with the mean and variance of the projected classes given by

$$\mu_k = \langle \psi, \bar{\Lambda}_k \rangle = \frac{1}{N_k} \sum_{i=1}^{N_k} \langle \psi, \Lambda_{s_i^k} \rangle \tag{8.37a}$$

$$\sigma_k^2 = \frac{1}{N_k} \sum_{i=1}^{N_k} \left(\langle \psi, \Lambda_{s_i^k} \rangle - \mu_k \right)^2 = \frac{1}{N_k} \sum_{i=1}^{N_k} \langle \psi, \Lambda_{s_i^k} \rangle^2 - \mu_k^2 \tag{8.37b}$$

where $\bar{\Lambda}_k = \frac{1}{N_k} \sum_{i=1}^{N_k} \Lambda_{s_i^k}$ is the mean, in the RKHS, of the kth class.

By the representer theorem (Kimeldorf and Wahba, 1971; Schölkopf et al., 2001), ψ can be written as a linear combination of the spike trains mapped onto the RKHS $\{\Lambda_{s_i} \in \mathcal{H} : i = 1, \dots, N, s_i \in C\}$. That is, there exist coefficients $c_1, \dots, c_N \in \mathbb{R}$ such that

$$\psi = \sum_{j=1}^{N} c_j \Lambda_{s_j} = \mathbf{c}^T \Lambda \tag{8.38}$$

where $\mathbf{c}^T = [b_1, \dots, b_N]$ and $\Lambda = \left[\Lambda_{s_1}, \dots, \Lambda_{s_N} \right]^T$. Substituting in Eq. (8.37a) yields

$$\mu_k = \frac{1}{N_k} \sum_{i=1}^{N_k} \sum_{j=1}^{N} c_j \langle \Lambda_{s_j}, \Lambda_{s_i^k} \rangle = \frac{1}{N_k} \mathbf{c}^T \mathbf{P}_k \mathbf{1}_{N_k} = \mathbf{c}^T \mathbf{M}_k \tag{8.39}$$

with $\mathbf{M}_k = \frac{1}{N_k} \mathbf{P}_k \mathbf{1}_{N_k}$ and with $\mathbf{1}_{N_k}$ as the $N_k \times 1$ vector of all ones. Likewise, substituting in Eq. (8.37b) results in

$$\sigma_k^2 = \frac{1}{N_k} \sum_{i=1}^{N_k} \sum_{j=1}^{N} \sum_{n=1}^{N} c_j c_n \langle \Lambda_{s_j}, \Lambda_{s_i^k} \rangle \langle \Lambda_{s_i^k}, \Lambda_{s_n} \rangle - \mu_k^2$$

$$= \frac{1}{N_k} \mathbf{c}^T \mathbf{P}_k \mathbf{P}_k^T \mathbf{c} - \frac{1}{N_k^2} \mathbf{c}^T \mathbf{P}_k \mathbf{1}_{N_k} \mathbf{1}_{N_k}^T \mathbf{P}_k^T \mathbf{c} \tag{8.40}$$

$$= \frac{1}{N_k} \mathbf{c}^T \mathbf{P}_k \left(\mathbf{I}_{N_k} - \frac{1}{N_k} \mathbf{1}_{N_k} \mathbf{1}_{N_k}^T \right) \mathbf{P}_k^T \mathbf{c}$$

where $\mathbf{P}_k$ is the $N \times N_k$ matrix with the ij^{th} element given by the inner product between the ith spike train of C and the jth spike train of C_k, and $\mathbf{I}_{N_k}$ is the $N_k \times N_k$ identity matrix. Consequently, the Fisher criterion in Eq. (8.36) can be written as

$$J_F(\mathbf{c}) = \frac{\mathbf{c}^T \mathbf{S}_b \mathbf{c}}{\mathbf{c}^T \mathbf{S}_w \mathbf{c}} \tag{8.41}$$

with between-class scatter matrix, $\mathbf{S}_b$, and within-class scatter matrix, $\mathbf{S}_w$, given by

$$\mathbf{S}_b = (\mathbf{M}_1 - \mathbf{M}_2)(\mathbf{M}_1 - \mathbf{M}_2)^T \tag{8.42a}$$

$$\mathbf{S}_w = \sum_{k=1,2} \mathbf{P}_k \left(\mathbf{I}_{N_k} - \frac{1}{N_k} \mathbf{1}_{N_k} \mathbf{1}_{N_k}^T \right) \mathbf{P}_k^T \tag{8.42b}$$

As in the usual Fisher linear discriminant result (Duda et al., 2000), the vector $\mathbf{c}$ that maximizes J_F can be obtained as

$$\mathbf{c} = \mathbf{S}_w^{-1}(\mathbf{M}_1 - \mathbf{M}_2) \tag{8.43}$$

Clearly, this solution is ill posed in general since $\mathbf{S}_w$ is not guaranteed to be invertible. Hence, in practice $\mathbf{S}_w + \epsilon \mathbf{I}_N$ is used instead. Basically, the term $\epsilon \mathbf{I}_N$ implements Tikhonov regularization, with multiplier ϵ.

To compute the projection of a given spike train onto the Fisher linear discriminant, in the training set (i.e., the set C) or in testing, we need to compute the inner product, in the RKHS, between the transformed spike train and ψ, given as

$$\begin{aligned} \text{Proj}_\psi(\Lambda_s) &= \langle \Lambda_s, \psi \rangle \\ &= \sum_{j=1}^{N} c_j \langle \Lambda_s, \Lambda_{s_j} \rangle \\ &= \sum_{j=1}^{N} c_j P(s, s_j) \end{aligned} \tag{8.44}$$

After projection onto the Fisher linear discriminant, one still needs to find the optimum threshold for classification. However, this is a relatively simple task and many methods exist in the literature (Duda et al., 2000).

To summarize, the algorithm can be implemented in the following steps:

1. Compute the $\mathbf{P}_k$ matrices, $k = 1, 2$.
2. Compute the matrices $\mathbf{M}_k$ and $\mathbf{S}_w$ according to Eqs. (8.39) and (8.42b).
3. Compute the vector $\mathbf{c}$ which defines the Fisher linear discriminant in the RKHS, according to Eq. (8.43).
4. Find the optimum classification threshold in the training set.
5. Classify the testing spike trains by computing their projection onto the Fisher linear discriminant (Eq. (8.44)) and thresholding.

8.4.2.2 *Results*

In this section, we demonstrate the algorithm for synthetic spike trains generated from two homogeneous renewal point processes. Specifically, we utilized the same data sets used to evaluate the PCA algorithm in Section 8.4.1.2, shown in Figure 8.3. In this example, classification is primarily a task of discriminating between spike trains with different inter-spike interval distributions.

As before, the Fisher linear discriminant algorithm was compared using three different spike train inner products, defined in Eqs. (8.17), (8.18) and (8.20). The same parameters were used. As with the PCA algorithm, the Fisher linear discriminant algorithm is independent of the inner product used; the difference is only in the computation of the inner product matrices, $\mathbf{P}_k$. The classification threshold was the value that minimized the probability of error in the training set.

For the inner product of Eq. (8.18) one must carefully choose the appropriate scaling parameter g_{max} of the f nonlinearity. For simplicity, in this case we tested the Fisher discriminant classifier using a range of values between 0.5 and 50, and reported the best result on the testing set, obtained with $g_{max} = 2$. Of course, in a practical scenario the optimum g_{max} value would have to be found by cross-validation. The probability of error in the testing set is shown in Figure 8.7. The figure can be intuitively understood in light of the role of nonlinearity in the inner product. For large values of g_{max}, the nonlinearity operates in its linear range, and therefore the result is similar to using the inner product in Eq. (8.17). As g_{max} is decreased, the classification improves because the nonlinearity allows the inner product to utilize more information from the higher-order moments (see Eq. (8.46)). However, for very small values the performance decreases because the nonlinearity obfuscates the dynamics of the linear component of the synapse model.

The classification results using each of the inner products are summarized in Table 8.1, with mean $\pm$ standard deviation statistics estimated over 100 Monte Carlo runs. To help understand the results, the distribution of

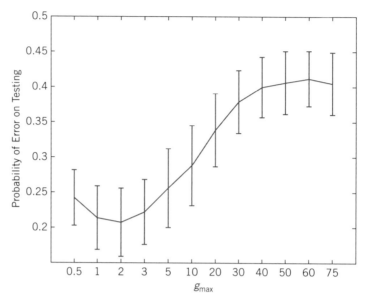

FIGURE 8.7

Probability of error in the testing set of the Fisher discriminant classifier for different values of g_{max}. The statistics, mean and standard deviation, were estimated over 100 Monte Carlo runs.

Table 8.1 Probability of Error in the Testing Set of the Fisher linear Discriminant Classifier Using Three Different Inner Products.

Inner Product	Probability of Error
Equation (8.17)	0.401 ± 0.040
Equation (8.18)	0.207 ± 0.048
Equation (8.20)	0.025 ± 0.013

the projections onto the Fisher discriminant for one of the Monte Carlo runs is shown in Figure 8.8. Overall, these results clearly show the importance of using inner products that can capture memory information. For these data sets, the inner product defined in Eq. (8.17) performs clearly worse (notice in Figure 8.8(a) the overlap in the projections onto the Fisher discriminant). Because the discriminating characteristic is the different inter-spike interval distributions, but since this inner product is sensitive only to differences in intensity, the results are only slightly above chance level. Using the inner product in Eq. (8.18) greatly improves the probability of error because of the use of memory information from the higher-order interactions, but only if an

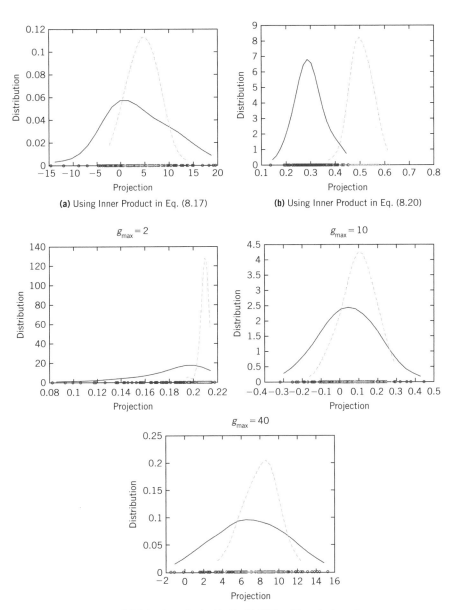

(a) Using Inner Product in Eq. (8.17)

(b) Using Inner Product in Eq. (8.20)

$g_{max} = 2$

$g_{max} = 10$

$g_{max} = 40$

(c) Using Inner Product in Eq. (8.18) for different values of g_{max}

FIGURE 8.8

Distribution of the projections onto the Fisher discriminant of spike trains in the testing set. For the inner product in Eq. (8.18), distributions are shown for three values of g_{max}.

appropriate value of g_{max} is chosen, as shown in Figure 8.7 and noticed in Figure 8.8(c). Finally, using the inner product in Eq. (8.20) yields the best results. As with the PCA these results are considerably better than the previous ones because the latter inner product nonlinearly measures the similarity between functional representations. This avoids the ambiguity in the inner product in equation (8.18),[8] adding to the expressiveness of the inner product (Figure 8.8(b)). Also, in the latter inner product we have to choose the bandwidth parameter σ of the nonlinear kernel $\mathcal{K}_\sigma$. However, unlike the former, the sensitivity on this parameter is very low. We tried values between 0.1 and 10, but the results changed only $\pm 0.1\%$.

8.5 DISCUSSION

A general framework to develop spike train machine learning methods was presented. The most important contribution of this work is in defining spike train inner products for building the framework. Unlike the top-down approach often taken in kernel methods, the approach emphasized here builds the inner products bottom up from basic spike train descriptors. As argued in Section 8.3, inner products are the fundamental operators needed for machine learning. Then, utilizing the mathematical structure of the reproducing kernel Hilbert space induced by a spike train inner product, it is easy to derive machine learning algorithms from first principles, as demonstrated in Section 8.4.

Two approaches to defining spike train inner products were presented. The first, introduced in Section 8.3.1.1, builds on spike times as basic elements. Although it allows for a priori knowledge of the spike domain to be easily incorporated in the inner product definition, these definitions per se lack a more in-depth understanding of both biological and/or statistical implications. Hence, the approach emphasized in this chapter is based on functional representations of spike trains, which were motivated either as a biological modeling problem or as statistical descriptors. One of the key advantages of this approach is the explicit construction of the inner product. In this way, the properties of the inner product, induced RKHS, and derived constructs are brought forth. The biological modeling perspective highlights the role of synaptic models, and thus aims to approximate the computational mechanisms in a neuron. On the other hand, by defining the inner products from conditional intensity functions, one focuses on the characteristics of the underlying

---

[8]Specifically, we are referring to the ambiguity arising from the fact that the functional representations of pairs of spike trains with the same geometric mean in time and/or the same arithmetic mean over time have the same inner product value.

point processes. Perhaps most important, verifying the duality between the two representations can provide valuable insight into the statistical limitations of a synaptic model. Therefore, these inner products (or their derived constructs) can be used to analyze neurophysiological data, and they hint at the underlying principles of information encoding from these perspectives.

The inner product with the form given in Eq. (8.16) is statistically, completely general but, in practice, its usefulness is limited by the current methods for estimation of the conditional intensity function. Although methods for point processes with memory have been recently proposed in the literature, currently effective intensity function estimators exist only for Poisson process.[9] In this regard, however, the inner products defined in Eqs. (8.18) and (8.20) show considerable promise. Through the use of a nonlinearity, these inner products circumvent these limitations and show sensitivity to spike time statistics, albeit with limited expressiveness. In our experiments using renewal point processes, both inner products clearly outperform the inner product in Eq. (8.17). It must be remarked that the inner product in Eq. (8.17) is in essence a continuous version of the widely used cross-correlation of binned spike trains (Paiva et al., 2008; Park et al., 2008), which shows the fundamental limitations of many spike train methods currently in use.

Appendix A HIGHER-ORDER SPIKE INTERACTIONS THROUGH NONLINEARITY

In this section we prove the statement that the nonlinearity in Eq. (8.7) reflects higher-order interactions in the spike train.

As an example, consider the nonlinear function

$$f(x) = g_{max} \left[1 - \exp\left(-\frac{x^2}{2g_{max}^2} \right) \right] \tag{8.45}$$

That is, f is an inverted Gaussian, appropriately scaled by g_{max}. The MacLaurin series expansion of f (i.e., Taylor series around the origin) is

$$f(x) = \frac{1}{2} \frac{x^2}{g_{max}} - \frac{1}{8} \frac{x^4}{g_{max}^3} + \frac{1}{48} \frac{x^6}{g_{max}^5} - \frac{1}{384} \frac{x^8}{g_{max}^7} + \cdots \tag{8.46}$$

---

[9]By "effective" it is meant estimation with a reasonably small amount of data, such as 1-second spike trains.

In the nonlinear model in Eq. (8.7), the argument of the nonlinearity is the evoked potential according to the linear model, given by Eq. (8.3). Hence, the powers on x in the MacLaurin series extend to

$$x^2 = \left(\sum_{i=1}^{N} h(t-t_i) \right)^2 = \sum_{i=1}^{N}\sum_{j=1}^{N} h(t-t_i)h(t-t_j)$$

$$x^4 = \sum_{i=1}^{N}\sum_{j=1}^{N}\sum_{k=1}^{N}\sum_{l=1}^{N} h(t-t_i)h(t-t_j)h(t-t_k)h(t-t_l)$$

$$\cdots$$

(8.47)

This shows that the membrane potential of the nonlinear model, Eq. (8.7), is a weighted sum of the moments of the linear potential. In turn, these moments depend on the interactions, smoothed by h, of a number of spikes of the same order as the moment.

The moments accounted for and their weighting in the nonlinear membrane potential depend on the choice of f. In the case just shown, it depends only on even-order moments. However, if the tanh nonlinearity is utilized in f instead,

$$f(x) = g_{max} \tanh\left(\frac{x}{g_{max}} \right), \qquad x \in \mathbb{R}^+$$

(8.48)

the nonlinear membrane potential depends only on odd-order moments.

Appendix B PROOFS

This section presents the proofs for properties 8.3.2, 8.3.3, and 8.3.5, given in Section 8.3.2.

Proof of Property 8.3.2. *The symmetry follows immediately from property 8.3.1. In the following we will use the general form of the inner products in Eqs. (8.15) and (8.16), denoted P:*

$$P(s_i, s_j) = \langle \phi_{s_i}, \phi_{s_j} \rangle = \int_{\mathcal{T}} \phi_{s_i}(t)\phi_{s_j}(t)dt$$

(8.49)

where ϕ denotes either the membrane potential or the conditional intentity function.

By definition, a kernel on spike trains, $P : \mathcal{S}(\mathcal{T}) \times \mathcal{S}(\mathcal{T}) \to \mathbb{R}$, is positive definite if and only if (Aronszajn, 1950; Parzen, 1959)

$$\sum_{i=1}^{n}\sum_{j=1}^{n} a_i a_j P(s_i, s_j) \geq 0 \qquad (8.50)$$

for any $n > 0$, $s_1, s_2, \ldots, s_n \in \mathcal{S}(\mathcal{T})$ and $a_1, a_2, \ldots, a_n \in \mathbb{R}$. Using the general form yields

$$
\begin{aligned}
\sum_{i=1}^{n}\sum_{j=1}^{n} a_i a_j P(s_i, s_j) &= \sum_{i=1}^{n}\sum_{j=1}^{n} a_i a_j \int_{\mathcal{T}} \phi_{s_i}(t)\phi_{s_j}(t)dt \\
&= \int_{\mathcal{T}} \left(\sum_{i=1}^{n} a_i \phi_{s_i}(t) \right) \left(\sum_{j=1}^{n} a_j \phi_{s_j}(t) \right) dt \\
&= \left\langle \sum_{i=1}^{n} a_i \phi_{s_i}, \sum_{j=1}^{n} a_j \phi_{s_j} \right\rangle \\
&= \left\| \sum_{i=1}^{n} a_i \phi_{s_i} \right\| \geq 0
\end{aligned}
\qquad (8.51)
$$

since the norm is non-negative by definition.

Proof of Property 8.3.3. *Again, symmetry follows immediately from property 8.3.1. A matrix* $\mathbf{P}$ *is non-negative definite if and only if* $\mathbf{a}^T \mathbf{P} \mathbf{a} \geq 0$, *for any* $\mathbf{a}^T = [a_1, \ldots, a_n]$ *with* $a_i \in \mathbb{R}$. *For an inner product matrix, we have that (using the general inner product form)*

$$\mathbf{a}^T \mathbf{P} \mathbf{a} = \sum_{i=1}^{n}\sum_{j=1}^{n} a_i a_j P(s_i, s_j) \geq 0 \qquad (8.52)$$

as shown in the previous proof (Eq. (8.51)). Hence, the inner product matrices are symmetric and non-negative definite.

Proof of Property 8.3.5. *Consider the* 2×2 *inner product matrix associated with any of the defined inner products,*

$$\mathbf{P} = \begin{bmatrix} P(s_i, s_i) & P(s_i, s_j) \\ P(s_i, s_j) & P(s_j, s_j) \end{bmatrix} \qquad (8.53)$$

From property 8.3.3, this matrix is symmetric and non-negative definite. Hence, its determinant is non-negative (Harville, 1997, p. 245). That is,

$$\det(\mathbf{P}) = P(s_i, s_i)P(s_j, s_j) - P(s_i, s_j)^2 \geq 0 \tag{8.54}$$

which proves the result in Eq. (8.21).

Appendix C BRIEF INTRODUCTION TO RKHS THEORY

In this appendix, we briefly introduce some basic concepts of reproducing kernel Hilbert space (RKHS) theory necessary for understanding the concepts in this chapter. The presentation here is meant to be as general and introductory as possible.

The fundamental result in RKHS theory is the well-known *Moore-Aronszajn theorem* (Moore, 1916; Aronszajn, 1950). Let K denote a symmetric and positive definite function of two variables defined on some space E. That is, a function $K(\cdot, \cdot) : E \times E \to \mathbb{R}$, which verifies

1. Symmetry: $K(x, y) = K(y, x)$, $\forall x, y \in E$.
2. Positive definiteness: for any finite number of $n \in \mathbb{N}$ points $x_1, x_2, \ldots, x_n \in E$, and any corresponding n coefficients $c_1, c_2, \ldots, c_n \in \mathbb{R}$,

$$\sum_{i=1}^{l} \sum_{j=1}^{l} c_i c_j K(x_i, x_j) \geq 0 \tag{8.55}$$

These are sometimes called Mercer conditions (Mercer, 1909). The Moore-Aronszajn theorem (Moore, 1916; Aronszajn, 1950) guarantees that there exists a unique Hilbert space $\mathcal{H}$ of real-valued functions on E such that, for every $x \in E$,

1. $K(x, \cdot) \in \mathcal{H}$.
2. For any $f \in \mathcal{H}$,

$$f(x) = \langle f(\cdot), K(x, \cdot) \rangle_{\mathcal{H}} \tag{8.56}$$

The identity in Eq. 8.56 is called the *reproducing property* of K, and, for this reason, $\mathcal{H}$ is said to be a reproducing kernel Hilbert space with reproducing kernel K. Note Eq. (8.55) is equivalent to saying that, for a set of points

$\{x_i \in E : i = 1, \ldots, n\}$, the matrix

$$\mathbf{K} = \begin{bmatrix} K(x_1, x_1) & K(x_1, x_2) & \ldots & K(x_1, x_n) \\ K(x_2, x_1) & K(x_2, x_2) & \ldots & K(x_2, x_n) \\ \vdots & \vdots & \ddots & \vdots \\ K(x_n, x_1) & K(x_n, x_2) & \ldots & K(x_n, x_n) \end{bmatrix} \tag{8.57}$$

is positive *semi*-definite. $\mathbf{K}$ is often referred to as the *Gram matrix*.

A very important result of this theorem is that K evaluates the inner product in the RKHS. This can readily verified using the reproducing property,

$$K(x, y) = \langle K(x, \cdot), K(y, \cdot) \rangle_{\mathcal{H}} \tag{8.58}$$

for any $x, y \in E$. This identity is the *kernel trick*, well known in kernel methods, and the main tool for computation in the RKHS because most algorithms can be formulated using only inner products, as discussed in the introduction of Section 8.3.

Another important concept is that of congruent spaces. Two spaces, say E and F, are said to be congruent if there exists an isometric isomorphism between them. This means that there exists a one-to-one mapping (i.e., an isomorphism) $\mathscr{G} : E \to F$, such that, for any $x, y \in E$,

$$\langle x, y \rangle_E = \langle \mathscr{G}(x), \mathscr{G}(y) \rangle_F \tag{8.59}$$

That is, if there exists a one-to-one inner product–preserving mapping between the two spaces. The mapping $\mathscr{G}$ is called a congruence. By two congruent inner products it is meant that there exists a mapping between the spaces where the inner products are defined, such that Eq. (8.59) holds.

Acknowledgments

A. R. C. Paiva was supported by Fundação para a Ciência e a Tecnologia (FCT), Portugal, under grant SRFH/BD/18217/2004. This work was partially supported by NSF grants ECS-0422718 and CISE-0541241.

References

Aronszajn, N., 1950. Theory of reproducing kernels. Trans. Am. Math. Soc. 68 (3), 337–404.
Baccalá, L.A., Sameshima, K., 1999. Directed coherence: a tool for exploring functional interactions among brain structures. In: Nicolelis, M.L.A. (E.d.,) Methods for Neural Ensemble Recordings. CRC Press, 179–192.

Barbieri, R., Quirk, M.C., Frank, L.M., Wilson, M.A., Brown, E.N., 2001. Construction and analysis of non-Poisson stimulus-response models of neural spiking activity. J. Neurosci. Methods 105 (1), 25–37. doi: 10.1016/S0165-0270(00)00344-7.

Bohte, S.M., Kok, J.N., Poutré, H.L., 2002. Error-backpropagation in temporally encoded networks of spiking neurons. Neurocomputing 48 (1–4), 17–37. doi: 10.1016/S0925-2312(01)00658-0.

Brown, E.N., Nguyen, D.P., Frank, L.M., Wilson, M.A., Solo, V., 2001. An analysis of neural receptive field plasticity by point process adaptive filtering. Proc. Natl. Acad. Sci. 98, 12261–12266.

Carnell, A., Richardson, D., 2005. Linear algebra for time series of spikes. In: Proceedings of the European Symposium on Artificial Neural Networks. Apr. 2005, Bruges, Belgium, pp. 363–368.

Chi, Z., Margoliash, D., 2001. Temporal precision and temporal drift in brain and behavior of zebra finch song. Neuron 32 (1–20), 899–910.

Chi, Z., Wu, W., Haga, Z., Hatsopoulos, N.G., Margoliash, D., 2007. Template-based spike pattern identification with linear convolution and dynamic time warping. J. Neurophysiol. 97 (2), 1221–1235. doi: 10.1152/jn.00448.2006.

Christen, M., Kohn, A., Ott, T., Stoop, R., 2006. Measuring spike pattern reliability with the Lempel-Ziv-distance. J. Neurosci. Methods 156 (1-2), 342–350. doi: 10.1016/j.jneumeth.2006.02.023.

Cox, D.R., Isham, V., 1980. Point Processes. Chapman and Hall.

Daley, D.J., Vere-Jones, D., 1988. An Introduction to the Theory of Point Processes. Springer-Verlag, New York, NY.

Dayan, P., Abbott, L.F., 2001. Theoretical Neuroscience: Computational and Mathematical Modeling of Neural Systems. MIT Press, Cambridge, MA.

Diamantaras, K.I., Kung, S.Y., 1996. Principal Component Neural Networks: Theory and Applications. John Wiley & Sons.

Dowling, J.E., 1992. Neurons and Networks: An Introduction to Behavioral Neuroscience. Harvard University Press, Cambridge, MA.

Duda, R.O., Hart, P.E., Stork, D.G., 2000. Pattern Classification, second ed. Wiley Interscience.

Eden, U.R., Frank, L.M., Barbieri, R., Solo, V., Brown, E.N., 2004. Dynamic analysis of neural encoding by point process adaptive filtering. Neurocomputing 16 (5), 971–998.

Freeman, W.J., 1975. Mass Action in the Nervous System. Academic Press, New York. http://sulcus.berkeley.edu/MANSWWW/MANSWWW.html.

Gerstner, W., Kistler, W., 2002. Spiking Neuron Models. MIT Press.

Harville, D.A., 1997. Matrix Algebra from a Statistician's Perspective. Springer.

Houghton, C., 2009. Studying spike trains using a van Rossum metric with a synapse-like filter. J. Comp. Neurosci. 26, 149–155. doi: 10.1007/s10827-008-0106-6.

Houghton, C., Sen, K., 2008. A new multineuron spike train metric. Neural Comp. 20 (6), 1495–1511. doi: 10.1162/neco.2007.10-06-350.

Hurtado, J.M., Rubchinsky, L.L., Sigvardt, K.A., 2004. Statistical method for detection of phase-locking episodes in neural oscillations. J. Neurophysiol. 91 (4), 1883–1898.

Jabri, M.A., Coggins, R.J., Flower, B.G. (Eds.), 1996 Learning in Graphical Models. Springer. ISBN 0-412-61630-0, 978-0-412-61630-3.

Jensen, F.V., 2001. Bayesian Networks and Decision Graphs. Springer, New York, NY. ISBN 0-387-95259-4.

Jordan, M.I. (Ed.), 1998. Learning in Graphical Models. MIT Press. ISBN 0-262-60032-3, 978-0-262-60032-3.

Kailath, T., 1971. RKHS approach to detection and estimation problems–part I: Deterministic signals in gaussian noise. IEEE Trans. Inform. Theory 17 (5), 530–549.

Kailath, T., Duttweiler, D.L., 1972. An RKHS approach to detection and estimation problems–part III: Generalized innovations representations and a likelihood-ratio formula. IEEE Trans. Inform. Theory 18 (6), 730–745.

Kailath, T., Weinert, H.L., 1975. An RKHS approach to detection and estimation problems–part II: Gaussian signal detection. IEEE Trans. Inform. Theory 21 (1), 15–23.

Karr, A.R., 1986. Point Processes and their Statistical Inference. Marcel Dekker, New York, NY.

Kass, R.E., Ventura, V., 2001. A spike-train probability model. Neural Comp. 13 (8), 1713–1720.

Kass, R.E., Ventura, V., Cai, C., 2003. Statistical smoothing of neuronal data. Netw. Comp. Neural Syst. 14, 5–15.

Kass, R.E., Ventura, V., Brown, E.N., 2005. Statistical issues in the analysis of neuronal data. J. Neurophysiol. 94, 8–25. doi: 10.1152/jn.00648.2004.

Kimeldorf, G., Wahba, G., 1971. Some results on Tchebycheffian spline functions. J. Math. Anal. Appl. 33 (1), 82–95.

Kreuz, T., Haas, J.S., Morelli, A., Abarbanel, H.D.I., Politi, A., 2007. Measuring spike train synchrony. J. Neurosci. Methods 165 (1), 151–161. doi: 10.1016/j.jneumeth.2007.05.031.

Maass, W., Bishop, C.M. (Eds.), 1998. Pulsed Neural Networks. MIT Press.

Maass, W., Natschläger, T., Markram, H., 2002. Real-time computing without stable states: a new framework for neural computation based on perturbations. Neural Comp. 14 (11), 2531–2560. doi: 10.1162/089976602760407955.

Machens, C.K., Prinz, P., Stemmler, M.B., Ronacher, B., Herz, A.V.M., 2001. Discrimination of behaviorally relevant signals by auditory receptor neurons. Neurocomputing 38–40, 263–268.

Machens, C.K., Schütze, H., Franz, A., Kolesnikova, O., Stemmler, M.B., Ronacher, B., et al., 2003. Single auditory neurons rapidly discriminate conspecific communication signals. Nature Neurosci. 6 (4), 341–342.

Marmarelis, V.Z., 2004. Nonlinear Dynamic Modelling of Physiological Systems. IEEE Press Series in Biomedical Engineering. John Wiley & Sons.

Mercer, J., 1909. Functions of positive and negative type, and their connection with the theory of integral equations. Philos. Trans. R. Soc. Lond. A 209, 415–446.

Mika, S., Ratsch, G., Weston, J., Schölkopf, B., Müller, K.-R., 1999. Fisher discriminant analysis with kernels. In: Proceedings of International Workshop on Neural Networks for Signal Processing. August 1999, pp. 41–48. doi: 10.1109/NNSP.1999.788121.

Moore, E.H., 1916. On properly positive Hermitian matrices. Bull. Am. Math. Soc. 23, 59.

Narayanan, N.S., Kimchi, E.Y., Laubach, M., 2005. Redundancy and synergy of neuronal ensembles in motor cortex. J. Neurosci. 25 (17), 4207–4216. doi: 10.1523/JNEUROSCI.4697-04.2005.

Paiva, A.R.C., 2008. Reproducing Kernel Hilbert Spaces for Point Processes, with applications to neural activity analysis. PhD thesis, University of Florida.

Paiva, A.R.C., Park, I., Príncipe, J.C., 2008. Reproducing kernel Hilbert spaces for spike train analysis. In: Proceedings of the IEEE International Conference on Acoustics, Speech, and Signal Processing, ICASSP-2008. April 2008, Las Vegas, NV, USA.

Paiva, A.R.C., Park, I., Príncipe, J.C., 2009. A reproducing kernel Hilbert space framework for spike train signal processing. Neural Comp. 21 (2), 424–449. doi: 10.1162/neco.2008.09-07-614.

Park, I., Paiva, A.R.C., DeMarse, T.B., Príncipe, J.C., 2008. An efficient algorithm for continuous-time cross correlation of spike trains. J. Neurosci. Methods 128 (2), 514–523. doi: 10.1016/j.jneumeth.2007.10.005.

Parzen, E., 1959. Statistical inference on time series by Hilbert space methods. Technical Report 23, Applied Mathematics and Statistics Laboratory, Stanford University, Stanford, California.

Pesaran, B., Pezaris, J.S., Sahani, M., Mitra, P.P., Andersen, R.A., 2002. Temporal structure in neuronal activity during working memory in macaque parietal cortex. Nature Neurosci. 5 (8), 805–811. doi: 10.1038/nn890.

Ramsay, J.O., Silverman, B.W., 1997. Functional Data Analysis. Springer-Verlag.

Reiss, R.-D., 1993. A Course on Point Processes. Springer-Verlag, New York, NY.

Richmond, B.J., Optican, L.M., Spitzer, H., 1990. Temporal encoding of two-dimensional patterns by single units in primate primary visual cortex. I. Stimulus-response relations. J. Neurophysiol. 64 (2), 351–369.

Schneidman, E., Bialek, W., Berry, I.M.J., 2003. Synergy, redundancy, and independence in population codes. J. Neurosci. 23 (37), 11539–11553.

Schölkopf, B., Smola, A., Müller, K.-R., 1998. Nonlinear component analysis as a kernel eigenvalue problem. Neural Comp. 10 (5), 1299–1319.

Schölkopf, B., Burges, C.J.C., Smola, A.J. (Eds.), 1999. Advances in Kernel Methods: Support Vector Learning. MIT Press.

Schölkopf, B., Herbrich, R., Smola, A.J., 2001. A generalized representer theorem. In: Proceedings of the 14th Annual Conference on Computational Learning Theory and and 5th European Conference on Computational Learning Theory. July 2001.

Schrauwen, B., Campenhout, J.V., 2007. Linking non-binned spike train kernels to several existing spike train distances. Neurocomputing 70 (7–8), 1247–1253. doi: 10.1016/j.neucom.2006.11.017.

Schreiber, S., Fellous, J.M., Whitmer, D., Tiesinga, P., Sejnowski, T.J., 2003. A new correlation-based measure of spike timing reliability. Neurocomputing 52–54, 925–931. doi: 10.1016/S0925-2312(02)00838-X.

Snyder, D.L., 1975. Random Point Processes. John Viley & Sons, New York.

Song, D., Chan, R.H.M., Marmarelis, V.Z., Hampson, R.E., Deadwyler, S.A., Berger, T.W., 2007. Nonlinear dynamic modeling of spike train transformations for hippocampal-cortical prostheses. IEEE Trans. Biomed. Eng. 54 (6), 1053–1066. doi: 10.1109/TBME.2007.891948.

Truccolo, W., Eden, U.T., Fellows, M.R., Donoghue, J.P., Brown, E.N., 2005. A point process framework for relating neural spiking activity to spiking history, neural ensemble, and extrinsic covariate effects. J. Neurophysiol. 93, 1074–1089. doi: 10.1152/jn.00697.2004.

Tsodyks, M., Markram, H., 1997. The neural code between neocortical pyramidal neurons depends on neurotransmitter release probability. Proc. Natl. Acad. Sci. 94 (2), 719–723.

van Rossum, M.C.M., 2001. A novel spike distance. Neural Comp. 13 (4), 751–764.

Vapnik, V.N., 1995. The Nature of Statistical Learning Theory. Springer.

Victor, J.D., 2002. Binless strategies for estimation of information from neural data. Phys. Rev. E 66 (5), 051903, 2002. doi: 10.1103/PhysRevE.66.051903.

Victor, J.D., Purpura, K.P., 1996. Nature and precision of temporal coding in visual cortex: A metric-space analysis. J. Neurophysiol. 76 (2), 1310–1326.

Victor, J.D., Purpura, K.P., 1997. Metric-space analysis of spike trains: theory, algorithms, and application. Netw.: Comp. Neural Syst. 8, 127–164.

Wahba, G., 1990. Spline Models for Observational Data, vol. 59. In: CBMS-NSF Regional Conference Series in Applied Mathematics. SIAM.

Wang, Y., Paiva, A.R.C., Príncipe, J.C., Sanchez, J.C., 2009. Sequential Monte Carlo point process estimation of kinematics from neural spiking activity for brain machine interfaces. Neural Comp. 21 (10), 2894–2930. doi: 10.1162/neco.2009.01-08-699.

Signal Processing and Machine Learning for Single-trial Analysis of Simultaneously Acquired EEG and fMRI

Paul Sajda*[†], **Robin I. Goldman***, **Mads Dyrholm*** **, **Truman R. Brown***[†]

*Department of Biomedical Engineering, [†]Department of Radiology Columbia University, New York, New York, **Department of Psychology, University of Copenhagen, Denmark

9.1 INTRODUCTION

The simultaneous acquisition of electroencephalography (EEG) and functional magnetic resonance imaging (fMRI) is a potentially powerful multimodal imaging technique for measuring the functional activity of the human brain. Given that EEG measures the electrical activity of neural populations while fMRI measures hemodynamics via a blood oxygenation-level–dependent (BOLD) signal related to neuronal activity, simultaneous EEG/fMRI (hereafter referred to as EEG/fMRI) offers a modality to investigate the relationship between these two phenomena within the context of noninvasive neuroimaging. Though fMRI is widely used to study cognitive and perceptual function, there is still substantial debate regarding the relationship between local neuronal activity and hemodynamic changes (Logothetis and Wandell (2004); Sirotin and Das (2009)). That is, there is a need for a more comprehensive understanding of the specific mechanisms underlying neurovascular coupling.

Another rationale for EEG/fMRI is that, despite the fact that the individual modalities measure markedly different physiological phenomena, in terms of spatial and temporal resolution they are quite complementary. EEG offers millisecond temporal resolution; however, the spatial sampling density and ill-posed nature of the inverse model problem limit its spatial resolution. On the other hand, fMRI provides millimeter spatial resolution, but

Statistical Signal Processing for Neuroscience and Neurotechnology. DOI: 10.1016/B978-0-12-375027-3.00009-0

because of scanning rates and the low-pass nature of the BOLD response, the temporal resolution is limited. One approach that has been adopted to take advantage of this complementarity is to use fMRI activations to seed EEG source localization (Menon et al. (1997); Obitz et al. (1999); Wang and Zhou (1999)). Though in this case simultaneous acquisition may be convenient, for many experimental paradigms it may not add substantially to separate acquisitions, particularly if there are no significant adaptation or learning effects that would change the fMRI activation maps. Similarly, approaches that link trial-averaged ERP features (such as amplitude and latency) to BOLD activation (e.g., Philiastides and Sajda (2007)) could in principle benefit from EEG/fMRI, though once again true simultaneous acquisition may be more for convenience than for providing additional information, unless the trial-averaged ERPs are significantly affected via adaptation/learning.

Perhaps the most compelling reason for EEG/fMRI is that temporally specific variations in the EEG's trial-to-trial fluctuations can be captured and correlated with the fMRI BOLD signal. Many aspects of perception and cognition, including decision making, attentional shifting, and memory encoding, are essentially characterized by trial-to-trial fluctuations, and there is the potential that many of the associated neural correlates can be measured via scalp EEG. Several groups have investigated single-trial analysis of EEG/fMRI. Some have considered the frequency domain, measuring trial-to-trial power variations in subbands of single electrodes (Mantini et al., 2007) while others have considered single-trial variations in features of the time domain signal, such as amplitude and latency, extracted from individual-channel ERPs (Bénar et al., 2007; Muler et al., 2008).

Multivariate analysis, both unsupervised and supervised, has been used to identify meaningful linear projections of the EEG and subsequently correlate single-trial variations along these projections with the BOLD activity. For example, independent component analysis (ICA), an unsupervised approach, has been used to exploit statistical correlations between electrodes, particularly in high-density arrays, to identify "meaningful" variations in component amplitude—the source separation problem, which in neuroscience has been termed the "neural cocktail party problem" (Makeig et al., 1999). Several groups have applied ICA to EEG acquired simultaneously with fMRI (Debener et al., 2005; Eichele et al., 2008; Eichele et al., 2009). The ICA method typically requires visual inspection and can thus introduce significant bias when choosing components.

In the following sections we describe the hardware design, signal processing and machine learning for a system we developed for single-trial EEG/fMRI analysis. We address issues related to the hardware and software of data acquisition, MR-induced artifacts in the EEG, lead-induced artifacts in the MR images, MR gradient and RF artifacts in the EEG, and the ballistocardiogram (BCG) artifact. We then focus on a specific method for single-trial analysis that uses supervised learning to extract task-relevant discriminant components. It is the single-trial variability of the projection onto these task-specific components that we correlate with the BOLD signal. We present results for an auditory detection paradigm and describe the significance of the results relative to standard fMRI analyses. We then conclude by discussing challenges and future directions, particularly within the context of statistical analysis and machine learning.

9.2 HARDWARE DESIGN AND SETUP: CHALLENGES IN EEG/fMRI ACQUISITION

Acquiring EEG and fMRI simultaneously poses a significant technical challenge since the large magnetic field gradients and radio frequency (RF) pulses used to produce the MR images are major noise sources in the EEG. EEG leads are by definition a conducting loop, and the time-varying magnetic fields required to form the MR images thus induce an electromotive force (EMF) in the lead wires by Faraday's law. On present-day scanners, the time rate of change in the scanning gradients, or slew rate, is quite high (e.g. $200\,mT/m/ms$ on our scanner) and therefore induces large artifacts in the EEG every time the gradients fire. The RF pulses are also a large source of noise, for even though they are typically only about $10\,\mu T$ in amplitude, they have a fundamental frequency of $42.6\,MHz/T$ ($128\,MHz$ on a 3T scanner) and are turned on rapidly, at about $4000\,T/s$. These very high magnetic fluxes through the EEG lead wires can completely obscure the EEG signals (Ives et al., 1993; Huang-Hellinger et al., 1995; Goldman et al., 2000).

Conventional EEG amplifiers use AC-coupled inputs to high-pass-filter the signal of interest (typically microvolts) before it is digitized in order to remove the DC component resulting from the potential differences across the scalp (typically millivolts). This is usually accomplished by placing a simple RC filter with a long time constant between the output of the first-stage amplifiers and the input to the analog-to-digital converters. This AC filter is designed to pass frequencies in the EEG range, above about 0.3 Hz, and thus has a time constant on the order of several seconds. In a conventional EEG setting, saturation of the input stage rarely occurs. However, when the MR gradients are active, the induced EMF is large enough to saturate the initial input stage for several milliseconds, leading to long recovery times for the EEG signal following each MR acquisition (Huang-Hellinger et al., 1995; Bonmassar et al.,

1999; Krakow et al., 1999; Lovblad et al., 1999). For simultaneous EEG and fMRI studies, this means that the EEG amplifier must be specially designed to remove the DC component without allowing the gradients to saturate the input stage.

As mentioned, an EMF is induced in the EEG leads with any time variation in B field through the lead wire loop. The imaging gradients are one such time-varying field, but EMF is also induced by motion of the leads in the large (1.5 to 3 T in most scanners) static field of the magnet. For this reason, motion artifacts in the EEG are greatly amplified when a subject is placed in the MR scanner (Hill et al., 1995).

Perhaps not so obvious, however, is another prominent artifact caused by motion of the leads in the static field. Every time the heart beats, the electrodes and leads move with the small translations of the scalp caused by circulation of the blood. This pulsatile motion of the scalp is enough to induce an EMF in the leads. Further, motion of a conductive fluid (e.g. blood) in a magnetic field can cause voltage differences. These two effects give rise to an artifact in the EEG called the ballistocardiogram (Ives et al., 1993; Muri et al., 1998; Goldman et al., 2000). This artifact differs from lead pair to lead pair because of the differing orientations of the loops in the static field and the slight differences in motion across the scalp. It also varies rather widely from subject to subject—in some people, the ballistocardiogram is in the microvolt range and can be twice as high in amplitude as the EEG itself; in others, the artifact is barely noticeable in the EEG traces.

An important issue when recording EEG in an MR environment is subject safety. Currents induced in the EEG leads from MR gradients, RF pulses, or even motion of the leads in the static field could potentially cause harm to the subject. The risks include ulcers due to electrolysis from DC currents, electric shock or stimulation from AC currents below 100 kHz, and tissue heating or burns from AC currents above 100 kHz (Lemieux et al., 1997). The problem is not from current passing through the subject *per se* (current is induced in subjects during MR scanning under normal circumstances and is not harmful), but instead from the concentration of current in low impedance components that then yield high current density in tissue they come in contact with. The largest risk for such high currents to be induced comes from low impedance loops. These can be formed by exposed lead wire touching the subject, two leads in direct electrical contact, or even a single lead wire crossing back on itself and current flowing through the insulation. Long lead wires can also act as dipole antennas and pick up the electrical component of the RF, yielding current in the leads even without such loops. Filters at the front end of the pre-amp can also be low impedance at RF frequencies, making the entire circuit through the subject a burn risk. In addition, eddy currents can form in metallic components, and thus, standard EEG electrodes made from silver or gold can be a problem as well. For these reasons, great care must be taken to design the amplifiers and leads used in EEG/fMRI studies to ensure that the circuits are high impedance, such that there is little risk of large currents flowing through the subject. Lemieux calculated impedance thresholds to ensure subject safety from the currents induced in their EEG/fMRI scanning environment (Lemieux et al. (1997)), and we used these thresholds in our design.

Introduction of standard EEG equipment into the scanner potentially causes significant artifacts in the MR images. When highly conductive materials such as electrodes, EEG paste, and leads are brought into the field of view, they can greatly disrupt the homogeneity of the magnetic field and cause susceptibility artifacts in the images (Ives et al., 1993; Bonmassar et al., 2001). Such artifacts can be greatly diminished by using nonferrous materials (Lufkin et al., 1988). In addition, the EEG electronics can introduce RF noise. Such problems, however, are relatively easy to overcome by using specialized magnet-compatible hardware (Goldman et al., 2000, 2009; Sajda et al., 2007).

9.2.1 EEG Cap Design

Our EEG cap includes features that address many of the above issues. It is designed both to minimize induced EMF in the leads and to ensure subject safety. To minimize lead loop area, and thus induced EMF, the leads for each electrode are twisted with the lead of a neighboring electrode for its entire length. In this way, the flux through the leads is self-canceling for all but the small area at the electrode pair (Figure 9.1(a)). So that all leads can be twisted with a neighboring electrode, each electrode is connected to either two or three lead wires creating a chained bipolar montage (Figure 9.1(b)). The signal from each twisted pair is fed to a separate differential amplifier. Our current cap consists of 36 Ag/AgCl electrodes and 43 bipolar pairs. This oversampling makes the cap robust to any given electrode failing, as rereferencing of the electrode pairs can follow algebraically over multiple lead paths. It also enables a novel signal separation method for removing artifacts (see Section 9.4). To ensure subject safety from lead heating, there is a 10K-Ohm current-limiting surface mount resistor in series with each lead wire where it terminates at the electrode.

9.2.2 EEG Amplifier

To remove MR-induced artifacts, EEG recorded during MR scanning must be kept in the linear range. Because the signal generated in the EEG traces by MR gradients and RF pulses can be on the order of 100 times that of the EEG signal itself, scanner-induced voltages must be attenuated before the

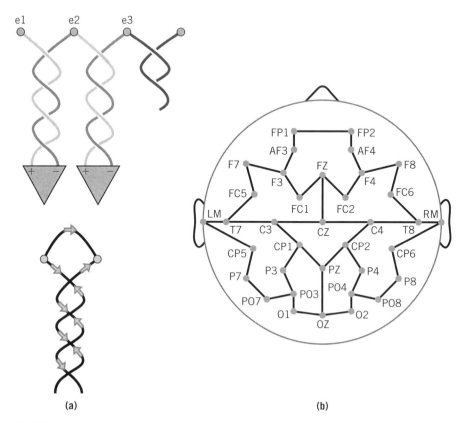

FIGURE 9.1

Bipolar electrode and EEG cap montage design. (a) A pair of EEG electrodes (small circles) are connected to a differential amplifier to form a bipolar channel. Leads from the electrodes to the differential amplifiers are twisted so as to minimize the inductive loop area and cancel induced currents. (b) In our current design, 34 electrodes are connected to form 43 bipolar channels. Electrodes are labeled using the standard 10–20 labeling and each line between electrodes represents a bipolar channel. The montage for the cap is designed so as to minimize the loop area as well as be robust to any given electrode failing (i.e., if one electrode fails, all channels in its loop will be affected). In addition, the overcomplete sampling of the montage enables a novel signal separation method for removing BCG artifacts (see Section 9.4). Please see this figure in color at the companion web site: www.elsevierdirect.com/companions/9780123750273

final amplification stage. This can be achieved with sharp low-pass filtering, because the gradients and RF signals are well above the 100 Hz clinical EEG frequency range.

We designed a special amplifier for this purpose (Goldman et al., 2000, 2009; Sajda et al., 2007). In each of its 64 channels, the RF contamination is attenuated passively using a balanced LC low-pass filter and a pair of crossed diodes provide input protection against high-voltage signals. Baseline drifts that bring the signal near the maximum or minimum voltage of the amplifier are detected and automatically reset as needed to ensure that the amplifier does not saturate. The input is differentially amplified with a user-defined gain (10 k or 2 k), and filtered with a 24 dB/octave Butterworth low pass filter tuned to either 125 or 300 Hz (set by the user), which limits the signal bandwidth to attenuate gradient artifacts with spectral energy out of the EEG bands of interest. The signal is then digitized using the synchronized method described below.

9.2.3 Synchronized Sampling

Identically repeating signals, such as the artifacts caused by the MR gradients and RF with each fMRI volume acquisition (TR), can be removed from the EEG traces using simple averaging and subtraction if they are well characterized. In our system, analog-to-digital conversion of the EEG is synchronized to the MR scanner clock by sending a transistor-transisitor logic (TTL) trigger pulse at the start of each image TR to a field-programmable array (FPGA) card (National Instruments, Austin, TX). This card is programmed to emit a pulse train that resets with each TTL trigger pulse. A simple schematic of the hardware and a photo of the amplifier and cap are shown in Figure 9.2.

9.3 SIGNAL PROCESSING AND REMOVAL OF BCG ARTIFACTS

Removal of the BCG artifact is perhaps the greatest challenge for obtaining a high-fidelity EEG signal during simultaneous acquisition of fMRI. As mentioned earlier, the BCG artifact is believed to arise via electromagnetic induction from heart beat-related movement of the head/electrodes and from flow-related phenomena when a subject is situated inside the magnetic field of an MRI scanner (see for example Debener et al., 2007). Methods for removing the BCG can provide cleaner-looking EEG with reduced power at the BCG frequencies; at the same time, however, they attenuate or distort that part of the underlying neural signal informative for single-trial classification (Dyrholm et al., 2009). As a result, our goal has been to develop a BCG removal method that does not suffer from this problem.

Roughly speaking, BCG removal methods can be grouped in two main categories: component analysis for identifying a linear BCG subspace based on temporal assumptions (see for example Srivastava et al. (2005)), and template fitting using ECG related information (see for example Goldman (2000);

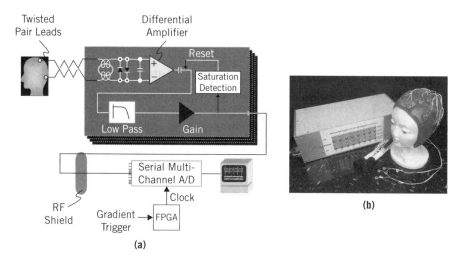

FIGURE 9.2

Overview of the system. (a) Leads from a pair of electrodes from the EEG cap (only one pair shown) are twisted to form a bipolar channel, feeding to a differential amplifier that includes RF filtering at its inputs. The channel is low-passed-filtered and then amplified by a selectable gain stage. There is also circuitry to reset the amplifiers in case of saturation. The amplifiers, all of which are co-located in the MR scanner room (outside 5 Gauss line), pass their outputs through the RF shield to the acquisition computer located outside the scanner room. The A/D conversion is triggered by the gradient pulse from the MR scanner, via an FPGA card. (b) A photograph of our current 43-channel MR-compatible EEG system. Please see this figure in color at the companion web site: www.elsevierdirect.com/companions/9780123750273

Bonmassar et al. (2002); Niazy et al. (2005)). The method that we developed belongs to the first category, but includes a *non-linear* pre-processing step such that the BCG subspace is identified, not within the raw data, but within a suitable representation of it.

ICA has been proposed for removing BCG artifacts from EEG by con-catenating data epochs along the temporal dimension and performing a decomposition assuming temporal independence between BCG and EEG (Srivastava et al., 2005). Although this method is appealing because it does not rely on an ECG regressor signal, the artifact is extracted by linear unmixing; hence, the rank of the cleaned EEG matrix plus the rank of the BCG matrix must not exceed the rank of the data matrix. Consequently, as BCG compo-nents are subtracted from the ICA decomposition, the rank of the remaining EEG matrix decreases; in other words, the neural signal is distorted or attenuated.

The method we have developed relies on the multipath properties of our EEG cap and provides an overcomplete basis that can then be used to infer a matrix of latent variables. This matrix has extra dimensions that are spanned if the data cannot be accounted for using the simplified set of Maxwell's equations known as Kirchhoff's laws (Dyrholm et al., 2009). An ICA decomposition is then performed on the matrix, and the cleaned EEG can be produced without rank deficiency.

9.3.1 The Kirchhoffian Account

The key idea in our method is to identify more components than there are sensors available. With such "overcomplete representation," components can be subtracted but the remaining representation suffers no rank deficiency. Overcomplete representations can in some cases be estimated blindly from data (see for example Bofill and Zibulevsky (2000); Lewicki and Sejnowski (2000)); however, it turns out that our multi-path cap can provide the necessary redundancy for building an overcomplete basis. The remainder of the problem thus boils down to inferring a matrix of latent variables assuming appropriate priors.

To see how the overcomplete basis can be derived, let u_p denote the differential voltage of channel p at an arbitrary point in time. Now, split the differential measurement in two parts $u_p = e_p + n_p$ where e_p is a voltage that sums to zero over closed loops and n_p is everything else. Voltages always sum to zero over closed loops if the system is in a conservative magnetic field—the assumption behind Kirchhoff's laws—and we say that the e-terms are Kirchhoffian while the n-terms are not. Adding Kirchhoffian contributions along any path connecting the two electrodes, say of channel p, always adds up to exactly e_p—that is, $\pm e_x \pm \ldots \pm \ldots \pm e_y = e_p$. The sum can be taken over any path that connects the electrodes of channel p, and the individual $\pm$ can be either $+$ or $-$ depending on the amplifier polarities in that path relative to the amplifier polarity of e_p. For any non-Kirchhoffian component, this does not hold; hence, a non-zero n_p would indicate that u_p has a non-Kirchhoffian component. The BCG artifact is non-Kirchoffian because it involves deformation and rotation of the twisted-pair loops (within the magnetic field), so our method tries to estimate n_p such that it can be used to subtract the BCG.

Consider all possible routes on the wiring diagram in Figure 9.1, leading from the electrodes of channel p without adding the same measurement twice:

$$u_p = e_p + n_p$$

$$\pm u_x \pm \ldots \pm u_y = e_p \pm n_x \pm \ldots \pm n_y \qquad (9.1)$$

$$\vdots$$

where the pattern of +'s and −'s for the u-terms is the same as for the n-terms. These equations all include one single e-term, namely e_p, while adding different combinations of n-terms. The resulting set of equations can be formulated as a matrix (of zeros, ones, and minus ones), denoted by $\mathbf{M}_p$, so that in matrix notation

$$\mathbf{M}_p \mathbf{u} = \begin{bmatrix} \mathbf{1}_p & \mathbf{M}_p \end{bmatrix} \begin{bmatrix} e_p \\ \mathbf{n} \end{bmatrix} \tag{9.2}$$

where $\mathbf{u}$ is a 43-dimensional vector of all the differential amplifier measurements, $\mathbf{n}$ the corresponding vector of n-terms, and $\mathbf{1}_p$ is a column vector of implicit length filled with ones. The matrix $\mathbf{M}_p$ can be determined by exhaustive search. For instance, given our cap, there are 70 different routes between frontal electrodes FP1 and FP2. Figure 9.3 shows some of the rows of $\mathbf{M}_{\mathrm{FP1\text{-}FP2}}$. The joint linear system involving all measurements and latent variables is found by consolidating Eq. (9.2) for all channels p, and the over-complete basis is then the joint system matrix that forwards the latent variables

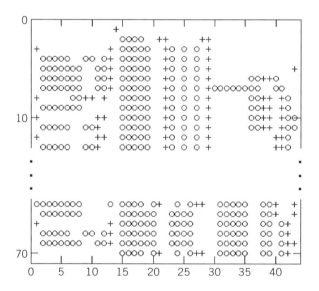

FIGURE 9.3

Matrix representation of the 70 different routes between electrodes FP1 and FP2. Plus signs represent a+1 element; circles represent a−1. The first row represents the shortest route, which is simply the measurement $u_{\mathrm{FP1,FP2}}$ itself. The differential amplifier is connected with the + terminal on FP1, so the matrix element is +1. The second row represents the path going through electrodes AF3–F3–FC1–FZ–FC2–F4–AF4. The signs are set according to the differential amplifier polarity.

Source: Reproduced/adapted from Dyrholm et al. (2009).

n and **e** to form **u**. Note that the overcomplete basis matrix is rectangular, and isolation of **n** and **e** requires a pseudo-inverse; in other words, the problem of inferring the latent variables becomes that of maximizing the posterior assuming an appropriate prior. If we were to assume a Gaussian prior, the latent variable inference would boil down to the Moore-Penrose pseudo-inverse, which simply produces various linear combinations of the 43 input channels. However, linear mixing of 43 inputs can never produce an output of rank higher than 43; thus, we conclude that a Gaussian prior is too vague for our purpose. We expect the BCG to be somewhat sparse, compared to a Gaussian distribution, which might lead us more in the direction of the Laplace distribution.

The Laplace distribution has been suggested as a generically useful prior in overcomplete representations, and the inference can be carried out by a linear program that minimizes the 1-norm of the latent variables (Lewicki and Sejnowski (2000)). However, it is clear that minimizing the 1-norm of $\{\hat{\mathbf{e}}, \hat{\mathbf{n}}\}$ does not uniquely determine the balance between e_p and n_p. In particular, there is the trivial solution $(\hat{\mathbf{e}}, \hat{\mathbf{n}}) = (\mathbf{0}, \mathbf{u})$ from which we learn nothing about the Kirchhoffian component. In fact, by our initial definition of e_p, we seek a solution that is as far from this trivial solution as possible. It can be achieved by minimizing the 1-norm of **n** without regard to the norm of **e**. It is beneficial to consider this problem in the electrode domain:

$$\hat{\mathbf{v}} = \arg\min \|\mathbf{n}\|_1 = \arg\min \|\mathbf{u} - \mathbf{A}\mathbf{v}\|_1 \qquad (9.3)$$

where **v** represents electrode potentials, and **A** represents the electrode-amplifier connections. We solve Eq. (9.3) using Eq. (9.1) and finally let $\hat{\mathbf{e}} = \mathbf{A}\hat{\mathbf{v}}$ and $\hat{\mathbf{n}} = \mathbf{u} - \mathbf{A}\hat{\mathbf{v}}$. This solution has the desired property that, in the event that **u** is strictly Kirchhoffian, the data will remain unaltered ($\hat{\mathbf{n}}$ becomes zero). Now, since $\hat{\mathbf{e}}$ and $\hat{\mathbf{n}}$ are distinguished by their Kirchhoffian–to–non-Kirchhoffian content ratio, the non-Kirchhoffian BCG activity in $\hat{\mathbf{n}}$ can be used to remove Kirchoffian BCG activity from **u**. To do so we use ICA in the latent space of $\{\hat{\mathbf{e}}, \hat{\mathbf{n}}\}$. We inspect the component time courses visually, and remove all components that are dominated by BCG activity. This leads to a varying number of components, which are then removed from the data by subtracting from **u** its contributions to $\hat{\mathbf{e}} + \hat{\mathbf{n}}$. Since the inference is carried out with a temporal IID assumption, the components are smoothed (lowpass-filtered at 15 Hz) before subtraction. This choice of transition frequency is simply based on visual inspection and suggests that $\sim$15 harmonics, having a fundamental frequency of 1 Hz, will reproduce a realistic BCG waveform.

Example raw EEG and cleaned epochs are shown in Figure 9.4. Looking at the raw versus the cleaned EEG, we see that the cleaned data has visibly reduced BCG artifacts. Besides producing clean-looking data, our new

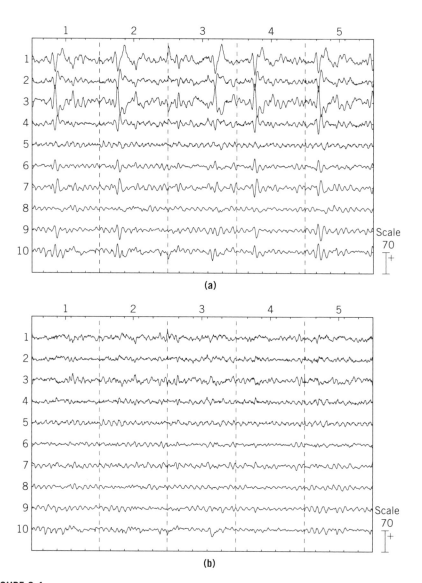

(a)

(b)

FIGURE 9.4

Epochs for the first five target or standard events, each showing 200 ms before to 1000 ms after each event, for channels 1–10 are shown for subject 11 in two cases (same scale). (a) The raw data with BCG artifact. One can clearly see the BCG artifact "bumps" in the signal. (b) The cleaned data using our new method to remove the BCG artifacts. The artifacts are visibly reduced.

method results in the best average single-trial classification performance (see Section 9.4.1) in comparison to two prominent methods: the optimal basis sets (OBS) (Niazy et al., 2005) and ICA (Srivastava et al., 2005). When using either OBS or ICA, the single-trial classification was significally worse than when BCG was not removed at all (12 subjects, two-sided, signed rank test, $p < 0.05$; see Dyrholm et al. (2009) for details). With our method, the average single-trial classification performance was higher in comparison to no BCG removal, but not significantly so. This result clearly indicates that the neural activity is not attenuated or distorted by our method.

9.4 LINKING SINGLE-TRIAL VARIATIONS OF TASK-RELEVANT EEG COMPONENTS TO THE BOLD SIGNAL

Traditionally, the analysis of EEG has relied on averaging across tens to hundreds of trials to uncover the neural signatures related to any given behavioral paradigm. The underlying assumption has always been that trial averaging increases signal-to-noise ratio (SNR). Although this assumption is generally valid, it has a major detriment—it conceals inter-trial response variability. Single-trial variability can arise from a number of factors, including changes in attention, adaptation, and habituation, as well as changes in the recording environment.

Single-trial methods usually exploit the multi-channel electrode arrays to integrate spatially across the scalp and determine spatial components that discriminate between experimental conditions. Spatial integration enhances signal quality without the loss of temporal precision common in trial averaging in event-related potential (ERP) analysis. Methods that have been used for such analysis include ICA (Makeig et al., 2002), common spatial patterns (CSP) (Guger et al., 2000), and, by our group, linear discrimination (LD) (Parra et al., 2002, 2005). Specifically, our group pioneered the use of supervised linear discriminant analysis to identify neural correlates of perceived error detection (Parra et al., 2003), target detection (Gerson et al. (2005)), and perceptual decision making (Philiastides et al., 2006; Philiastides and Sajda, 2006).

9.4.1 Identifying EEG Components Using Linear Discrimination

We use logistic regression to find an optimal basis for discriminating between the two conditions over a specific temporal window (Parra et al., 2002). Specifically, we define a training window starting at a post-stimulus onset time τ, with a duration of δ, and use logistic regression to estimate a spatial weighting vector $\mathbf{w}_{\tau,\delta}$ that maximally discriminates between electrode array signals $\mathbf{X}$ for two conditions (e.g., binary-labeled trials):

$$y = \mathbf{w}_{\tau,\delta}^T \mathbf{X} \tag{9.4}$$

where $\mathbf{X}$ is an $N \times T$ matrix (N electrodes and T time samples). The result is a "discriminating component" y that is specific to activity correlated with condition 1 but minimizes activity correlated with both task conditions. In practice, we choose a duration of the training window (δ) and vary the window onset (τ) across time. We use the reweighted least squares algorithm to learn the optimal discriminating spatial weighting vector $\mathbf{w}_{\tau,\delta}$ (Jordan and Jacobs, 1994). To provide a functional neuroanatomical interpretation of the resultant discriminating activity, and given the linearity of our model, we compute the electrical coupling coefficients for the linear model (see Parra et al., 2005 for derivation):

$$\mathbf{a} = \frac{\mathbf{X}y}{y^T y} \tag{9.5}$$

Equation (9.5) describes the electrical coupling $\mathbf{a}$ of the discriminating component y that explains most of the activity $\mathbf{X}$. Strong coupling indicates low attenuation of the component and can be visualized as the intensity of the "sensor projections" $\mathbf{a}$. Training the discriminator over several temporal windows allows us to visualize the temporal evolution of the discriminating component activity by displaying the coupling coefficients $\mathbf{a}$.

We quantify the performance of the linear discriminator by using a leave-one-out approach (Duda et al., 2001) and computing the area under the ROC curve, which we refer to as A_z. We use the A_z metric to characterize the discrimination performance and compute a significance level for A_z by performing the leave-one-out test after randomizing the truth labels. We repeat this randomization process 100 times to produce an A_z randomization distribution (i.e., bootstrapping) and identify the A_z value leading to a significance level of $p = 0.01$. Figure 9.5 provides a graphical representation of our single-trial analysis.

9.4.2 Constructing fMRI Regressors from Single-Trial Variability in EEG Components

To obtain EEG-derived fMRI activation maps that depend on task- and subject-specific electrophysiological source variability, we construct fMRI regressors based on the output of our single-trial EEG analysis. Specifically, we use the output of our linear discriminator for the different EEG components we identify to construct these regressors. Given an EEG component identified using a

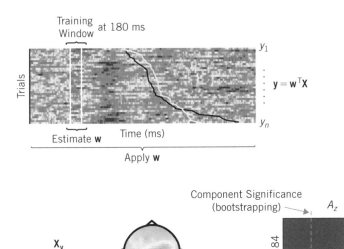

FIGURE 9.5

Summary of our single-trial analysis of the EEG. We construct discriminant component maps, aligning trials to the event onset (black vertical line) and sorting them by reaction time (sigmoidal curves). Each row of the discriminant component map represents a single trial across time. Discriminant components are represented by the **y** vectors. To construct this map we choose a training window, indicated by white vertical bars (for this example starting at 180 ms post-stimulus), during which we train the linear discriminator to estimate a weighting vector **w** across all sensors in **X**, such that **y** is maximally discriminating between the two experimental conditions (e.g., trial type 1 versus trial type 2). Once **w** is estimated using data derived only from the training window, we apply **w** to the data across all trials and all time. The resultant discriminant component map is shown here. Medium gray represents positive activity, and dark gray represents negative activity. For this example the resultant discriminating component appears $\approx$ 180 ms after the onset of the stimulus and is stimulus-locked. Response-locked components have a sigmoidal profile similar to the subject's actual response times. We use the forward model to project this discriminating component back to the sensors. The resultant scalp projections, **a** are shown and are used for interpreting the neuroanatomical significance of the components. To quantify the discriminator's performance we use ROC analysis and compute the area under the ROC curve (A_z value). Finally, in order to assess the significance of the resultant discriminating component we use a bootstrapping technique to compute an A_z value, leading to a significance level of $p = 0.01$ shown here by the dotted medium gray line. Please see this figure in color at the companion web site: www.elsevierdirect.com/companions/9780123750273

Source: Reproduced/adapted from Philiastides and Sajda (2006).

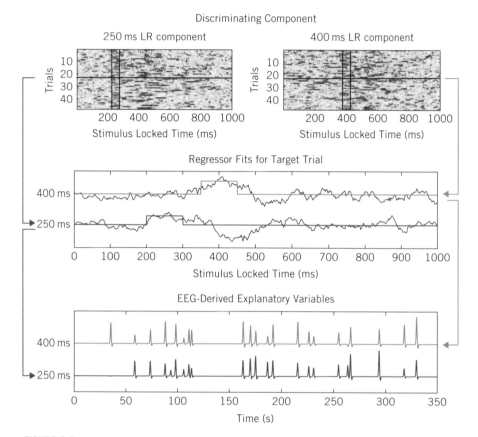

FIGURE 9.6

Graphical representation of how EEG can be used to construct fMRI regressors during a simultaneous EEG/fMRI acquisition. The EEG components shown are taken from the auditory oddball experiments discussed in Section 9.5. (a) Output for all trials of the single-trial EEG discriminator for two stimulus-locked 50 ms windows (data between black vertical bars) centered at 250 ms and 400 ms post-stimulus onset. The medium gray indicates positive and the dark gray negative values of the projection (i.e., positive and negative correlations). (b) EEG discriminator output for a single target trial for each of the two components (black curves), showing the fMRI event model amplitude as the average of the discriminator output for each window, 250 ms (dark gray) and 400 ms (medium gray), for the trial. (c) Single-trial fMRI model for target trials across the entire session for the 250-ms and 400-ms windows. Note that the event timing for each of the two windows is the same, but the event amplitudes are different. Please see this figure in color at the companion web site: www.elsevierdirect.com/companions/9780123750273

Source: Reproduced from Goldman et al. (2009).

linear discriminator trained with an N-ms training window, the discriminator output $\mathbf{y}$ has dimension $N \times T$, where T is the number of trials. We average across all training samples and compute:

$$\mathbf{y_j} = \frac{1}{N} \sum_{i=1}^{N} \mathbf{y_{ij}} \qquad (9.6)$$

where $\mathbf{j}$ is the trial number index. We use $\mathbf{y_j}$ to vary the amplitude of our fMRI regressor. The onset and duration of the events making up each regressor are determined by the temporal characteristics of the discriminating EEG discriminating components as identified by the single-trial EEG analysis. Figure 9.6 illustrates how two EEG components, identified for different time windows, can be used to construct fMRI regressors specific to each EEG discriminating component. To identify the unique contribution of each component, we must orthogonalize regressors with respect to one another.

9.5 RESULTS

We present results showing how our approach yields activations that are unique to the single-trial variability of EEG components (for a complete description see Goldman et al. (2009)). Using the system described, we tested our methodology for coupling EEG component variability with BOLD for an auditory oddball task. The oddball task generates the P300 (P3), which is the well-studied late positive complex in the EEG observed when subjects discriminate rare stimuli (targets) from frequent stimuli (standards) (Donchin and Coles, 1988; Picton, 1992; Polich, 2007). There is substantial evidence via analysis of the EEG that the P300 is in fact a superposition of subcomponents that reflect underlying constituent processes of the discrimination task, from stimulus evaluation to response execution (Makeig et al., 1999).

We analyzed the EEG data—both stimulus-locked and response-locked—and constructed fMRI regressors from the single-trial variability of both stimulus-locked and response-locked EEG components. We emphasize that our approach utilizes the trial-to-trial amplitude variability of temporally specific EEG components to differentiate hemodynamic activations associated with the different components. That is, we differentiate EEG components by their temporal onset and subsequently differentiate hemodynamic activations (via regressors) by the components' trial-to-trial variability. In this way, there is no issue with the temporal resolution of the hemodynamic changes given sufficient inter-stimulus interval (ISI), since the timing of regressor events is determined by the ISI.

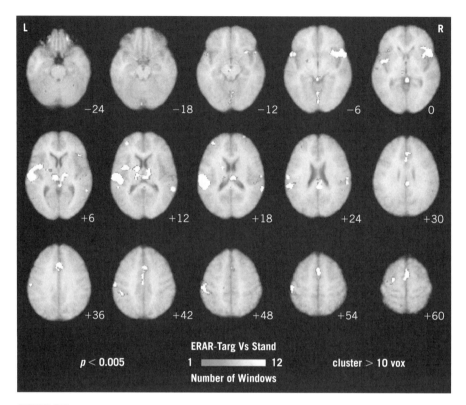

FIGURE 9.7

Areas of significant activation for targets versus standards using the traditional event-related model, ERAR-TargVsStand. These areas are activated on average to the presentation of target tones versus standard tones. Axial slices (MNI space) ranging from $z = -24$ mm to $z = +60$ mm are shown. The shading bar at the bottom displays the number of windows, or analyses (see methods in Goldman et al. (2009)), out of 12 total for which the ERAR-TargVsStand contrast passed the threshold of per-voxel $p < 0.005$, cluster > 10 voxels at that voxel (positive correlation). Please see this figure in color at the companion web site: www.elsevierdirect.com/companions/9780123750273
Source: Reproduced from Goldman et al. (2009).

Figure 9.7 shows the EEG/fMRI results using a purely stimulus-derived event-related analysis (i.e., contrasts constructed as target versus standard trials). The distributed activity of this event-related contrast is consistent with previous fMRI-only studies investigating the auditory oddball task (Linden et al., 1999; Kiehl et al., 2005; Friedman et al., 2009). We analyzed the data using the EEG-derived regressors, via the procedure outlined in Figure 9.6 (additional details can be found in Goldman et al. (2009)). These regressors were orthogonalized with respect to the event-related regressors as well as the

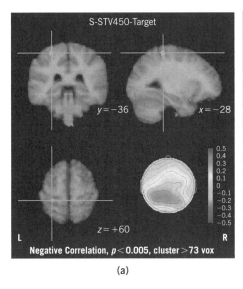

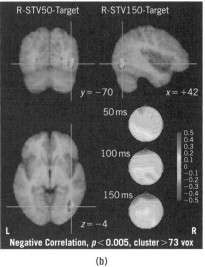

(a)   (b)

FIGURE 9.8

(a) fMRI activations for the stimulus-locked single-trial analysis showing regions with significant BOLD signal correlation ($p < 0.005$, cluster > 73 voxels, negative correlation) with single-trial variability to targets for the 450-ms window, S-STV450-Targ. For target tones, only the stimulus-locked 450-ms window passed both the EEG and fMRI thresholds. Shown also is the scalp topography of the corresponding 450-ms window stimulus-locked EEG discriminating component (arbitrary units). (b) fMRI activations for the response-locked single-trial analysis showing regions with significant BOLD signal correlation ($p < 0.005$, cluster >73 voxels, negative correlation) to response-locked single-trial variability to target tones. The response-locked 50-ms window, R-STV50-Targ (dark gray), and 150-ms window, R-STV150-Targ (light gray), passed both the EEG and fMRI thresholds. The scalp maps of the output of the EEG discriminator for the three windows from 50-150–ms response-locked are also shown (arbitrary units). Please see this figure in color at the companion web site: www.elsevierdirect.com/companions/9780123750273
Source: Reproduced from Goldman et al. (2009).

reaction times of the subjects. Activations seen with the regressors are thus due purely to the trial-to-trial variability of the EEG components. Figure 9.8 shows the results for EEG-derived regressors (for discussion of the criteria used to identify activations and the cluster criteria used to correct for multiple comparisons, see Goldman et al. (2009)). We found negative correlations in the somatosensory cortex (SMC) and the lateral occipital complex (LOC).

Given that one interpretation of the single-trial variability of our EEG components is as a surrogate for attentional engagement in the task, our results are consistent with the hypothesis that though the SMC and LOC are not activated by auditory stimuli, the ongoing waxing and waning of attention modulates a large number of areas, including those not directly participating in the information processing for the task. This in fact may also explain why we see our single-trial EEG components producing negative BOLD correlations. As attention shifts to the auditory task, it reduces attentional resources available for visual processing and somato-sensation (a push-pull effect) (Shomstein and Yantis (2004))–that is, an increase in the single-trial regressor for a trial represents an increase in auditory attention at the cost of a decrease in visual and somatosensory attention; the result is a negative correlation in those areas. Additional simultaneous EEG/fMRI experiments that explicitly probe auditory, visual, and somatosensory target detection within the context of push-pull attentional resources are needed to further substantiate this hypothesis.

Figure 9.8 also shows the resulting scalp plots for the EEG components constructed via Eq. (9.5). It is important to note that these highlighted areas show excellent spatial consistency with the scalp topographies independently derived from the forward models of the discriminating components. This is in spite of the fact that the spatial information in the scalp topographies is not used in any way for localizing the fMRI activations. This finding provides even greater confidence that the EEG variability that generates fMRI activations is in fact related to the task and the cortical areas localized via fMRI. In summary, these results provide compelling evidence that single-trial analysis of simultaneous EEG/fMRI can be used to measure the effects of latent brain states such as attention on areas not directly participating in the information processing of the task at hand.

9.6 FUTURE DIRECTIONS

EEG/fMRI is in its infancy as a neuroimaging tool, and there are many directions for future research. Specific to our system, we have yet to determine an optimal number of EEG electrodes and, perhaps more intriguing, an optimal number of bipolar channels—that is, how much oversampling should we do? With regard to the MRI methodology, we believe extending the acquisition to 3 Tesla will result in increased sensitivity and better spatial localization.

From the perspective of optimally utilizing the information in the simultaneously acquired EEG and fMRI data, more principled methods are needed for linking the multivariate signals of the EEG with the multi-voxel measurements of the fMRI. For instance, the approach we described in this chapter treats EEG as a multivariate signal, extracts linear components from it, and then links these components to the fMRI data using a conventional massive univariate GLM. More compelling would be to treat both EEG and fMRI as multivariate signals and identify components in a joint multivariate space.

Though there has been substantial effort focused on multivariate analyses of fMRI (for a review see Norman et al. (2006)), research on analysis of the joint space of EEG/fMRI has been limited because neuroimaging modalities are relatively new (for an exception, see Martinez-Montes et al. (2004)).

The EEG components we identify are modulated by task, stimulus, and/or the state of the user. One area for future work is to consider an alternative method of directly modulating EEG activity in a certain region of the brain through transcranial magnetic stimulation (TMS). This approach might enable us to directly identify causal changes between EEG patterns and BOLD signals, and may provide a powerful check on the results of these analyses as it would directly confirm the accuracy of the methods. Also, replacing BOLD image acquisition with a sequence sensitive to cerebral blood flow (CBF) using arterial spin labeling (ASL) should yield similar results, although we would expect the localized regions to be smaller since CBF studies based on ASL are known to be less affected by the "draining vein" problem that spatially spreads the BOLD response (Turner (2002)).

In summary, we believe that simultaneously acquired EEG/fMRI is a wonderful example of sophisticated imaging technology, signal processing, and machine learning ultimately enabling an entirely new type of multimodal imaging with the potential for increasing our understanding of how the mind, the brain, and behavior relate to one another.

Acknowledgments

This research was supported by funding from NIH (grant EB004730) and DARPA (contract NBCHC080029).

References

Abdelmalek, N.N., 1980. Algorithm 551: a Fortran subroutine for the L1 solution of overdetermined systems of linear equations. ACM Trans. Math. Softw. 6 (2), 228–230.

Bénar, C.G., Schön, D., Grimault, S., Nazarian, B., Burle, B., Roth, M., et al., 2007. Single-trial analysis of oddball event-related potentials in simultaneous EEG-fMRI. Hum. Brain Mapp. 28 (7), 602–613.

Bofill, P., Zibulevsky, M., 2000. Blind separation of more sources than mixtures using sparsity of their short-time fourier transform. In: Proceedings of ICA, pp. 87–92.

Bonmassar, G., Anami, K., et al., 1999. Visual evoked potential (VEP) measured by simultaneous 64-channel EEG and 3T fMRI. Neuroreport 19 (9), 1893–1897.

Bonmassar, G., Hadjikhani, N., et al., 2001. Influence of EEG electrodes on the BOLD fMRI signal. Hum. Brain Mapp. 14 (2), 108–115.

Bonmassar, G., Purdon, P.L., Jaaskelainen, I.P., Chiappa, K., Solo, V., Belliveau, J.W., et al., 2002. Motion and ballistocardiogram artifact removal for interleaved recording of EEG and EPs during MRI. Neuroimage 16 (4), 1127–1141.

Debener, S., Mullinger, K.J., Niazy, R.K., Bowtell, R.W., 2007. Properties of the ballisto-cardiogram artefact as revealed by EEG recordings at 1.5, 3 and 7 T static magnetic field strength. Int. J. Psychophysiol.

Debener, S., Ullsperger, M., Siegel, M., Fiehler, K., von Cramon, D.Y., Engel, A.K., 2005. Trial-by-trial coupling of concurrent electroencephalogram and functional magnetic resonance imaging identifies the dynamics of performance monitoring. J. Neurosci. 25 (50), 11730–11737.

Donchin, E., Coles, M.G., 1988. Is the p300 component a manifestation of context updating? Behav. Brain Sci. 11, 357–374.

Duda, R.O., Hart, P.E., Stork, D.G., 2001. Pattern Classification. Wiley, New York.

Dyrholm, M., Goldman, R., Sajda, P., Brown, T.R., 2009. Removal of BCG artifacts using a non-Kirchhoffian overcomplete representation. IEEE Trans. Biomed. Eng. 56 (2), 200–204.

Eichele, T., Calhoun, V.D., Moosmann, M., Specht, K., Jongsma, M.L.A., Quiroga, R.Q., et al., 2008. Unmixing concurrent eeg-fmri with parallel independent component analysis. Int. J. Psychophysiol. 67 (3), 222–234.

Eichele, T., Calhoun, V.D., Debener, S., 2009. Mining eeg-fmri using independent component analysis. Int. J. Psychophysiol. 73 (1), 53–61.

Friedman, D., Goldman, R.I., Stern, Y., Brown, T.R., 2009. The brain's orienting response: an event-related functional magnetic resonance imaging investigation. Hum. Brain Mapp. 30 (4), 1144–1154.

Gerson, A.D., Parra, L.C., Sajda, P., 2005. Cortical origins of response time variability during rapid discrimination of visual objects. Neuroimage 28 (2), 342–353.

Goldman, R.I., Stern, J.M., Engel, Jr., J., Cohen, M.S., 2000. Acquiring simultaneous EEG and functional MRI. Clin. Neurophysiol. 111 (11), 1974–1980.

Goldman, R.I., Stern, J.M., et al., 2000. Acquiring simultaneous EEG and functional MRI. Clin. Neurophysiol. 111 (11), 1974–1980.

Goldman, R.I., Wei, C.-Y., Philiastides, M.G., Gerson, A.D., Friedman, D., Brown, T.R., et al., 2009. Single-trial discrimination for integrating simultaneous EEG and fMRI: identifying cortical areas contributing to trial-to-trial variability in the auditory oddball task. Neuroimage 47 (1), 136–147.

Guger, C., Ramoser, H., et al., 2000. Real-time EEG analysis with subject-specific spatial patterns for a brain computer interface (BCI). IEEE Trans. Rehabil. Eng. 8 (4), 441–446.

Hill, R.A., Chiappa, K.H., et al., 1995. EEG during MR imaging: differentiation of movement artifact from paroxysmal cortical activity. Neurology 45 (10), 1942–1943.

Huang-Hellinger, F.R., Breiter, H.C., et al., 1995. Simultaneous functional magnetic resonance imaging and electrophysiological recording. Hum. Brain Mapp. 3 (1), 13–25.

Ives, J.R., Warach, S., Schmitt, F., Edelman, R.R., Schomer, D.L., 1993. Monitoring the patient's EEG during echo planar MRI. Electroencephalogr. Clin. Neurophysiol. 87 (6), 417–420.

Jordan, M.I., Jacobs, R.A., 1994. Hierarchical mixtures of experts and the EM algorithm. Neural Comput. 6, 181–214.

Kiehl, K.A., Stevens, M.C., Laurens, K.R., Pearlson, G., Calhoun, V.D., Liddle, P.F., 2005. An adaptive reflexive processing model of neurocognitive function: supporting evidence from a large scale ($n = 100$) fMRI study of an auditory oddball task. Neuroimage 25, 899–915.

Krakow, K., Woermann, F.G., Symms, M.R., Allen, P.J., Lemieux, L., Barker, G.J., et al., 1999. EEG-triggered functional MRI of interictal epileptiform activity in patients with partial seizures. Brain 122 (Pt 9), 1679–1688.

Lemieux, L., Allen, P.J., Franconi, F., Symms, M.R., Fish, D.R., 1997. Recording of EEG during fMRI experiments: patient safety. Magn. Reson. Med. 38 (6),943–952.

Lewicki, M.S., Sejnowski, T.J., 2000. Learning overcomplete representations. Neural Comput. 12 (2), 337–365.

Linden, D.E., Prvulovic, D., Formisano, E., Vollinger, M., Zanella, F.E., Goebel, R., et al., 1999. The functional neuroanatomy of target detection: an fMRI study of visual and auditory oddball tasks. Cereb. Cortex 9, 815–823.

Logothetis, N.K., Wandell, B.A., 2004. Interpreting the bold signal. Annu. Rev. Physiol. 66, 735–769.

Lovblad, K.O., Thomas, R., Jakob, P.M., Scammell, T., Bassetti, C., Griswold, M., et al., 1999. Silent functional magnetic resonance imaging demonstrates focal activation in rapid eye movement sleep. Neurology 53 (9), 2193–2195.

Lufkin, R., Jordan, S., Lylyck, P., Vinuela, F., 1988. MR imaging with topographic EEG electrodes in place. AJNR Am. J. Neuroradiol. 9 (5), 953–954.

Makeig, S., Westerfield, M., Jung, T.-P., Covington, J., Townsend, J., Sejnowski, T., et al., 1999. Independent components of the late positive response complex in a visual spatial attention task. J. Neurosci. Methods 19, 2665–2680.

Makeig, S., Westerfield, M., Jung, T.P., Enghoff, S., Townsend, J., Courchesne, E., et al., 2002. Dynamic brain sources of visual evoked responses. Science 295, 690–694.

Mantini, D., Perrucci, M.G., Del Gratta, C., Romani, G.L., Corbetta, M., 2007. Electrophysiological signatures of resting state networks in the human brain. Proc. Natl. Acad. Sci. USA 104 (32), 13170–13175.

Martinez-Montes, E., Valdes-Sosa, P.A., Miwakeichi, F., Goldman, R.I., Cohen, M.S., 2004. Concurrent EEG/fMRI analysis by multiway partial least squares. Neuroimage 22 (3), 1023–1034.

Menon, V., Ford, J.M., et al. 1997. Combined event-related fMRI and EEG evidence for temporal-parietal cortex activation during target detection. Neuroreport 8 (14), 3029–3037.

Mulert, C., Seifert, C., Leicht, G., Kirsch, V., Ertl, M., Karch, S., et al., 2008. Single-trial coupling of EEG and fMRI reveals the involvement of early anterior cingulate cortex activation in effortful decision making. Neuroimage 42 (1), 158–168.

Muri, R.M., Felblinger, J., Rosler, K.M., Jung, B., Hess, C.W., Boesch, C., 1998. Recording of electrical brain activity in a magnetic resonance environment: distorting effects of the static magnetic field. Magn. Reson. Med. 39 (1), 18–22.

Niazy, R.K., Beckmann, C.F., Iannetti, G.D., Brady, J.M., Smith, S.M., 2005. Removal of fMRI environment artifacts from EEG data using optimal basis sets. Neuroimage 28 (3), 720–737.

Norman, K.A., Polyn, S.M., Detre, G.J., Haxby, J.V., 2006. Beyond mind-reading: multi-voxel pattern analysis of fmri data. Trends Cogn. Sci. 10 (9), 424–430.

Obitz, B., Mecklinger, A., Friederici, A.D., Von Cramon, D.Y., 1999. The functional neuroanatomy of novelty processing: integrating ERP and fMRI results. Cereb. Cortex 9, 379–391.

Parra, L., Alvino, C., Tang, A., Pearlmutter, B., Young, N., Osman, A., et al., 2002. Linear spatial integration for single-trial detection in encephalography. Neuroimage 17, 223–230.

Parra, L.C., Spence, C.D., Gerson, A.D., Sajda, P., 2003. Response error correction: A demonstration of improved human-machine performance using real-time eeg monitoring. IEEE Trans. Neural Syst. Rehabil. Eng. 11 (2), 173–177.

Parra, L.C., Spence, C.D., Gerson, A.D., Sajda, P., 2005. Recipes for the linear analysis of EEG. Neuroimage 28 (2), 326–341.

Philiastides, M.G., Ratcliff, R., Sajda, P., 2006. Neural representation of task difficulty and decision making during perceptual categorization: a timing diagram. J. Neurosci. 26 (35), 8965–8975.

Philiastides, M.G., Sajda, P., 2006. Temporal characterization of the neural correlates of perceptual decision making in the human brain. Cereb. Cortex 16 (4), 509–518.

Philiastides, M.G., Sajda, P., 2007. EEG-informed fMRI reveals spatiotemporal characteristics of perceptual decision making. J. Neurosci. 27 (48), 13082–13091.

Picton, T. W., 1992. The P300 wave of the human event-related potential. J. Clin. Neurophysiol. 9 (4), 456–479.

Polich, J., 2007. Updating P300: an integrative theory of P3a and P3b. Clin. Neurophysiol. 118 (10), 2128–2148.

Sajda, P., Goldman, R.I., Philiastides, M.G., Gerson, A.D., Brown, T.R., 2007. A system for single-trial analysis of simultaneously acquired EEG and fMRI. In: Proceedings of 3rd International IEEE/EMBS Conference on Neural Engineering, pp. 287–290, 2–5 May 2007.

Shomstein, S., Yantis, S., 2004. Control of attention shifts between vision and audition in human cortex. J. Neurosci. 24, 10702–10706.

Sirotin, Y.B., Das, A., 2009. Anticipatory haemodynamic signals in sensory cortex not predicted by local neuronal activity. Nature 457 (7228), 475–479.

Srivastava, G., Crottaz-Herbette, S., Lau, K.M., Glover, G.H., Menon, V., 2005. ICA-based procedures for removing ballistocardiogram artifacts from EEG data acquired in the MRI scanner. Neuroimage 24 (1), 50–60.

Turner, R. 2002. How much cortex can a vein drain? downstream dilution of activation-related cerebral blood oxygenation changes. Neuroimage 16 (4), 1062–1067.

Wang, J., Zhou, T., et al., 1999. Relationship between ventral stream for object vision and dorsal stream for spatial vision: an fMRI and ERP study. Hum. Brain Mapp. 8 (4), 170–181.

Statistical Pattern Recognition and Machine Learning in Brain–Computer Interfaces

10

Rajesh P. N. Rao, Reinhold Scherer

Department of Computer Science and Engineering, University of Washington, Seattle, Washington

Brain–computer interfaces (BCIs) allow a subject to directly control objects such as a cursor or a robot using brain signals. BCIs depend crucially on the ability to reliably identify behaviorally induced changes (or "cognitive states") in the brain signals being recorded. Statistical pattern recognition and machine learning algorithms play an important role in identifying these changes in brain signals and mapping them to appropriate control signals. We focus in this chapter on brains signals recorded from the scalp (EEG) and from the surface of the brain (ECoG) in humans. We provide an overview of some of the major feature extraction, pattern recognition, and machine learning techniques that have been successfully applied to EEG- and ECoG-based BCIs. The techniques reviewed include methods to enhance the signal-to-noise ratio of EEG and ECoG signals (e.g., Laplacian rereferencing and common spatial patterns (CSP)), feature extraction techniques, classification methods (e.g., linear discriminant analysis (LDA) and support vector machines (SVMs)), and methods for evaluating BCI performance. We conclude by illustrating the application of these methods in two BCI systems, one for controlling a humanoid robot and another for operating the popular global navigation program Google Earth.

10.1 INTRODUCTION

Brain–computer interfaces (BCIs) augment the human ability to communicate and interact with the external world by directly linking the brain to computers

Statistical Signal Processing for Neuroscience and Neurotechnology. DOI: 10.1016/B978-0-12-375027-3.00010-7

and robotic devices. BCIs bypass the normal neuromuscular output pathways for translating brain signals into action. Instead, physiological brain signals are processed in real time by digital signal processing methods to allow a novel form of communication and interaction with the environment.

A major goal of BCIs has been to improve of the quality of life of physically impaired individuals, including those paralyzed because of degenerative neurological diseases such as amyotrophic lateral sclerosis, spinal cord injury, or stroke. BCIs have been developed to aid communication (e.g., spelling devices (Birbaumer et al., 1999; Obermaier et al., 1999; Donchin et al., 2000; Millàn and Mourino, 2003; Scherer et al., 2004; Müller and Blankertz, 2006; Pfurtscheller et al., 2006) and Internet browsing (Tomori and Moore, 2003; Karim et al., 2006; Bensch et al., 2007; Mugler et al., 2008)); to restore motor control (e.g., grasp function for individuals with spinal cord injury) (Pfurtscheller et al., 2003; Müller-Putz et al., 2005b, 2006; Hochberg et al., 2006); to control wheelchairs and other robotic devices (Millàn et al., 2002, 2004; Moore, 2003; Rebsamen et al., 2007; Vanacker et al., 2007; Bell et al., 2008) to facilitate rehabilitation of neurological conditions (e.g., stroke (Birbaumer et al., 2006; Pfurtscheller and Neuper, 2006; Birbaumer and Cohen, 2007; Dobkin, 2007; Tecchio et al., 2007; Scherer et al., 2007); and to enhance entertainment (e.g. performing arts or computer games (Pineda et al., 2003; Lalor et al., 2005; Leeb et al., 2007; BrainPlay, 2007; Krepki et al., 2007a,b; Scherer et al., 2007, 2008)). Since an additional hands-free communication channel is useful in many other areas, the field has been steadily growing. Recent developments include the manipulation of topographic maps (Trejo et al., 2006; Scherer et al., 2007), person authentication (Marcel and Millán, 2007), and enhancement of astronaut capabilities (Rossini et al., 2009).

In this chapter, we introduce the reader to basic concepts in BCI research. We review some of the most common statistical pattern recognition and machine learning methods used in EEG/ECoG-based BCIs. We illustrate the application of these methods using two BCI systems, one for controlling a humanoid helper robot and another for navigating in a virtual world. Given the limited space available, we have been forced to omit some important alternative approaches; we hope that the extensive reference list compensates for some of these omissions. The chapter emphasizes approaches that have been evaluated by our group and demonstrated in practical implementations, along with some new trends and ideas. We hope that the overview it provides helps stimulate further research in this exciting new discipline.

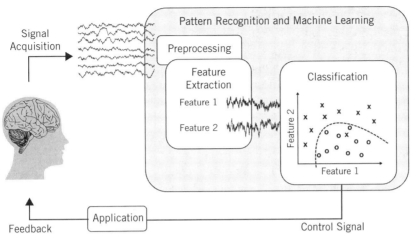

FIGURE 10.1

Basic components of a brain–computer interface.

10.2 SIGNAL PROCESSING AND PATTERN RECOGNITION IN BCI SYSTEMS

Figure 10.1 illustrates the typical components found in a brain–computer interface. The aim is to translate brain activity into messages (control commands) for devices. This translation typically involves a series of processing stages—signal acquisition, preprocessing, feature extraction, and classification (or regression)—to generate a control signal, followed by feedback to the user from the application being controlled. We review each of these stages in the following sections.

10.2.1 Signal Acquisition and Major Signal Types

The noninvasive electroencephalogram (EEG) (Birbaumer et al., 1999; Millàn and Mourino, 2003; Pfurtscheller et al., 2003, 2006; Pineda et al., 2003; Blankertz et al., 2006; Piccione et al., 2006; Vaughan et al., 2006) and the invasive electrocorticogram (ECoG) (Crone et al., 1998; Graimann et al., 2004; Leuthardt et al., 2004; Engel et al., 2005; Miller et al., 2007; Schalk et al., 2007) have emerged as important sources of brain signals for BCI in humans. EEG and ECoG signals are potential differences recorded from electrodes placed on the scalp or on the surface of the cortex, respectively. The measured signals are thought to reflect the temporal and spatial summation of post-synaptic potentials from large groups of cortical neurons located beneath the electrode (Purpura, 1959). EEG and ECoG have similar time resolution (in the millisecond range). EEG has a spatial resolution in the range of cm^2 with amplitudes in the range of a few microvolts ($\pm 100 \mu V$). ECoG, in contrast, has a spatial resolution in the mm^2 range with amplitudes up to 100 times greater than EEG amplitudes (Walter, 1936; Cooper et al, 1965). The poor spatial resolution of EEG is caused primarily by the different layers of tissue (meninges, cerebrospinal fluid, skull, scalp) interposed between the source of the signal (neural activity in the cortex) and the sensor placed on the scalp. These layers act as a volume conductor and effectively low-pass-filter and smear the original signal (Adrian and Matthews, 1934). Because of their small amplitude, EEG signals are very sensitive to noise and artifacts. The most frequent artifacts are muscle activities (electromyogram (EMG)) and eye movements (electro-oculogram (EOG)) generated by the user and signal contamination due to nearby electrical devices (e.g., 60-Hz power line interference). Additional noise sources include changing electrode impedance and the user's varying psychological states due to boredom, distraction, stress, or frustration (e.g., caused by BCI mistranslation). The muscle and movement artifacts commonly seen in EEG are minimized in ECoG because signals are measured directly from the brain surface. However, being invasive, ECoG has an increased medical risk. Another drawback is its availability: ECoG is typically recorded in patients who are being monitored prior to epilepsy surgery. Electrodes usually remain implanted for less than a week, and BCI-related experiments are only performed if the patient is willing and able.

Two major types of EEG/ECoG signals that have been used in BCIs are event-related potentials (ERPs) and event-related desynchronization/synchronization (ERD/ERS). ERPs are electrical potential shifts which are time-locked to perceptual, cognitive, and motor events; thus they represent the temporal signature of macroscopic brain electrical activity. Typical ERPs include the P300, so named because it is characterized by a positive potential shift occurring about 300 ms after stimulus presentation (Donchin et al., 2000; Allison and Pineda, 2003; Bayliss, 2003; Serby et al., 2005; Sellers and Donchin, 2006; Piccione et al., 2006; Bell et al., 2008), and the steady-state visually evoked potential (SSVEP) (Middendorf et al., 2000; Cheng et al., 2002; Lalor et al., 2005; Müller-Putz et al., 2005a; Scherer et al., 2007; Müller-Putz and Pfurtscheller, 2008) Steady-state refers to the fact that by presenting external stimuli in a rapid sequential order, one can evoke overlapping brain responses, thereby reflecting the actual stimulation frequency. ERD and ERS (Pfurtscheller and Aranibar, 1977; Pfurtscheller and Lopes da Silva, 1999a,b; Pfurtscheller and Neuper, 2001) are time-locked but have phase jitter—that is, a varying onset time. They can be initiated, for example, through motor imagery and involve a decrease (ERD) or increase (ERS) in power in particular frequency bands. Further details regarding the use of ERPs and ERD/ERS in BCIs are given in Section 10.3.

10.2.2 Pattern Recognition and Machine Learning

Two major problems confronting BCI developers are the nonstationarity and inherent variability of EEG and ECoG brain signals. Data from the same experimental paradigm but recorded on different days (or even in different sessions on the same day) are likely to exhibit significant differences. Additionally, the noisy superposition of the electrical activity of large populations of neurons as measured on the scalp (EEG) or on the brain surface (ECoG) can mask the underlying neural patterns and hamper their detection. Besides the variability in neuronal activity, the user's current mental state may affect the measured signals as well: Stress, excessive workload, boredom, or frustration may cause temporary distortions of EEG/ECoG activity. User adaptation to the BCI, as well as the changing impedance of EEG sensors during recording, contribute to making the recorded signal statistically nonstationary.

Pattern recognition and machine learning algorithms play a crucial role in recognizing ERP/ERD/ERS patterns in noisy EEG/ECoG signals and translating them into control signals. Pattern recognition methods are sometimes broadly categorized as performing classification or regression. Classification involves assigning one of N labels to new brain signals, given a labeled training set consisting of previously seen signals and their corresponding labels. Regression typically involves mapping brain signals directly to a continuous output signal, such as position or velocity of a prosthetic device.

The standard training of a classification or regression algorithm consists of collecting data from a user prior to BCI use and analyzing it offline to find user-specific parameters. A major drawback to this procedure is that the brain signal patterns of a user typically change over time as a result of feedback during BCI use, making parameters learned offline suboptimal. This problem is an active research area and is currently being addressed through the development of online adaptation methods (Shenoy et al., 2006; Vidaurre et al., 2007) as well as research on features that remain stable over users and time (Shenoy et al., 2008).

In the next few sections, we review the three main components of signal processing in EEG/ECoG BCIs: preprocessing, feature extraction, and classification. We conclude by summarizing research on feature subset selection and visualization.

10.2.2.1 *Preprocessing*

Preprocessing methods are typically applied to EEG/ECoG signals to detect, reduce, or remove noise and artifacts. The goal is to increase the signal-to-noise ratio (SNR) and isolate the signals of interest. One very common but also very subjective approach is to visually screen and score the recordings. In this setting, experts manually mark artifacts and exclude this data from further analysis. While this may work well for offline analysis, visual inspection is not practical during online BCI use.

A number of online preprocessing methods have been utilized in BCIs to date. These include notch filtering (e.g., to filter out 60-Hz power line interference), regression analysis for the reduction of eye movement (EOG) artifacts, and spatial filtering (e.g., bipolar, Laplacian, or common average rereferencing). Spatial filtering aims at enhancing local EEG activity over single channels while reducing common activity visible in several channels (McFarland et al., 1997; Wolpaw et al., 2002). The more complex method of common spatial patterns (CSP) transforms the EEG signal in such a way that the resulting variance is maximally discriminative between different classes.

We discuss some commonly used spatial filtering methods below. A comprehensive review of methods for detecting and removing EMG/EOG artifacts can be found in Fatourechi et al. (2007).

Spatial Filtering: Bipolar, Laplacian, and Common Average Rereferencing

A simple way to perform spatial filtering is to rereference EEG recordings as follows. Let s_i denote the EEG or ECoG signal from channel i. One can then extract bipolar signals $\tilde{s}_{i,j} = s_i - s_j$ to enhance the electrical potential differences between the two electrodes of interest (i and j).

A second spatial filter method, orthogonal source Laplacian filtering (Hjorth, 1975), highlights local activity at electrode i by subtracting the average activity present in the four orthogonal nearest-neighbor electrodes Θ:

$$\tilde{s}_i = s_i - \frac{1}{4} \sum_{i \in \Theta} s_i$$

This causes common activity such as EMG activity to be subtracted away from the electrode of interest. A closely related type of spatial filtering, common average rereferencing, enhances local activity at electrode i by subtracting the average over all electrodes:

$$\tilde{s}_i = s_i - \frac{1}{N} \sum_{i=1}^{N} s_i$$

Spatial Filtering: Common Spatial Patterns

The method of common spatial patterns (CSP) designs spatial filters in such a way that the variances in the filtered time series data are optimal (in the least squares sense) for discrimination (Koles, 1991; Müller-Gerking et al., 1999; Guger et al., 2000; Ramoser et al., 2000).

We are given input data $\{\mathbf{S}_c^i\}_{i=1}^{K}$ denoting EEG/ECoG data from trial i for class $c \in \{1, 2\}$ (e.g., hand versus foot motor imagery). Each $\mathbf{S}_c^i$ is an $N \times T$ matrix, where N is the number of EEG channels and T the number of samples in time per channel. We assume that the $\mathbf{S}_c^i$ are centered and scaled.

The goal of CSP is to find M spatial filters, given by an $N \times M$ matrix $\mathbf{W}$ (each column is a spatial filter), that linearly transform the input signals according to

$$\mathbf{s}_{CSP}(t) = \mathbf{W}^T \mathbf{s}(t)$$

where $\mathbf{s}(t)$ is the vector of input signals at time t from all the channels. In order to find the filters, the class-conditional covariances are first estimated as

$$\mathbf{R}_c = \frac{1}{K} \sum_i \mathbf{S}_c^i \left(\mathbf{S}_c^i\right)^T$$

for $c \in \{1, 2\}$. The CSP technique involves determining a matrix $\mathbf{W}$ such that

$$\mathbf{W}^T \mathbf{R}_1 \mathbf{W} = \Lambda_1$$
$$\mathbf{W}^T \mathbf{R}_2 \mathbf{W} = \Lambda_2$$

where the Λ_i are diagonal matrices and $\Lambda_1 + \Lambda_2 = I$ (I is the identity matrix). This can be done by solving a generalized eigenvalue problem given by

$$\mathbf{R}_1 w = \lambda \mathbf{R}_2 w$$

The generalized eigenvectors w_j that satisfy the above equation form the columns of $\mathbf{W}$ and represent the CSP spatial filters. The generalized eigenvalues $\lambda_1^j = w_j^T \mathbf{R}_1 w_j$ and $\lambda_2^j = w_j^T \mathbf{R}_2 w_j$ form the diagonal elements of Λ_1 and Λ_2, respectively. Since $\lambda_1^j + \lambda_2^j = 1$, a high value for λ_1^j means that the filter output based on filter w_j yields a high variance for input signals in class 1 and a low variance for signals in class 2 (and vice versa); spatial filtering with such filters can thus significantly enhance discrimination ability. Typically, a small number of eigenvectors (e.g., six) are used as CSP filters in BCI applications. A more detailed overview of the CSP method can be found in Blankertz et al. (2008), and various enhancements to boost robustness and applicability are given in Lemm et al. (2005), Dornhege et al. (2006), and Grosse-Wentrup and Buss (2008). Although CSP has not been extensively studied in the context of ECoG, there have been some reports of poor generalization performance for CSP when applied to ECoG signals (Hill et al., 2006).

10.2.2.2 *Feature Extraction*

The goal of feature extraction is to transform the acquired and preprocessed EEG/ECoG signal into a set of p features $\mathbf{x} = [x_1,\ldots,x_p]^T \in X^p$ that are suitable for subsequent classification or regression. In some cases, the preprocessed signal may already be in the form of an appropriate set of features (e.g., the outputs of CSP filters). More typically, some form of time or frequency domain transform is used.

The most commonly used features for EEG/ECoG are frequency domain–based. Overt and imagined movements typically activate premotor and primary sensorimotor areas, resulting in amplitude/power changes in the mu (7–13 Hz), central beta (13–30 Hz) and gamma (>30 Hz) rhythms in EEG/ECoG. These changes can be characterized using classical power spectral density estimation methods such as the digital bandpass filter, the short-term Fourier transform, and wavelets. The resulting feature vectors are usually high dimensional because the features are extracted from several EEG/ECoG channels and from several time segments prior to, during, and after the movement. The computation of a large number of spectral components can be computationally demanding. Another common and often faster way to estimate the power spectrum is to use (adaptive) auto-regressive ((A)AR) (Schlögl, 2000) models, although the model parameters have to be selected a priori.

Features that are extracted in the time domain include (A)AR parameters (Schlögl, 2000), (smoothed) signal amplitude Hjorth parameters (Hjorth, 1975), the fractal dimension (FD), and common spatial patterns (CSP; discussed previously). Hjorth parameters are a set of parameters that characterize wideband-filtered signal power, mean frequency, and change in frequency (Hjorth, 1975). The FD quantifies the complexity and self-similarity of the input signal. The temporal average from movement trials has also been used as a template to detect movement-related potentials through correlation/template matching (Levine et al., 2000).

Finally, features estimated in the phase domain have not received much attention in EEG/ECoG-based BCIs, although coherence and phase-locking values (Brunner et al., 2006) have been explored as potential features.

10.2.2.3 *Classification*

Classification is the problem of assigning one of N labels to a new input signal, given labeled training data of inputs and their corresponding output labels. Regression is the problem of mapping input signals to a continuous output signal. Given the limited spatial resolution of EEG/ECoG, most BCIs based on EEG/ECoG rely on classification to generate discrete control outputs (e.g., move a cursor up or down by a small amount by detecting left- versus right-hand motor imagery). BCIs based on neuronal recordings, on the other hand, have utilized regression to generate continuous output signals, such as position or velocity signals for a prosthetic device. Given our emphasis on EEG/ECoG BCIs in this chapter, we focus primarily on classification methods used in BCIs.

A useful categorization of classifiers into various types (not necessarily mutually exclusive) was suggested in Lotte et al. (2007):

- *Generative* classifiers learn a statistical model for each class of inputs. For classification of an input, the likelihood of the input within each class is computed and the most likely class is selected (e.g., hidden Markov models (HMMs), Bayes classifiers).

- *Discriminative* classifiers attempt to learn a boundary that best discriminates between input classes (e.g., linear discriminant analysis (LDA)).

- *Linear* classifiers use a linear function to approximate the boundary between classes. This is the most commonly used type of classifier in BCI applications (e.g., LDA, linear support vector machine (SVM)).

- *Nonlinear* classifiers attempt to learn a nonlinear decision boundary between classes (e.g., k-nearest neighbors (k-NN), learning vector quantization (LVQ), kernel SVMs, HMMs).

- *Dynamic* classifiers capture the temporal dynamics of the input classes and use this information for classifying a new input time series of data (e.g., HMMs, recurrent neural networks).

- *Stable* classifiers are those that remain robust to small variations in the training data (e.g., LDA, SVM).

- *Unstable* classifiers are those whose performance can be significantly affected by small variations in the training data (e.g., neural networks).

- *Regularized* classifiers incorporate methods to prevent overfitting to the training data, allowing better generalization and greater robustness to outliers (e.g., *regularized linear discriminant analysis* (RDA)).

We now briefly review four of the most common classification methods used in BCI applications: LDA, RDA, quadratic discriminant analysis, and

SVM. We begin by noting that the task of a classifier is to assign class labels $y \in Y$ to a p-dimensional feature vector $\mathbf{x}$. The most simple case is $Y = [-1, +1]$—that is, to discriminate between two classes (binary classification). Note, however, that these methods can be applied to multiclass classification as well (see the subsection Multiclass Classification in Section 10.2.2.3).

Linear Discriminant Analysis

Linear discriminant analysis (LDA; sometimes also called Fisher's linear discriminant) is a linear classifier that projects a p-dimensional feature vector onto a hyperplane that divides the space into two half-spaces (Duda et al., 2000). Each half-space represents a class ($+1$ or -1). The decision boundary

$$\left[w_1, \ldots, w_p\right]^T \left[x_1, \ldots, x_p\right] + w_0 = \mathbf{w}^T \mathbf{x} + w_0 = 0$$

is characterized by the hyperplane's normal vector $\mathbf{w}$ and the threshold w_0.

Given a new input vector $\mathbf{x} \in X^p$, classification is achieved by computing

$$y = sign(\mathbf{w}^T \cdot \mathbf{x} + w_0)$$

and assigning the resulting class label $y = -1$ or $y = +1$ to the input $\mathbf{x}$. During online BCI experiments, the (signed) distance to the hyperplane, given by $d(\mathbf{x}) = \mathbf{w}^T \mathbf{x} + w_0$ (assuming $\|\mathbf{w}\| = 1$), is sometimes also used to provide feedback to the user.

To compute $\mathbf{w}$, LDA assumes that the class-conditional distributions $P(\mathbf{x}|c = 1)$ and $P(\mathbf{x}|c = 2)$ are normal distributions with mean μ_c and covariance Σ_c for $c \in \{1, 2\}$. It can be shown that the optimal classification strategy is to assign inputs to the first class if the log likelihood ratio $\log[P(\mathbf{x}|c = 1)/P(\mathbf{x}|c = 2)]$ is above a threshold (and to assign them to the second class if below). Since the two distributions are Gaussian, this reduces to the comparison

$$(\mathbf{x} - \mu_1)^T \Sigma_1^{-1} (\mathbf{x} - \mu_1) - (\mathbf{x} - \mu_2)^T \Sigma_2^{-1} (\mathbf{x} - \mu_2) > T \quad (10.1)$$

where T is a threshold. If we now make the assumption that the class covariances are equal (i.e., $\Sigma_1 = \Sigma_2 = \Sigma$) and have full rank, we obtain

$$\mathbf{w}^T \mathbf{x} > c \text{ where } \mathbf{w} = \Sigma^{-1} (\mu_1 - \mu_2) \quad (10.2)$$

The threshold c is often defined to be in the middle of the projection of the two class means:

$$c = \mathbf{w}^T \left(\boldsymbol{\mu}_1 + \boldsymbol{\mu}_2 \right) / 2$$

It can be shown that the above choice for $\mathbf{w}$ defines a decision boundary that maximizes the distance between the means of the projected data $\tilde{\mathbf{y}} = \mathbf{w}^T \mathbf{x}$ from each class while minimizing its within-class variance. Further details can be found in Duda et al. (2000).

LDA has been a popular classifier in BCI research because it is simple to implement and can be computed fast enough for online use. In general, it has been found to produce good results, although, because of the strong assumptions made in its derivation, factors such as non-Gaussian data distributions, outliers, and noise can adversely affect LDA performance (Mülller et al., 2003).

Regularized Linear Discriminant Analysis

Regularization techniques are typically used to promote generalization and avoid overfitting, especially when the number of parameters to be estimated is large and the number of available observations is small. For example, in the case of LDA we might have insufficient data to accurately estimate the class mean $\boldsymbol{\mu}_c$ and the class covariance $\boldsymbol{\Sigma}_c$. In particular, $\boldsymbol{\Sigma}_c$ can become singular. Regularized linear discriminant analysis (RLDA) (Friedman, 1989) is a simple variant of LDA where the common covariance $\boldsymbol{\Sigma}$ (see earlier) is replaced by its regularized form:

$$\boldsymbol{\Sigma}_\lambda = (1 - \lambda)\boldsymbol{\Sigma} + \lambda\mathbf{I}$$

where $\lambda \in (0,1)$ denotes the regularization parameter and $\mathbf{I}$ is the identity matrix. By adding small constant values to the diagonal elements of $\boldsymbol{\Sigma}$, one can ensure nonsingularity and the existence of $\boldsymbol{\Sigma}_\lambda^{-1}$, which is needed to compute $\mathbf{w}$ as in Eq. (10.2). The regularization parameter λ can be chosen via model selection techniques to allow better generalization.

Although more robust in general, RDA has not been extensively used in BCI applications. Comparisons suggest that the classification results using RDA are similar to those achieved by LDA (Vidaurre et al., 2007).

Quadratic Discriminant Analysis

Quadratic discriminant analysis (QDA) begins with the same assumptions as in LDA—that is, the class-conditional distributions $P(\mathbf{x}|c = 1)$ and $P(\mathbf{x}|c = 2)$ are normal with mean $\boldsymbol{\mu}_c$ and covariance $\boldsymbol{\Sigma}_c$ for $c \in \{1,2\}$, assuming multivariate normal distribution $N(\mu_c, \Sigma_c)$. It differs from LDA in allowing different covariance matrices (Σ_1 and Σ_2) for the two classes. This results in a quadratic decision boundary (see Eq. (10.1)) based on (the square of) the Mahalanobis distance (Mahalanobis, 1936) between the new observation $\mathbf{x}$ and the class mean μ_c:

$$m_c(\mathbf{x}) = (\mathbf{x} - \mu_c)^T \Sigma_c^{-1}(\mathbf{x} - \mu_c)$$

Classification is performed as in Eq. (10.1) by comparing the difference between the two distances with a predetermined threshold T:

$$y = sign(m_1(\mathbf{x}) - m_2(\mathbf{x}) - T)$$

Support Vector Machine

Recall that LDA selects a hyperplane $\mathbf{w}^T\mathbf{x} + w_0 = 0$ based on the assumption of Gaussian class-conditional distributions with equal covariance. This hyperplane is only one among a potentially infinite number of hyperplanes that could separate the two input classes. It can be shown (Vapnik, 1995) that among such hyperplanes, the best generalization is achieved by selecting the one with the largest separation ("margin") between the two separable classes.

The support vector machine (SVM) (Vapnik, 1995) is a classifier that finds the separating hyperplane for which the margin between the samples of the two classes is maximized. Since the width of the margin is inversely proportional to $\|\mathbf{w}\|_2^2$ (Duda et al., 2000),[1] the search for the optimal $\mathbf{w}$ can be framed as a quadratic optimization problem, subject to the constraint that each training data point is correctly classified. However, because of the nature of EEG and ECoG data, one cannot assume that the data will be linearly separable. In this case, one might try to separate the training data with a minimal number of errors. To allow for misclassifications and possible outliers, the *soft margin* variant of the SVM (Cortes and Vapnik, 1995) uses slack variables ξ_i to measure the degree of misclassification of an input i. The resulting optimization problem for the linear soft margin SVM is given by Cortes and Vapnik (1995):

$$\min_{\mathbf{w}, \xi, w_0} \left\{ \frac{1}{2} \|\mathbf{w}\|_2^2 + \frac{C}{K} \|\xi\|_1 \right\}$$

---

[1] We use $\|\cdot\|_2$ to represent the Euclidean (or L2) norm and $\|\cdot\|_1$ to represent the L1 norm (e.g., $\|\mathbf{w}\|_1 = \sum |\mathbf{w}|$).

subject to

$$y_i\left(\mathbf{w}^T\mathbf{x}_i + w_0\right) \geq 1 - \xi_i$$

with

$$\xi_i \geq 0 \text{ for } i = 1,\dots,K$$

Here, $\mathbf{x}_i$ denotes input feature vector i, K is the number of inputs, and $y_i \in \{-1,+1\}$ is the class membership.

Linear SVMs have been successfully applied in a large number of BCI applications (e.g., Blankertz et al. (2008) and Shenoy et al. (2008)). In cases where linear SVMs are not sufficient, it is possible to utilize the "kernel trick" (Boser et al., 1992) to achieve a nonlinear mapping of the data to a sufficiently high-dimensional space where the two classes are linearly separable. The most commonly used kernel in BCI applications is the Gaussian or radial basis function kernel (Lotte et al., 2007). Further information regarding nonlinear SVMs can be found in Burges (1998).

Multiclass Classification

The classifiers discussed thus far were designed for assigning data to one of two classes. In BCI applications, the number of desired output signals is frequently greater than two, requiring methods for multiclass classification. There are several strategies for applying binary classifiers to the multiclass problem. One is to apply majority voting to pairwise classification. Given N_Y classes, a total of $N_Y(N_Y - 1)/2$ binary classifiers are trained, one for each binary combination of classes. For classification, a given input is fed to each of these classifiers and the class with the most votes (i.e., the one detected most often) is selected as the output.

An alternate strategy for multiclass classification using binary classifiers is the one-versus-the-rest approach: For each class, an individual classifier is trained to separate the data belonging to it from the data in the other classes. Classification is achieved by running each of these N_Y classifiers on the given input and choosing the class with the highest output value (the distance of the input to the hyperplane in the case of LDA).

Learning Vector Quantization and Distinction-Sensitive Learning Vector Quantization

Multiclass classification can also be accomplished directly through methods that are true multiclass classifiers. One such method that has successfully been applied to EEG and ECoG BCIs is learning vector quantization (LVQ) and its variant, distinction sensitive learning vector quantization (DSLVQ) (Pregenzer et al., 1994; Pregenzer, 1997). LVQ and DSLVQ differ from the methods discussed above in that they combine feature selection and classification within a single framework.

In learning vector quantization (LVQ), classification is based on a small set of labeled feature vectors $\{\mathbf{m}_i, Y_i\}_{i=1}^N$ (also known as codebook vectors) with labels $Y_i \in [1, \ldots, N_Y]$. Classification of a new sample is achieved by assigning to it the label Y_k of its closest codebook vector $\mathbf{m}_k$. How close a codebook vector is to an input sample is determined using, for example, the Euclidean distance between vectors. The codebook (or feature) vectors $\mathbf{m}_i$ and their labels are initialized randomly. Learning proceeds by adapting the codebook vectors to the training data. The closest codebook vector is selected for each training sample. If it correctly classifies the sample, the vector is moved closer to it; otherwise, it is moved away to make it less similar to the sample.

Note that in LVQ each codebook or feature vector contributes equally. A more common scenario in BCI is one in which we are given a fixed set of features $\mathbf{f}_i$ (e.g., power spectral features) but want to weight them differently in terms of their discriminative ability. A variant of the LVQ algorithm, DSLVQ (Pregenzer, 1997), can be used in this case. DSLVQ employs a weighted distance function:

$$\mathbf{d}_{\mathbf{w},\mathbf{x},\mathbf{m}} = \sqrt{\sum_{n=1}^{N}(w_n \cdot (x_n - m_n))^2}$$

to differentially weight features in classification. The weights vectors $\mathbf{w}$ are adapted in a manner similar to that for adapting codebook vectors in LVQ (see Pregenzer (1997) for details). An application of DSLVQ to a BCI system is illustrated in Section 10.3.

Evaluation of Classification Performance
In BCI applications, as in other applications in which classifiers are used, it is important to evaluate the accuracy and generalization performance of the (feature/classifier) combination selected for use. We briefly review some of these evaluation techniques and refer the reader to Schlögl et al. (2007), Bianchi et al. (2007), and Mason et al. (2006) for more comprehensive reviews of BCI performance measures.

A useful measure of classifier performance is the $N_Y \times N_Y$ "confusion" matrix $\mathbf{M}$, where N_Y denotes the number of classes, with the rows representing the true class labels and the columns representing the classifier's output. Binary classification ($N_Y = 2$) is considered in Table 10.1. The resulting four

Table 10.1 Confusion Matrix for 2-Class Problems

"True" class	Classification	
	Positive	Negative
Positive	TP	FN
Negative	FP	TN

entries provide information on the number of *true positives* (TP), or correct positive classifications; the number of *false negatives* (FN), or missed positive detections; the number of *false positives* (FP), or incorrect positive detections; and the number of *true negatives* (TN), or correct rejections. The diagonal elements $m_{i,i}$ of the matrix represent the number of correctly classified samples. The off-diagonal elements $m_{i,j}$ show how many samples of class i have been misclassified as class j.

Classification accuracy, ACC, is defined as the ratio between the number of correctly classified samples and the total number of samples. It can be derived from the confusion matrix $\mathbf{M}$ as follows:

$$ACC = \frac{TP + TN}{TP + FN + FP + TN} = \frac{\sum_{i=1}^{N} m_{i,i}}{\sum_{i=1}^{N}\sum_{j=1}^{N} m_{i,j}}$$

We can then define the error rate as $err = 1 - ACC$. When the number of samples for each class is the same, the chance level is $ACC_0 = 1/N_Y$.

Another useful performance measure is the kappa coefficient (Cohen's κ):

$$\kappa = \frac{ACC - ACC_0}{1 - ACC_0}$$

By definition, the kappa coefficient (which lies between -1 and $+1$) is independent of the number of samples per class and the number of classes, so $\kappa = 0$ means chance-level performance and $\kappa = 1$, perfect classification. $\kappa = -1$ means that all samples are assigned to the wrong class.

Perhaps the most commonly reported performance measure for BCIs is the information transfer rate (ITR):

$$B = \log_2(N_Y) + p_0 \cdot \log_2(p_0) + (1 - p_0)\log_2(1 - p_0)/(N_Y - 1)$$

measured in bits/trial (dividing B by the trial duration in minutes gives the rate in bits/min) (Wolpaw et al., 2002). Here, N_Y denotes the number of classes and $p_0 = ACC$. Since a BCI can be regarded as a communication channel, quantifying its ITR makes sense in an information-theoretic context. The equation for B is derived from the classification error rate, but the assumptions made are not always fulfilled; thus B only provides an upper limit on the performance that can be achieved (Schlögl et al., 2007).

A final but important issue that we briefly discuss here is the error rate, err. To obtain a true estimate of err, classifiers are typically tested on "test data," which are different from the data used to train the classifier. One approach is to simply partition a given input data set into two subsets, one for training and one for testing, but this strategy is sensitive to how the data is split. A more sophisticated method is K-fold cross-validation, in which the data set is split into K subsets of approximately equal size: $K - 1$ subsets are used to train the classifier and the remaining subset is used for testing. The classifier is trained and tested K times, resulting in K different error rates, err_k. The final "true" error rate is computed by averaging the individual err_k:

$$err = \frac{1}{K} \sum_{k=1}^{K} err_k$$

Variations of this procedure exist. For example, leave-one-out cross-validation is an extreme form of K-fold cross-validation where K is set equal to the number of training samples. Also, to minimize the effects of specific partitions of the data, K-fold cross-validation can be repeated N times, yielding $N \cdot K$ individual error rates, err_i, with the final one being the average over these $N \cdot K$ values.

In many applications, it is common to split the training data set into three subsets: a training subset to find the parameters of the classifier, a validation subset to tune these parameters, and a test subset to test the optimized classifier. Although these procedures are computationally costly, they play an important role in improving the classifier's ability to generalize.

10.3 APPLICATIONS

To illustrate the applications of the methods reviewed in this chapter, we discuss two EEG-based BCI systems developed by the authors and their collaborators. The first system is based on the detection of an event-related potential (ERP) known as the P300. The second is based on detecting event-related desynchronization/synchronization (ERD/ERS) caused by different types of motor imagery. These two systems exemplify the two major paradigms in EEG/ECoG BCI research:

Cue-guided BCIs. The BCI detects the change in brain activity caused by a stimulus or cue generated by the BCI—for example, an unexpected presentation of a visual stimulus.

Self-paced BCIs. After a period of training, the user autonomously and voluntarily modulates brain activity each time a message needs to be sent to the BCI.

From a machine learning point of view, cue-guided BCIs are easier to train because the brain patterns generated are typically robust and stereotypical and, since the pattern is locked to stimulus onset, the time window of analysis is clearly defined. Self-paced BCIs, on the other hand, require analysis and interpretation of EEG/ECoG signals throughout the entire period of BCI operation. The on-demand mode of control offered by self-paced BCIs in general makes them more versatile than their cue-guided counterparts. These advantages and disadvantages of the two paradigms become more apparent in the context of actual applications such as the two BCI systems discussed next.

10.3.1 P300-Based Control of a Humanoid Robot

We discuss in this section an EEG-based cue-guided BCI system for controlling a humanoid helper robot (Bell et al., 2008). The system utilizes the P300 visually evoked potential to select high-level commands, which the robot executes autonomously. The robot has the ability to autonomously move and pick up/release objects. It also possesses some computer vision capabilities, such as being able to segment objects on a table and use vision to navigate to a destination. The BCI was designed to command the robot to navigate to a specific (known) location and transmit images of objects it sees at that location, to allow the user to select an object for pickup, and, finally, to command the robot to bring the object to the user or transport it to a different location (see Figure 10.2 for an example).

EEG signals were used to select the two main types of commands for the robot: which object to select among the ones whose images were transmitted by it, and which location to choose as the destination from among a set of known locations. The images of the possible choices (objects or destination locations) were scaled and arranged as a grid on the user's computer screen. Figure 10.2 illustrates the case of two objects, one medium gray and one light gray, and two locations (two dark gray tables, one with a white square in the

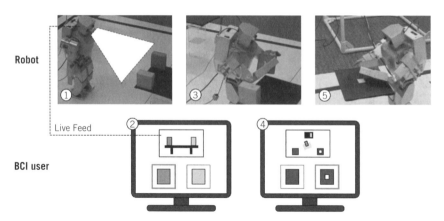

Robot

Live Feed

BCI user

FIGURE 10.2

BCI for controlling a humanoid helper robot. The top row shows images of the humanoid robot in action. The bottom row depicts the user's computer screen. The user receives a live feed from the robot's cameras (image 1), thereby immersing her in the robot's environment and allowing her to select actions based on objects seen in the robot's cameras (computer screen in image 2). Objects are found using computer vision techniques. The robot transmits the segmented images of the objects (light gray and medium gray colored boxes in these images) and queries the user about which one to pick up. The selection is made by the user using a P300 BCI as described in the text. After picking up the object selected by the user (image 3), the robot asks the user which location to bring the selected object to. Images of possible locations (dark gray tables on the left and right sides) from an overhead camera are presented to the user (image 4). Again, the selection of the destination is made by the user by means of the P300. Finally, the robot walks to the destination selected by the user and places the object on the table at that location (image 5). Please see this figure in color at the companion web site: www.elsevierdirect. com/companions/9780123750273

center). According to the oddball paradigm used to evoke the P300 response (Figure 10.3), the user focuses his or her attention on the image of choice while the border of a randomly selected image is flashed every 250 ms. When the flash occurs on the attended object, a P300 can be expected; this response is then detected by the BCI and employed to infer the user's choice. To focus their attention, users were asked to mentally count the number of flashes on their image of choice.

Thirty-two EEG channels were recorded from electrodes placed according to the standard 10-20 system of EEG electrode placement. A linear soft-margin classifier (Section 10.2.2.3) was trained to discriminate between the P300 response generated by a flash on a desired object and EEG responses due to the other flashes. The feature vectors used in classification were based on a set of spatial filters similar to CSP filters. As with LDA (see Section 10.2.2.3),

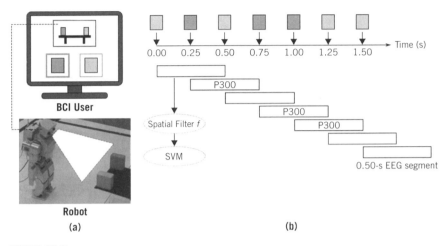

BCI User

Robot

(a)

Time (s)
0.00   0.25   0.50   0.75   1.00   1.25   1.50

P300

P300

P300

Spatial Filter *f*

SVM

0.50-s EEG segment

(b)

FIGURE 10.3

Using the P300 response to infer the user's choice. (a) When the robot finds objects of interest (in this experiment a medium gray and a light gray box), segmented images are sent to the user and arranged in a grid format in the lower part of the user screen. (b) The oddball paradigm is used to evoke the P300 response. The dark and medium gray boxes at the top show a random temporal order of flashed images. EEG segments of 0.5-s duration from flash onset are spatially filtered and classified by a soft-margin SVM into segments containing a P300 and those not containing a P300. After a fixed number of flashes, the object associated with the most P300 segments is selected as the user's choice. Please see this figure in color at the companion web site: www.elsevierdirect.com/companions/9780123750273

these spatial filters were chosen to maximize the distance between the means of the filtered data from each class while minimizing the within-class variance (see Bell et al. (2008) for details). The first three such filters were selected (Figure 10.4). They were applied to data segments of 500-ms duration from the onset of each flash.

To learn the filters and train the classifier for a given user, a 10-minute data collection protocol was used prior to BCI operation. Four different images, arranged in a 2×2 grid, were presented on the computer screen. To collect labeled data, the user was instructed to attend to a preselected image while the borders around the four images were flashed every 250 ms in random order, with ten flashes per image. This procedure was repeated four times for each of the four image positions.

After training on the labeled data, the BCI was used to infer the user's choice regarding an object or destination location. The choices (objects or locations) were presented in a grid format (e.g., a 2×2 grid for four object images) and the borders were flashed in random order. The EEG data for the 500-ms duration after each image flash was classified as a P300 response

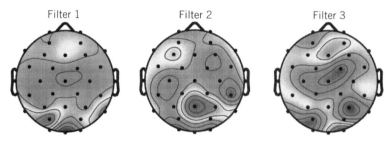

FIGURE 10.4

Spatial filters used in the P300 BCI. Shown are the first three spatial filters learned from the data across all subjects (medium gray denotes high positive values; dark gray denotes high negative values). The spatial filters are smoothly varying and are focused on the occipital and parietal regions, which are appropriate for a visuospatial attention task such as the one utilized by the P300 BCI discussed in the text. Please see this figure in color at the companion web site: www.elsevierdirect.com/companions/9780123750273

or a non-P300 response (see Figure 10.3, right panel). The image with the highest number of P300 classifications after all flashes was selected as the user's choice.

The results, based on nine able-bodied subjects, show that an accuracy of 95% can be achieved for discriminating between four classes (note that the theoretical chance classification level for four classes is 25%). With the implemented rate of four flashes per second, the selection of one out of four options takes 5 seconds, yielding an ITR of 24 bits per minute, which is comparable to rates achieved by other EEG-based cue-guided BCIs.

10.3.2 Motor Imagery–Based Control of Virtual Environments

We next describe a self-paced BCI based on sensorimotor rhythms that can discriminate between three motor imagery classes from ongoing EEG and use this output to navigate in virtual worlds (Scherer et al., 2008). In particular, the user operates the BCI by imagining left/right hand, foot, or tongue movements: It is well-known that such motor imagery causes ERD/ERS events (Pfurtscheller and Aranibar, 1977; Pfurtscheller and Neuper, 2006), which can in turn be detected by a classifier. The system was designed using a two-step process: First, a 3-class classifier was learned from cue-guided motor imagery data; the classifier was then enhanced to support self-paced navigation based on three commands: *rotate left*, *rotate right*, and *move forward*.

Features used to quantify the EEG activity were computed by band-pass filtering, squaring, and averaging the samples collected over the prior one second. To achieve 3-class classification, a committee of three binary LDA classifiers with majority voting was used (see the subsection Multiclass

Classification in Section 10.2.2.3). Each LDA discriminated between two of three motor imagery classes and the class chosen by the majority of LDAs was selected as the output class (e.g., in Figure 10.5, "Right Hand" imagery was chosen by both LDA 1.1 and LDA 1.3). To make the distribution of the feature values more Gaussian and thus better suited for linear methods, the logarithm of the feature value was used.

For the initial calibration of the 3-class BCI, 22-channel EEG signals (Figure 10.5) were recorded during cue-guided left-hand, right-hand, feet, and tongue motor imagery (MI). In each trial, the subject was instructed to start imagining, upon presentation of a cue, the movement corresponding to it. For each class, data from 144 trials, each of four seconds duration, were collected. For each subject, individual 4-class DSLVQ classifiers were learned, with band power features extracted at different time lags from cue presentation. For each time lag and each EEG channel, 15 nonoverlapping band power features between 6 and 36 Hz with a bandwidth of 2 Hz were computed. The most relevant BP features were selected by examining the weights for them computed by the DSLVQ algorithm (Section 10.2.2.3). From among the four possible MI types, the three MI tasks with the highest classification accuracies were selected for each subject using cross-validation (see the subsection Evaluation of Classification Performance in Section 10.2.2.3) and employed for online experiments.

The first BCI experiments conducted were cue-guided feedback experiments in which subjects learned to reliably generate the three MI patterns to control the BCI. The plot in the upper right panel of Figure 10.5 shows the average accuracy of 120 MI feedback trials (40 per class) as a function of time from cue onset. Figure 10.6 shows four log band power features during single trials of left-hand, right-hand, and tongue MI, respectively. Note that the features exhibit examples of both ERD (decrease in power) and ERS (increase in power); these characteristic changes in power are detected by the classifier. The final output from majority voting is shown at the bottom of Figure 10.6.

Results from the cue-guided BCI experiments were leveraged to build a self-paced BCI. The first problem faced was detecting when the subject was engaged in motor imagery in order to issue a command. After the online accuracy for each subject reached 80% in the cue-guided experiments, an additional LDA classifier was trained for each subject to discriminate between MI and non-MI patterns (LDA 2 in Figure 10.5). Thirty-one band power features (1-Hz overlap) of 6–36 Hz with a bandwidth of 2 Hz were extracted from each channel and analyzed via DSLVQ. The six most relevant were selected as input to the LDA. The LDA was trained using data from the last cue-guided feedback training session (all MI classes were pooled). The non-MI samples for training were obtained from a 2-min EEG segment recorded at the beginning

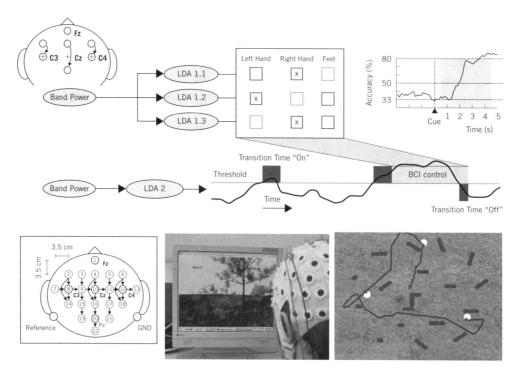

FIGURE 10.5

A self-paced BCI for navigation in a virtual environment. The 22-channel EEG montage used in the initial calibration experiments is shown at the lower left (top down view of the head). Three bipolar channels (top left) are selected from these channels and band power features are computed on a sample-by-sample basis from the ongoing EEG recorded from the channels. The MI class (left hand, right hand, or feet in this example) is classified by a committee of three LDAs using majority voting. Each LDA discriminates between two classes (symbolized by the black squares; e.g., LDA 1.1: left-hand versus right-hand imagery). The output of each is marked with an X. The class detected by the majority of the LDAs (right-hand imagery in this example) is the detected class. The curve in the upper right plot shows the average 3-class online accuracy during a cue-guided experiment. Self-paced operation is achieved by using a different LDA (middle, LDA 2) to detect the onset or end of motor imagery. A state transition from non-MI to MI (or vice versa) is triggered only after the threshold (medium gray line) is exceeded for predefined transition times (dark gray). The user employs the self-paced BCI to navigate through a virtual environment (middle photograph). An example trajectory is shown in the photograph on the right. The user's task is to navigate through the environment and collect a set of coins (white objects) distributed within the environment. Please see this figure in color at the companion web site: www.elsevierdirect.com/companions/9780123750273

of the session, during which subjects were asked to sit relaxed without moving. After training, a state switch between non-MI and MI (and vice versa) was triggered only when the LDA distance exceeded a subject-specific threshold for a subject-specific time. This threshold was chosen to be the value that

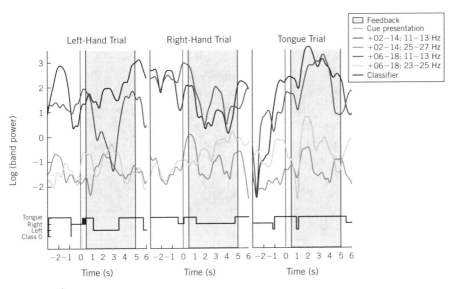

FIGURE 10.6

Classification of left-hand, right-hand, and tongue motor imagery by a committee of three LDA classifiers using majority voting. The upper plots show the logarithmic band power features during example trials involving cue-guided left-hand, right-hand, and tongue motor imagery. The lightest gray area indicates the period during which the user receives feedback. The output of the classifier is shown at the bottom for each example trial. In each case, about one second after cue onset, the classifier correctly classified the EEG samples as left-hand, right-hand, and tongue MI, respectively. Please see this figure in color at the companion web site: www.elsevierdirect.com/companions/9780123750273

would maximize the number of TP detections (see the subsection Evaluation of Classification Performance in Section 10.2.2.3) within the feedback period and, at the same time, minimize the number of FP detections.

The self-paced BCI was built by combining (1) the LDA classifier for detecting MI/non-MI and (2) the committee classifier for determining the imagery class. Both were applied to the recorded EEG signals on a sample-by-sample basis—that is, for each new EEG sample, a classification was performed (the sampling rate was 250 Hz). When the MI/non-MI LDA detected motor imagery, the output from the committee classifier was used as the BCI output; otherwise, the committee output was ignored.

To evaluate the self-paced BCI, three subjects participated in a study where the task was to use the BCI to navigate in a virtual environment (VE) and collect coins that were scattered within it (see Figure 10.5). Left-hand, right-hand, and foot (or tongue) MI were translated into the navigation commands *rotate left*, *rotate right*, and *move forward*, respectively. After about five hours of cue-guided feedback training, during which the classifier was trained to the subject's brain patterns, all three subjects successfully navigated through the VE and collected the coins (see Figure 10.5 for an example trajectory). Computation of classification accuracy (Section 10.2.2.3) for self-paced BCIs is difficult because the true intent of the subject at any point in time is unknown. However, in post-study interviews, all three subjects reported that the BCI correctly detected the MI and non-MI periods and allowed satisfactory control of navigation within the virtual environment.

10.4 DISCUSSION

This chapter introduced basic concepts pertaining to brain–computer interfaces (BCIs) based on EEG/ECoG. We reviewed some of the most common statistical signal processing and pattern recognition methods used in these interfaces. In particular, we provided an overview of the different signal processing stages typically used in EEG/ECoG BCIs: signal preprocessing, feature extraction, classification, and performance evaluation. The preprocessing step helps in increasing the signal-to-noise ratio and removes signals that do not originate in the brain, such as muscle-related artifacts. Feature extraction transforms the preprocessed signal into a low-dimensional set of features that characterize the underlying brain signal of interest. These features are subsequently classified using linear or nonlinear classifiers. We also briefly discussed basic performance measures for BCIs and methods used to estimate generalization performance.

The noisy nature of EEG and the fact that brain activity patterns are typically subject-specific means that signal processing and subject-specific optimization are essential. Statistical methods are well suited to tackling this problem. They can be used to derive both filters (e.g., CSP) and classifiers tailored to a given user. The nonstationarity and inherent variability of the EEG, along with limited sample size and limited knowledge about the underlying signal, makes this a challenging domain for statistical methods. Linear classifiers such as LDA or linear SVM have been popular given their simplicity and the availability of off-the-shelf software packages. The SVM is the most frequently used classifier in the field (Lotte et al., 2007).

The two example systems we presented reflect the two major trends in EEG-based BCIs:

- Using a large number of channels and sophisticated signal processing methods such as CSP and SVM to achieve good performance in a limited amount of time; the drawback here is usability, practicality, and the cost of using a large number of electrodes.

- Using only a few electrodes (e.g., three bipolar electrodes in Section 10.3.2) and very basic signal processing methods. The drawback here is a longer training time because users have to learn to more reliably generate the brain patterns to be detected.

Compared to EEG, the ECoG signal appears to be more robust, not only with respect to artifacts but also with respect to the induced brain activity patterns. We have evidence that in ECoG, band power features in the high-frequency band ("high gamma") produce the best results (compared to the low-frequency band features used in EEG) and remain stable over time (Shenoy et al., 2008; Blakely et al., 2009). Methods that achieve good results with EEG, such as common spatial patterns (CSP), may not necessarily work well with ECoG (Hill et al., 2006). These results suggest that consideration of the physiological properties of the recorded signals is important in choosing the appropriate signal processing and pattern recognition methods for BCI construction.

Much past research focused on cue-based BCIs, where mental states are more or less well defined. An important challenge for the future is the design and implementation of self-paced BCIs, where a number of distinct patterns must be reliably detected in ongoing brain activity. Although there have been several prototype systems (Millàn and Mourino, 2003; Bashashati et al., 2007; Scherer et al., 2007) based on statistical classifiers (one such system was described in Section 10.3.2), there is room for improvement. A particularly important open issue is developing useful performance measures for quantifying the performance of self-paced systems.

An important challenge for statistical pattern recognition and machine learning methods for BCIs is the co-adaptation of brain and machine. Promising preliminary results have been obtained using adaptive classifiers or "general" features for BCIs, but an overarching co-adaptation theory remains to be developed. Such a theory would entail finding statistical methods that can predict changes in brain activity and adapt in sync with the human user to achieve the common goal of brain-based control. Developing such methods remains an active area of BCI research.

Acknowledgments

We thank the two anonymous referees for their comments and suggestions. This work was supported by NSF grants 0622252 and 0930908, a Microsoft External Research grant, and the Packard Foundation.

References

Adrian, E.D., Matthews, B.H.C., 1934. The Berger rhythm: potential changes from the occipital lobes in man. Brain 57 (4), 355–385.

Agresti, A., Caffo, B., 2000. Simple and effective confidence intervals for proportions and differences of proportions result from adding two successes and two failures. Am. Stat. 54 (4), 280–288.

Allison, B.Z., Pineda, J.A., 2003. ERPs evoked by different matrix sizes: implications for a brain computer interface (BCI) system. IEEE Trans. Neural. Syst. Rehabil. Eng. 11 (2), 110–113.

Allison, B.Z., Wolpaw, E.W., Wolpaw, J.R., 2007. Brain-computer interface systems: progress and prospects. Expert Rev. Med. Devices 4 (4), 463–474.

Ball, T., Demandt, E., Mutschler, I., Neitzel, E., Mehring, C., Vogt, K., et al., 2008. Movement related activity in the high gamma range of the human EEG. Neuroimage 41 (2), 302–310.

Bashashati, A., Ward, R.K., Birch, G.E., 2007. Towards development of a 3-state self-paced brain-computer interface. Comput. Intell. Neurosci. vol. 2007, Article ID 84386, 8 pages, 2007. doi:10.1155/2007/84386.

Bayliss, J.D., 2003. Use of the evoked potential P3 component for control in a virtual apartment. IEEE Trans. Neural. Syst. Rehabil. Eng. 11 (2), 113–116.

Bell, C.J., Shenoy, P., Chalodhorn, R., Rao, R.P.N., 2008. Control of a humanoid robot by a noninvasive brain-computer interface in humans. J. Neural. Eng. 5 (2), 214–220.

Bensch, M., Karim, A.A., Mellinger, J., Hinterberger, T., Tangermann, M., Bogdan, M., et al., 2007. Nessi: An EEG-Controlled Web Browser for Severely Paralyzed Patients. Comput. Intell. Neurosci. vol 2007, Article ID 71863, 5 pages, doi:10.1155/2007/71863.

Bianchi, L., Quitadamo, L.R., Garreffa, G., Cardarilli, G.C., Marciani, M.G., 2007. Performances evaluation and optimization of brain computer interface systems in a copy spelling task. IEEE Trans. Neural. Syst. Rehabil. Eng. 15 (2), 207–216.

Birbaumer, N., 2006. Brain-computer-interface research: coming of age. Clin. Neurophysiol. 117 (3), 479–483.

Birbaumer, N., Cohen, L.G., 2007. Brain-computer interfaces: communication and restoration of movement in paralysis. J. Physiol. 579 Pt 3, 621–636.

Birbaumer, N., Ghanayim, N., Hinterberger, T., Iversen, I., Kotchoubey, B., Kübler, A., et al., 1999. A spelling device for the paralysed. Nature 398 (6725), 297–298.

Birbaumer, N., Weber, C., Neuper, C., Buch, E., Haapen, K., Cohen, L., 2006. Physiological regulation of thinking: brain-computer interface (BCI) research. Prog. Brain Res. 159, 369–391.

Blakely, T., Miller, K.J., Zanos, S.P., Rao, R.P.N., Ojemann, J.G., 2009. Robust, long-term control of an electrocorticographic brain-computer interface with fixed parameters. Neurosurg. Focus 27 (1), E13.

Blankertz, B., Dornhege, G., Krauledat, M., Müller, K.R., Kunzmann, V., Losch, F., et al., 2006. The Berlin brain-computer interface: EEG-based communication without subject training. IEEE Trans. Neural. Syst. Rehabil. Eng. 14 (2), 147–152.

Blankertz, B., Tomioka, R., Lemm, S., Kawanabe, M., Müller, K.R., 2008. Optimizing spatial filters for robust EEG single-trial analysis. IEEE Signal Process. Mag. 25 (1), 41–56.

Boser, B.E., Guyon, I.M., Vapnik, V.N., 1992. A training algorithm for optimal margin classifiers. In: D. Haussler (Ed.), COLT '92: Proceedings of the Fifth Annual Workshop on Computational Learning Theory, ACM, New York, 144–152.

BrainPlay'07: Playing with Your Brain, Brain-Computer Interfaces and Games, Workshop of the International Conference on Advances in Computer Entertainment Technology, ACE 2007.

Brunner, C., Scherer, R., Graimann, B., Supp, G., Pfurtscheller, G., 2006. Online control of a brain-computer interface using phase synchronization. IEEE Trans. Biomed. Eng. 53 (12 Pt 1), 2501–2506.

Burges, C.J.C., 1998. A tutorial on Support Vector Machines for Pattern Recognition. Data Min. Knowl. Discov. 2, 121–167.

Cheng, M., Gao, X., Gao, S., Xu, D., 2002. Design and implementation of a brain-computer interface with high transfer rates. IEEE Trans. Biomed. Eng. 49 (10), 1181–1186.

Chernick, M.R., 1999. Bootstrap Methods: A Practitioners's Guide. Wiley, New York.

Cooper, R., Winter, A.L., Crow, H.J., Walter, W.G., 1965. Comparison of subcortical, cortical and scalp activity using chronically indwelling electrodes in man. Electroencephalogr. Clin. Neurophysiol. 18, 217–228.

Cortes, C., Vapnik, V., 1995. Support-vector networks. Mach. Learn. 20 (3), 273–297.

Crone, N.E., Miglioretti, D.L., Gordon, B., Sieracki, J.M., Wilson, M.T., Uematsu, S., et al., 1998. Functional mapping of human sensorimotor cortex with electrocorticographic spectral analysis. I. Alpha and beta event-related desynchronization. Brain 121 Pt 12, 2271–2299.

Dobkin, B.H., 2007. Brain-computer interface technology as a tool to augment plasticity and outcomes for neurological rehabilitation. J. Physiol. 579 Pt 3, 637–642.

Donchin, E., Spencer, K.M., Wijesinghe, R., 2000. The mental prosthesis: assessing the speed of a P300-based brain-computer interface. IEEE Trans. Rehabil. Eng. 8 (2), 174–179.

Donoghue, J.P., 2002. Connecting cortex to machines: recent advances in brain interfaces. Nat. Neurosci. 5 (Suppl), 1085–1088.

Dornhege, G., Blankertz, B., Krauledat, M., Losch, F., Curio, G., Müller, K.R., 2006. Combined optimization of spatial and temporal filters for improving brain-computer interfacing. IEEE Trans. Biomed. Eng. 53 (11), 2274–2281.

Duda, R., Hart, P., Stork, D., 2000. Pattern Classification, second ed. Wiley Interscience, New York.

Engel, A.K., Moll, C.K.E., Fried, I., Ojemann, G.A., 2005. Invasive recordings from the human brain: clinical insights and beyond. Nat. Rev. Neurosci. 6 (1), 35–47.

Fatourechi, M., Bashashati, A., Ward, R.K., Birch, G.E., 2007. EMG and EOG artifacts in brain computer interface systems: a survey. Clin. Neurophysiol. 118 (3), 480–494.

Fawcett, T., 2006. An introduction to ROC analysis. Pattern Recognit. Lett. 27 (8), 861–874.

Flotzinger, D., Kalcher, J., Pfurtscheller, G., 1992. EEG classification by learning vector quantization. Biomed. Tech. (Berl) 37 (12), 303–309.

Friedman, J.H., 1989. Regularized discriminant analysis. J. Am. Stat. Assoc. 84 (405), 165–175.

Goldberg, D.E., 1989. Genetic Algorithms in Search, Optimization, and Machine Learning. Addison-Wesley Professional, Boston.

Goncharova, I.I., Barlow, J.S., 1990. Changes in EEG mean frequency and spectral purity during spontaneous alpha blocking. Electroencephalogr. Clin. Neurophysiol. 76 (3), 197–204.

Graimann, B., Huggins, J.E., Levine, S.P., Pfurtscheller, G., 2002. Visualization of significant ERD/ERS patterns in multichannel EEG and ECoG data. Clin. Neurophysiol. 113 (1), 43–47.

Graimann, B., Huggins, J.E., Levine, S.P., Pfurtscheller, G., 2004. Toward a direct brain interface based on human subdural recordings and wavelet-packet analysis. IEEE Trans. Biomed. Eng. 51 (6), 954–962.

Graimann, B., Pfurtscheller, G., 2006. Quantification and visualization of event-related changes in oscillatory brain activity in the time-frequency domain. Prog. Brain Res. 159, 79–97.

Grosse-Wentrup, M., Buss, M., 2008. Multiclass common spatial patterns and information theoretic feature extraction. IEEE Trans. Biomed. Eng. 55 (8), 1991–2000.

Guger, C., Ramoser, H., Pfurtscheller, G., 2000. Real-time EEG analysis with subject-specific spatial patterns for a brain-computer interface (BCI). IEEE Trans. Rehabil. Eng. 8 (4), 447–456.

Hill, N.J., Lal, T.N., Schröder, M., Hinterberger, T., Widman, G., Elger, C.E., et al., 2006. *Lecture Notes in Computer Science*, chapter Classifying Event-Related Desynchronization in EEG, ECoG and MEG Signals, pages 404–413. Springer Berlin / Heidelberg, 2006.

Hjorth, B., 1975. An on-line transformation of EEG scalp potentials into orthogonal source derivations. Electroencephalogr. Clin. Neurophysiol. 39 (5), 526–530.

Hochberg, L.R., Serruya, M.D., Friehs, G.M., Mukand, J.A., Saleh, M., Caplan, A.H., et al., 2006. Neuronal ensemble control of prosthetic devices by a human with tetraplegia. Nature 442 (7099), 164–171.

Holland, J.M., 1975. Adaptation in Natural and Artificial Systems. University of Michigan Press, Ann Arbor.

Jain, A.K., Duin, R.P.W., Mao, J., 2000. Statistical pattern recognition: a review. IEEE Trans. Pattern Anal. Mach. Intell. 22 (1), 4–37.

Jeannerod, M., 1994. The representing Brain: Neural correlates of motor imagery and intention. Behav. Brain Sci. 17 (2), 187–245.

Kalcher, J., Flotzinger, D., Neuper, C., Gölly, S., Pfurtscheller, G., 1996. Graz brain-computer interface II: towards communication between humans and computers based on online classification of three different EEG patterns. Med. Biol. Eng. Comput. 34 (5), 382–388.

Karim, A.A., Hinterberger, T., Richter, J., Mellinger, J., Neumann, N., Flor, H., et al., 2006. Neural internet: Web surfing with brain potentials for the completely paralyzed. Neurorehabil. Neural Repair 20 (4), 508–515.

Koles, Z.J., 1991. The quantitative extraction and topographic mapping of the abnormal components in the clinical EEG. Electroencephalogr. Clin. Neurophysiol. 79 (6), 440–447.

Krepki, R., Blankertz, B., Curio, G., Müller, K.R., 2007a. The Berlin Brain-Computer Interface (BBCI) – towards a new communication channel for online control in gaming applications. J. Multimedia Tools Applications 33 (1), 73–90.

Krepki, R., Curio, G., Blankertz, B., Müller, K.R., 2007b. Berlin brain-computer interface – the HCI communication channel for discovery. Int. J. Hum. Comp. Studies 65, 460–477. Special Issue on Ambient Intelligence.

Kübler, A., Kotchoubey, B., 2007. Brain-computer interfaces in the continuum of consciousness. Curr. Opin. Neurol. 20 (6), 643–649.

Lalor, E.C., Kelly, S.P., Finucane, C., Burke, R., Smith, R., Reilly, R.B., et al., 2005. Steady-state VEP-based brain-computer interface control in an immersive 3D gaming environment. EURASIP J. Appl. Signal Processing 19, 3156–3164.

Lebedev, M.A., Nicholelis, M.A.L., 2006. Brain-machine interfaces: past, present and future. Trends Neurosci. 29 (9), 536–546.

Leeb, R., Friedman, D., Müller-Putz, G.R., Scherer, R., Slater, M., Pfurtscheller, G., 2007. Self-paced (asynchronous) BCI control of a wheelchair in Virtual Environments: a case study with a tetraplegic. Comput. Intell. Neurosci. Article ID 79642.

Leeb, R., Lee, F., Keinrath, C., Scherer, R., Bischof, H., Pfurtscheller, G., 2007. Brain-computer communication: motivation, aim, and impact of exploring a virtual apartment. IEEE Trans. Neural. Syst. Rehabil. Eng. 15 (4), 473–482.

Lemm, S., Blankertz, B., Curio, G., Müller, K.R., 2005. Spatio-spectral filters for improving the classification of single trial EEG. IEEE Trans. Biomed. Eng. 52 (9), 1541–1548.

Leuthardt, E.C., Schalk, G., Wolpaw, J.R., Ojemann, J.G., Moran, D.W., 2004. A brain-computer interface using electrocorticographic signals in humans. J. Neural. Eng. 1 (2), 63–71.

Levine, S.P., Huggins, J.E., BeMent, S.L., Kushwaha, R.K., Schuh, L.A., Rohde, M.M., et al., 2000. A direct brain interface based on event-related potentials. IEEE Trans. Rehabil. Eng. 8 (2), 180–185.

Lopes da Silva, F.H., van Hulten, K., Lommen, J.G., Storm van Leeuwen, W., van Veelen, C.W., Vliegenthart, W., 1977. Automatic detection and localization of epileptic foci. Electroencephalogr. Clin. Neurophysiol. 43 (1), 1–13.

Lotte, F., Congedo, M., Lécuyer, A., Lamarche, F., Arnaldi, B., 2007. A review of classification algorithms for EEG-based brain-computer interfaces. J. Neural. Eng. 4 (2), R1–R13.

Lugger, K., Flotzinger, D., Schlögl, A., Pregenzer, M., Pfurtscheller, G., 1998. Feature extraction for on-line EEG classification using principal components and linear discriminants. Med. Biol. Eng. Comput. 36 (3), 309–314.

Mahalanobis, P.C., 1936. On the generalised distance in statistics. Proc. Natl. Inst. Sci. India 2 (49).

Marcel, S., Millán, J Del R., 2007. Person authentication using brainwaves (EEG) and maximum a posteriori model adaptation. IEEE Trans. Pattern Anal. Mach. Intell. 29 (4), 743–752.

Mason, S.G., Bashashati, A., Fatourechi, M., Navarro, K. F., Birch, G.E., 2007. A comprehensive survey of brain interface technology designs. Ann. Biomed. Eng. 35 (2), 137–169.

Mason, S., Kronegg, J., Huggins, J., Fatourechi, M., Schlögl, A., 2006. Evaluating the Performance of Self-Paced Brain Computer Interface Technology. Technical report, available at http://www.bci-info.tugraz.at/Research_Info/documents/articles, 2006.

McFarland, D.J., McCane, L.M., David, S.V., Wolpaw, J.R., 1997. Spatial filter selection for EEG-based communication. Electroencephalogr. Clin. Neurophysiol. 103 (3), 386–394.

Middendorf, M., McMillan, G., Calhoun, G., Jones, K.S., 2000. Brain-computer interfaces based on the steady-state visual-evoked response. IEEE Trans. Rehabil. Eng. 8 (2), 211–214.

Millán, J del R., Franzé, M., Mouriño, J., Cincotti, F., Babiloni, F., 2002. Relevant EEG features for the classification of spontaneous motor-related tasks. Biol. Cybern. 86 (2), 89–95.

Millàn, Jdel R., Mourino, J., 2003. Asynchronous BCI and local neural classifiers: an overview of the Adaptive Brain Interface project. IEEE Trans. Neural. Syst. Rehabil. Eng. 11 (2), 159–161.

Millàn, Jdel R., Renkens, F., Mouriño, J., Gerstner, W., 2004. Noninvasive brain-actuated control of a mobile robot by human EEG. IEEE Trans. Biomed. Eng. 51 (6), 1026–1033.

Miller, K.J., denNijs, M., Shenoy, P., Miller, J.W., Rao, R.P.N., Ojemann, J.G., 2007. Real-time functional brain mapping using electrocorticography. Neuroimage 37 (2), 504–507.

Moore, M.M., 2003. Real-world applications for brain-computer interface technology. IEEE Trans. Neural. Syst. Rehabil. Eng. 11 (2), 162–165.

Mugler, E., Bensch, M., Halder, S., Rosenstiel, W., Bogdan, M., Birbaumer, N., et al., 2008. Control of an Internet Browser Using the P300 Event-Related Potential. Int. J. Bioelectromagnetism 10 (1), 56–63.

Müller, K.-R., Blankertz, B, 2006. Toward noninvasive Brain Computer Interfaces. Signal Process. Mag. 23 (5), 125–128.

Müller-Gerking, J., Pfurtscheller, G., Flyvbjerg, H., 1999. Designing optimal spatial filters for single-trial EEG classification in a movement task. Clin. Neurophysiol. 110 (5), 787–798.

Müller-Putz, G.R., Pfurtscheller, G., 2008. Control of an electrical prosthesis with an SSVEP-based BCI. IEEE Trans. Biomed. Eng. 55 (1), 361–364.

Müller-Putz, G.R., Scherer, R., Brauneis, C., Pfurtscheller, G., 2005a. Steady-state visual evoked potential (SSVEP)-based communication: impact of harmonic frequency components. J. Neural. Eng. 2 (4), 123–130.

Müller-Putz, G.R., Scherer, R., Brunner, C., Leeb, R., Pfurtscheller, G., 2008. Better than random? A closer look on BCI results. Int. J. Bioelectromagnetism 10 (1), 52–55.

Müller-Putz, G.R., Scherer, R., Pfurtscheller, G., Rupp, R., 2005b. EEG-based neuroprosthesis control: a step towards clinical practice. Neurosci. Lett. 382 (1–2), 169–174.

Müller-Putz, G.R., Scherer, R., Pfurtscheller, G., Rupp, R., 2006. Brain-computer interfaces for control of neuroprostheses: from synchronous to asynchronous mode of operation. Biomed. Tech. (Berl) 51 (2), 57–63.

Mülller, K.R., Anderson, C.W., Birch, G.E., 2003. Linear and nonlinear methods for brain-computer interfaces. IEEE Trans. Neural. Syst. Rehabil. Eng. 11 (2), 165–169.

Naeem, M., Brunner, C., Leeb, R., Graimann, B., Pfurtscheller, G., 2006. Seperability of four-class motor imagery data using independent components analysis. J. Neural. Eng. 3 (3), 208–216.

Neuper, C., Scherer, R., Reiner, M., Pfurtscheller, G., 2005. Imagery of motor actions: differential effects of kinesthetic and visual-motor mode of imagery in single-trial EEG. Cogn. Brain Res. 25 (3), 668–677.

Nicolelis, M.A., 2001. Actions from thoughts. Nature 409 (6818), 403–407.

Obermaier, B., Guger, C., Pfurtscheller, G., 1999. Hidden Markov models used for the offline classification of EEG data. Biomed. Tech. (Berl) 44 (6), 158–162.

Obermaier, B., Müller, G.R., Pfurtscheller, G., 2003. "Virtual keyboard" controlled by spontaneous EEG activity. IEEE Trans. Neural. Syst. Rehabil. Eng. 11 (4), 422–426.

Peters, B.O., Pfurtscheller, G., Flyvbjerg, H., 1998. Mining multi-channel EEG for its information content: an ANN-based method for a brain-computer interface. Neural Netw. 11 (7–8), 1429–1433.

Pfurtscheller, G., Aranibar, A., 1977. Event-related cortical desynchronization detected by power measurements of scalp EEG. Electroencephalogr. Clin. Neurophysiol. 42 (6), 817–826.

Pfurtscheller, G., Lopes da Silva, F.H., 1999a. Event-related EEG/MEG synchronization and desynchronization: basic principles. Clin. Neurophysiol. 110 (11), 1842–1857.

Pfurtscheller, G., Lopes da Silva, F.H., 1999b. Handbook of Electroencephalogaphy and Clinical Neurophysiology – Event Related Desynchronization. Elsevier, Amsterdam.

Pfurtscheller, G., Müller, G.R., Pfurtscheller, J., Gerner, H.J., Rupp, R., 2003. 'Thought'–control of functional electrical stimulation to restore hand grasp in a patient with tetraplegia. Neurosci. Lett. 351 (1), 33–36.

Pfurtscheller, G., Müller-Putz, G.R., Schlögl, A., Graimann, B., Scherer, R., Leeb, R., et al., 2006. 15 years of BCI research at Graz University of Technology: current projects. IEEE Trans. Neural. Syst. Rehabil. Eng. 14 (2), 205–210.

Pfurtscheller, G., Neuper, C., 1997. Motor imagery activates primary sensorimotor area in humans. Neurosci. Lett. 239 (2–3), 65–68.

Pfurtscheller, G., Neuper, C., 2001. Motor Imagery and Direct Brain-Computer Communication. Proc. IEEE 89, 1123–1134.

Pfurtscheller, G., Neuper, C., 2006. Future prospects of ERD/ERS in the context of brain-computer interface (BCI) developments. Prog. Brain Res. 159, 433–437.

Pfurtscheller, G., Neuper, C., Birbaumer, N., 2005. Human Brain-Computer Interface. In: Vaadia, E., Riehle, A. (Eds.), Motor Cortex in Voluntary Movements: A Distributed System for Distributed Functions of Methods and New Frontiers in Neuroscience, CRC Press, Boca Raton, FL, pp. 367–401.

Pfurtscheller, G., Neuper, C., Flotzinger, D., Pregenzer, M., 1997. EEG-based discrimination between imagination of right and left hand movement. Electroencephalogr. Clin. Neurophysiol. 103 (6), 642–651.

Piccione, F., Giorgi, F., Tonin, P., Priftis, K., Giove, S., Silvoni, S., et al., 2006. P300-based brain computer interface: reliability and performance in healthy and paralysed participants. Clin. Neurophysiol. 117 (3), 531–537.

Pineda, J.A., Silverman, D.S., Vankov, A., Hestenes, J., 2003. Learning to control brain rhythms: making a brain-computer interface possible. IEEE Trans. Neural. Syst. Rehabil. Eng. 11 (2), 181–184.

Pop-Jordanova, N., Pop-Jordanov, J., 2005. Spectrum-weighted EEG frequency ("brain-rate") as a quantitative indicator of mental arousal. Prilozi 26 (2), 35–42.

Pregenzer, M. *DSLVQ*. PhD thesis, Graz University of Technology, Graz, Austria, 1997.

Pregenzer, M., Pfurtscheller, G., Flotzinger, D., 1994. Selection of electrode positions for an EEG-based brain computer interface (BCI). Biomed. Tech. (Berl) 39 (10), 264–269.

Purpura, D.P., 1959. Nature of electrocortical potentials and synaptic organizations in cerebral and cerebellar cortex. Int. Rev. Neurobiol. 1, 47–163.

Ramoser, H., Müller-Gerking, J., Pfurtscheller, G., 2000. Optimal spatial filtering of single trial EEG during imagined hand movement. IEEE Trans. Rehabil. Eng. 8 (4), 441–446.

Rebsamen, B., Teo, C.L., Zeng, Q., Ang Jr., M.H., Burdet, E., Guan, C., et al., 2007. Controlling a Wheelchair Indoors Using Thought. IEEE Intell. Syst. 22 (2), 18–24.

Rossini, L., Dario, I., Summerer, L. (Eds). 2009. Brain Machine Interfaces for Space applications: Enhancing astronaut's capabilities. International Review of Neurobiology, vol 86, New York, NY: Academic Press.

Schalk, G., Kubánek, J., Miller, K.J., Anderson, N.R., Leuthardt, E.C., Ojemann, J.G., et al., 2007. Decoding two-dimensional movement trajectories using electrocorticographic signals in humans. J. Neural. Eng. 4 (3), 264–275.

Scherer, G., Müller-Putz, G.R., Pfurtscheller, G., 2007. Self-initiation of EEG-based brain-computer communication using the heart rate response. J. Neural. Eng. 4 (4), L23–L29.

Scherer, R., Graimann, B., Huggins, J.E., Levine, S.P., Pfurtscheller, G., 2003. Frequency component selection for an ECoG-based brain-computer interface. Biomed. Tech. (Berl) 48 (1–2), 31–36.

Scherer, R., Lee, F., Schlögl, A., Leeb, R., Bischof, H., Pfurtscheller, G., 2008. Towards self-paced brain-computer communication: navigation through virtual worlds. IEEE Trans. Biomed. Eng. 55 (2), 675–682.

Scherer, R., Mohapp, A., Grieshofer, P., Pfurtscheller, G., Neuper, C., 2007. Sensorimotor EEG patterns during motor imagery in hemiparetic stroke patients. Int. J. Bioelectromagnetism 9 (3), 155–162.

Scherer, R., Müller, G.R., Neuper, C., Graimann, B., Pfurtscheller, G., 2004. An asynchronously controlled EEG-based virtual keyboard: improvement of the spelling rate. IEEE Trans. Biomed. Eng. 51 (6), 979–984.

Scherer, R., Schlögl, A., Lee, F., Bischof, H., Janša, J., Pfurtscheller, G., 2007. The self-paced graz brain-computer interface: methods and applications. Comput. Intell. Neurosci. Article ID 79826:9 pages.

Schlögl, A., 2000. The electroencephalogram and the adaptive autoregressive model: theory and applications. PhD thesis, Graz University of Technology, 2000.

Schlögl, A., Flotzinger, D., Pfurtscheller, G., 1997. Adaptive autoregressive modeling used for single-trial EEG classification. Biomed. Tech. (Berl) 42 (6), 162–167.

Schlögl, A., Keinrath, C., Zimmermann, D., Scherer, R., Leeb, R., Pfurtscheller, G., 2007. A fully automated correction method of EOG artifacts in EEG recordings. Clin. Neurophysiol. 118 (1), 98–104.

Schlögl, A., Kronegg, J., Huggins, J.E., Mason, S.G., 2007. Toward Brain-Computer Interfacing, chapter Evaluation Criteria for BCI Research, MIT Press, Cambridge, MA, pp. 327–342.

Schlögl, A., Supp, G., 2006. Analyzing event-related EEG data with multivariate autoregressive parameters. Prog. Brain Res. 159, 135–147.

Schwartz, A.B., Cui, X.T., Weber, D.J., Moran, D.W., 2006. Brain-controlled interfaces: movement restoration with neural prosthetics. Neuron 52 (1), 205–220.

Sellers, E.W., Donchin, E., 2006. A P300-based brain-computer interface: initial tests by ALS patients. Clin. Neurophysiol. 117 (3), 538–548.

Serby, H., Yom-Tov, E., Inbar, G.F., 2005. An improved P300-based brain-computer interface. IEEE Trans. Neural. Syst. Rehabil. Eng. 13 (1), 89–98.

Shenoy, P., Krauledat, M., Blankertz, B., Rao, R.P.N., Mueller, K.-R., 2006. Towards Adaptive Classification for BCI. J. Neural. Eng. 3 (1), R13–23.

Shenoy, P., Miller, K.J., Ojemann, J.G., Rao, R.P.N., 2008. Generalized features for electrocorticographic BCIs. IEEE Trans. Biomed. Eng. 55 (1), 273–280.

Student, 1908. The probable error of a mean. Biometrika 6 (1), 1–25.

Sumida, B.H., Houston, A.I., McNamara, J.M., Hamilton, W.D., 1990. Genetic algorithms and evolution. J. Theor. Biol. 147 (1), 59–84.

Tecchio, F., Porcaro, C., Barbati, G., Zappasodi, F., 2007. Functional source separation and hand cortical representation for a brain-computer interface feature extraction. J. Physiol. 580 Pt 3, 703–721.

Tomori, O., Moore, M., 2003. The neurally controllable internet browser (BrainBrowser). CHI '03: CHI '03 extended abstracts on Human factors in computing systems, ACM, New York, pp. 796–797.

Townsend, G., Graimann, B., Pfurtscheller, G., 2004. Continuous EEG classification during motor imagery–simulation of an asynchronous BCI. IEEE Trans. Neural. Syst. Rehabil. Eng. 12 (2), 258–265.

Townsend, G., Graimann, B., Pfurtscheller, G., 2006. A comparison of common spatial patterns with complex band power features in a four-class BCI experiment. IEEE Trans. Biomed. Eng. 53 (4), 642–651.

Trejo, L.J., Rosipal, R., Matthews, B., 2006. Brain-computer interfaces for 1-D and 2-D cursor control: designs using volitional control of the EEG spectrum or steady-state visual evoked potentials. IEEE Trans. Neural. Syst. Rehabil. Eng. 14 (2), 225–229.

Vanacker, G., Del R Millàn, J., Lew, E., Ferrez, P.W., Moles, F.G., Philips, J., et al., 2007. Context-based Filtering for Assisted Brain-Actuated Wheelchair Driving. Comput. Intell. Neurosci. 2007 (119), Article ID 25130.

Vapnik, V., 1995. The Nature of Statistical Learning Theory. Springer Verlag, Berlin.

Vaughan, T.M., McFarland, D.J., Schalk, G., Sarnacki, W.A., Krusienski, D.J., Sellers, E.W., et al., 2006. The Wadsworth BCI Research and Development Program: at home with BCI. IEEE Trans. Neural. Syst. Rehabil. Eng. 14 (2), 229–233.

Vidaurre, C., Scherer, R., Cabeza, R., Schlögl, A., Pfurtscheller, G., 2007. Study of discriminant analysis applied to motor imagery bipolar data. Med. Biol. Eng. Comput. 45 (1), 61–68.

Vidaurre, C., Schlögl, A., Cabeza, R., Scherer, R., Pfurtscheller, G., 2007. Study of on-line adaptive discriminant analysis for EEG-based brain computer interfaces. IEEE Trans. Biomed. Eng. 54 (3), 550–556.

Wackermann, J., 1999. Towards a quantitative characterisation of functional states of the brain: from the non-linear methodology to the global linear description. Int. J. Psychophysiol. 34 (1), 65–80.

Walter, W.G., 1936. The location of cerebral tumors by electro-encephalography. Lancet 228, 305–308.

Wolpaw, J.R., Birbaumer, N., McFarland, D.J., Pfurtscheller, G., Vaughan, T.M., 2002. Brain-computer interfaces for communication and control. Clin. Neurophysiol. 113 (6), 767–791.

Prediction of Muscle Activity from Cortical Signals to Restore Hand Grasp in Subjects with Spinal Cord Injury

11

Emily R. Oby*, **Christian Ethier***, **Matthew J. Bauman***, **Eric J. Perreault**[†]**,
Jason H. Ko[††], **Lee E. Miller***[†]**

*Department of Physiology, **Department of Physical Medicine and Rehabilitation, [††]Division of Plastic and Reconstructive Surgery, Feinberg School of Medicine, Northwestern University, Chicago, Illinois, [†]Department of Biomedical Engineering, Northwestern University, Evanston, Illinois

11.1 INTRODUCTION

In the United States, Spinal cord injury (SCI) paralyzes approximately 11,000 individuals every year. SCI leaves patients partially or entirely unable to activate muscles that are innervated by nerves below the level of injury. Roughly half of SCIs occur above the sixth cervical vertebra, thereby affecting all four limbs and leaving the individuals unable to grasp objects effectively with their hands. Injuries as high as C2 also affect the muscles of respiration, requiring that these patients be artificially ventilated.

Although there are important changes to both the muscles and the brain that occur as a result of either the injury or subsequent disuse, the essential problem is that motor commands, still formulated in the brain, can no longer reach the muscles. Toward the end of the eighteenth century, Luigi Galvani demonstrated that muscles are electrically excitable and can be made to contract by the application of appropriate electrical current (Galvani, 1791; Piccolino, 1998). Following this basic principle, functional electrical stimulation (FES) has been used for nearly five decades in an attempt to restore some level of motor function to paralyzed individuals through electrical stimulation of muscles (Liberson et al., 1961; Peckham and Knutson, 2005). FES has successfully been applied to provide appropriately timed dorsiflexion for

Statistical Signal Processing for Neuroscience and Neurotechnology. DOI: 10.1016/B978-0-12-375027-3.00011-9

patients suffering foot drop as a result of SCI or stroke (Vodovnik et al., 1978; Weber et al., 2004), and other systems provide some support for stance or walking (Popovic et al., 1989; Solomonow et al., 1997; Graupe and Kohn, 1998). Stimulation of the phrenic nerve has been used in the case of high-level injuries to replace or augment the use of an external ventilator (Creasey et al., 1996), and systems have been devised to provide bowel and bladder control (Johnston et al., 2005).

Most patients with injuries at the C5 and C6 levels report that, more than any other function, regaining the ability to grasp objects would provide them with the greatest benefit (Anderson, 2004). For this reason, considerable effort has also been devoted to the development of FES systems to restore voluntary grasp. Probably the most clinically successful of these has been the Freehand prosthesis, which has been implanted in several hundred patients since 1986 (Keith et al., 1989). An earlier version was tested in 1980 (Peckham et al., 1980).

One of the primary limitations of the Freehand and related grasp prostheses is the very limited means of control available to patients. Even if it were possible to provide fully articulated and forceful muscle activation, dexterous voluntary control would not be possible through the existing control channels. The problem becomes even more severe for patients with higher-level injuries, who have greater needs for replaced function, with less available control. Fortunately, the advent and rapid development in this decade of the brain-machine interface (BMI) provide an exciting new means by which a prosthesis with more degrees of freedom might be controlled.

Currently, virtually all BMI applications extract kinematic information from the brain. The approach typically involves recording a training data set including both neural discharge and limb position as a monkey makes a series of normal movements. These data are used to compute a mapping between the neural data and hand position that can be used to transform subsequent neural recordings to produce "predictions" of movement intent. If these predictions are made in real time, they can be used to allow an experimental subject voluntary control of an external device.

However, there is considerable evidence that M1 also encodes information about the forces and dynamics of movement (Hepp-Reymond et al., 1978; Thach, 1978; Cheney and Fetz, 1980; Kalaska and Hyde, 1985; Georgopoulos et al., 1992; Riehle and Requin, 1995; Sergio et al., 2005). A small number of groups have begun to pursue the possibility of using this dynamics information as a real-time control signal, through the prediction of grip force (Carmena et al., 2003), joint torque (Fagg et al., 2009), and muscle activity (Pohlmeyer et al., 2007b). We have also shown preliminary evidence that monkeys can easily learn to modulate the magnitude of isometric wrist flexion torque produced through cortically controlled FES following peripheral nerve block (Pohlmeyer et al., 2007a, 2009). Similar results have been reported by another group that operantly conditioned monkeys to modulate the activity of individual neurons that directly controlled the FES (Moritz et al., 2008).

In this chapter we describe our continuing work on developing a neuroprosthesis that uses real-time predictions from an ensemble of recorded neurons to provide voluntary control of hand use following peripheral nerve block.

11.2 BACKGROUND

Several systems have been developed to provide rudimentary hand grasp through FES. Most of these rely on surface stimulation, and all require that the stimulus waveforms be preprogrammed to accommodate the very limited range of voluntary control available to a patient with a mid-cervical lesion. Clearly, even an unimpaired individual is unlikely to be able to modulate independently the large number of muscles needed to form a natural grasp through any but the normal means. Fortunately, a fairly wide range of grasping behaviors can be represented by a much lower-dimensional coordinate system. One study used principal component analysis to determine the dimensionality of hand shape as subjects grasped a series of objects having different size and shape (Santello et al., 1998). Only two components were needed to reconstruct 80% of the variance in hand shape. It should be noted that a later study found that two to three times as many components, or "movement synergies," were required to reconstruct more complex movements (Todorov and Ghahramani, 2004). These results suggest that at least simple functional grasps may be accomplished with a low-dimensional controller.

The current Freehand consists of an implanted stimulator and electrodes implanted in 12 muscles of the forearm and hand (Figure 11.1). The stimulator communicates via a short-range RF link to an external controller that delivers preprogrammed stimulus waveforms to the full set of muscles. These patterns are customized for each patient in a series of sessions with an occupational therapist, who determines the threshold level of stimulation necessary to produce force in each muscle and the maximal stimulation that avoids "spillover" to adjacent muscles with different mechanical actions. Having determined this mapping from stimulus intensity to muscle force, the therapist determines the pattern of stimulation necessary to produce either a power grasp or a lateral grasp. This process accomplishes a mapping from the high-order space of hand muscles to a single control variable.

Control of the FreeHand requires discrete signals for switching between grasp patterns and a continuous signal for controlling hand opening and

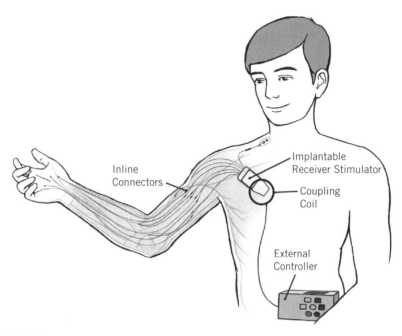

FIGURE 11.1

Freehand® neuroprosthesis developed to provide functional grasp to patients with C5/C6 tetraplegia due to spinal cord injury. Implanted stimulation electrodes cause the paralyzed muscles to contract in preprogrammed patterns, using signals sent from an external controller. The patient controls the system through residual proximal limb movement or muscle activity.
Source: Used with the permission of the Cleveland FES Center.

closing within a specific grasp (Scott et al., 1996; Hart et al., 1998). A number of discrete control signals have been used that depend on the user's abilities and preferences. These include switches mounted on the wheel chair or elsewhere, EMGs from one or more muscles, and voluntary movements such as those from a joystick mounted on the contralateral shoulder. Once a specific grasp is selected, continuous control of hand opening and closing is determined by a single degree-of-freedom (DOF) signal directly under the patient's control. Common sources of continuous control signals include external sensors that detect residual motion of the shoulder or wrist and implanted electrodes that sense EMG in muscles remaining under voluntary control. When the desired force level is achieved, the patient can use the discrete command signal to lock stimulation, thereby maintaining the grasp without additional conscious effort. He can then move his hand and the grasped object.

In a multi-center study to evaluate the safety, efficacy, and clinical impact of the FreeHand on fifty patients with spinal cord injuries (Peckham et al.,

2001), it was found that the neuroprosthesis produced increased pinch force in every patient. Additionally, in a test of grasp and release ability (Wuolle et al., 1994), 49 of the 50 participants moved at least one more object with the neuroprosthesis than they could without it. More than 90% were satisfied with the neuroprosthesis, and most used it regularly.

In addition to the Freehand, a number of other grasping FES systems are in use or have been tested (Popovic et al., 2002; Popovic, 2003; Peckham and Knutson, 2005). Although these systems share many underlying principles, they vary in several respects. Many of them rely on surface stimulation rather than implanted electrodes. This approach has the advantage that it can be adopted relatively soon after the injury. The Bionic Glove is designed simply to amplify the passive finger flexion movements caused by active wrist movements (tenodesis grip) for those patients with injuries at the C6 level. It uses surface stimulation to cause the fingers to flex when sensors in the glove detect wrist extension and to extend when it detects wrist flexion (Prochazka et al., 1997; Popovic et al., 1999). The Handmaster is another orthosis that uses surface stimulation (Snoek et al., 2000; Alon and McBride, 2003). However, it is intended for C5 spinal injuries in which there is limited or no voluntary control of the wrist. The patient pushes buttons to trigger preprogrammed stimulation patterns.

At least two systems have been designed in an effort to provide limited reaching functionality as well as grasping for patients with C4-level injuries. The Belgrade Grasping-Reaching System allows control of hand opening and closing simply by the push of a button. A goniometer is used to measure shoulder motion so triceps muscle stimulation can be modulated appropriately during reaching motions (Popovic et al., 2002; Popovic, 2003). The NEC FES-Mate, a multi-channel FES system that uses percutaneous electrodes, is also intended to control reaching as well as grasping. It uses amplitude modulation to control contractile force, modeled on averaged EMG signals recorded from able-bodied individuals. The predetermined stimulation patterns are triggered through a combination of voice control, head switches, sip-puff, and shoulder movement (Handa and Hoshimiya, 1987; Hoshimiya et al., 1989).

To provide an alternative control signal for severely paralyzed patients, several attempts have been made to employ electroencephalic (EEG) signals recorded noninvasively from the scalp. In one case, a patient using the Freehand device learned to modulate the amplitude of beta-band activity in order to control the rate of change in stimulus-driven grip force (Lauer et al., 1999). Modulation above one threshold caused gradually increasing flexion force, while extension was provided when the signal dropped below a minimum level. Achieving adequate control of EEG modulation required 20 sessions of practice over a period of several months. Subsequently, another group used bursts of beta-range signals generated by imagined foot movements to control the transition between four distinct phases of stimulation (Pfurtscheller et al., 2003). These phase transitions were quite slow (by design, greater than 5 seconds duration), but they did allow the patient to grasp a cylindrical object using lateral prehension.

Although even this 1-DOF control significantly improves a range of functional activities for most patients who have used it, it is clearly desirable to remove the requirement that grasp patterns be preprogrammed. This might allow the patient to grasp an irregularly shaped object more reliably, or to orient the hand through motion at the wrist rather than the shoulder or the entire trunk. The ability to predict the activity of individual muscles in real time provides a possible means by which this preprogramming stage might be completely avoided.

11.2.1 BMIs as a Potential Control Solution

The earliest demonstration of BMI technology actually used recordings from a human patient with amyotrophic lateral sclerosis who was nearly completely "locked in," retaining only the ability to move her eyes (Kennedy and Bakay, 1998). The patient could modulate the discharge of neurons recorded from the primary motor cortex (M1) adequately to control a simple binary switch. Several subsequent patients learned to move a cursor up or down and to slowly spell out words using a progressive menu selection system. The most successful patient eventually learned to communicate at a rate of approximately three letters per minute.

Research approaches to the study of BMIs can be divided into two major categories that we will call open-loop and closed-loop. Open-loop approaches refer to the prediction of a control signal based on recordings from the brain while the subject makes a series of normal limb movements. This signal may or may not be used to affect movement. The important point is that the subject remains unaware of the signal, such that it cannot affect subsequent movement commands. In many cases these predictions are made only offline using previously recorded data.

In contrast, closed-loop approaches use the predicted control signal to effect movement of a device. Information about that movement is fed back to the subject within tens of milliseconds (in nearly all cases through natural vision), so that it becomes possible for him to correct errors and guide the movement in something of a normal manner. Under these conditions, it is not necessary for the subject to move his own limb, and in many cases he does not.

Any useful control must be closed-loop, but the advantages of the open-loop approach to development include its relative simplicity and the fact that many different "decoding" algorithms can be tested and compared using a single data set. Quantitative comparisons of different approaches can be made by computing an error metric (e.g., the coefficient of determination, or mean square error) between the actual and the predicted movements. Such comparisons are not meaningful in the closed-loop state because the subject's natural limb movements (assuming they are made at all) may be influenced by the predictions.

In a sense, single-electrode recordings from the time of Evarts (1968) represent the open-loop approach to understanding how individual neurons in motor areas of the brain encode limb movement. The dichotomy between "muscle" and "movement" control has been explored in this manner for several decades (Evarts, 1968; Phillips, 1975; Georgopoulos et al., 1982; Scott and Kalaska, 1997). In 1970, Humphrey and his colleagues used multiple electrodes to record simultaneously from several neurons, thereby increasing the accuracy with which force- and movement-related signals could be predicted during wrist movements (Humphrey et al., 1970). In 1969, in what is often credited as the first closed-loop BMI, Fetz and Finocchio trained several monkeys to modulate the discharge of single neurons in M1 using auditory or visual feedback (Fetz and Finocchio, 1975).

In 2000, Nicolelis and colleagues chronically implanted large numbers of electrodes in several different motor areas of two owl monkeys (Wessberg et al., 2000). They used both a linear filter model and a nonlinear neural network model to fit ten minutes of data recorded from 100 neurons to single components of hand trajectory during a three-dimensional reaching task. The result was an R^2 as high as 0.5 or 0.6 between the actual and predicted trajectories of test data sets, with little difference between the linear and nonlinear approaches. Two years later, another group used similar linear filter methods to allow several monkeys to control the movement of a cursor viewed on a computer screen (Serruya et al., 2002). Although the movements were considerably slower than normal, the animals were able to make purposeful movements and detect and correct errors. The open-loop predictions in this experiment were similar to those achieved by Wessberg et al. (2000).

Rather than compute filters between input and output, Georgopoulos and his group used an approach based on the concept of neural preferred directions (the direction of hand movement yielding the greatest discharge) (Georgopoulos et al., 1982). They proposed the computation of a population vector that could be used to compute the mean direction of hand movement. Each neuron was characterized in terms of its preferred direction, and the preferred direction vectors were summed, weighted by the magnitude of discharge of the corresponding neuron during a movement. With this approach, several monkeys learned to control the movement of a cursor in three dimensions using a virtual reality display (Taylor et al., 2002).

As in the studies just described, the most prevalent BMI approach has been decoding of the entire movement trajectory, generally using recordings made in the primary motor cortex. An alternative approach, taken by several groups, has been to decode the final location of the monkey's intended reach. One study compared these two approaches using recordings made in either M1 or the dorsal premotor cortex (PMd). They concluded that the M1 recordings provided more trajectory-related information but that PMd was a better and earlier source of information about the intended movement target (Hatsopoulos et al., 2004).

One of the most successful examples of target prediction was demonstrated recently by Shenoy's group, which employed maximum likelihood methods using either Gaussian or Poisson models of the spiking statistics of neurons recorded from PMd to classify the intended reach target (Santhanam et al., 2006). While selecting among 8 or 16 targets, one monkey achieved information transfer rates above 6 bits per second (roughly equivalent to typing 15 words per minute) using signals recorded from 96 chronically implanted electrodes. Most groups use 500 to 1000 ms of spiking history to make continuous predictions of movement trajectory. However, to achieve such a high transfer rate in this experiment, the classifier was based on only 60 to 130 ms of spike data, making the result even more remarkable.

Recordings from the posterior parietal cortex have also been used to classify the intended reach target. The parietal reach region (PRR) is an area within the medial wall of the intraparietal sulcus and the dorsal aspect of the parietooccipital area that is thought to represent the goal of a reaching movement in visual coordinates (Batista et al., 1999). Discrimination accuracy among eight targets of about 65% was achieved using a relatively small number of neurons (Musallam et al., 2004). Interestingly, it was also possible to decode information about the monkeys' expectation of the likelihood or magnitude of the anticipated reward from the same signals. It is worth noting that PRR and, to a lesser extent, PMd, may undergo fewer functional changes following spinal cord injury than M1. For this reason, it may be a particularly attractive cortical site for clinical consideration.

11.2.2 BMIs for Control of Dynamics

A virtual device (a cursor for example) can simply be instantaneously repositioned based on the predicted endpoint position. Controlling a robotic limb is more difficult and typically accomplished with a PD (proportional-derivative) controller, which compares the intended state variables (position and/or velocity) to the current state of the limb. The PD controller drives the production of appropriate torques to reduce measured limb state errors. Its parameters typically are adjusted to optimize performance under particular dynamical situations. Consequently, if the mass of the plant changes significantly, as when the subject must grasp and apply force to an object in order to push it, the change in system dynamics can result in the controller behaving suboptimally. For the most part, this problem has not yet been faced in current BMI applications, which have been limited to a single dynamical condition. The purely kinematic BMI faces further limitations if the subject must apply controlled forces, as when grasping and manipulating an object.

There have been very few studies of M1 discharge and force-related variables using large numbers of chronically implanted electrodes. In one of these, monkeys learned to control the force of a gripper together with its position in two dimensions, first using actual hand movements and then using real-time force and endpoint velocity predictions (Carmena et al., 2003). The accuracy of offline prediction of grip force was significantly higher than that of either hand position or velocity. Another group showed that shoulder and elbow torque could be predicted during planar limb movements with an accuracy nearly as high as the corresponding kinematic signals (Fagg et al., 2009).

Likewise, we have shown that EMG signals for both arm and hand muscles can be predicted during reaching and grasping movements with an accuracy as high as or higher than that of most studies of kinematic signals (Pohlmeyer et al., 2007b). This was particularly surprising given the relatively noisy, stochastic nature of the EMG signal. We have since used these predictions as real-time inputs to a four-channel stimulator controlling the level of contraction of four flexor muscles. Two monkeys used the system to control isometric wrist flexion force despite temporary paralysis induced by a peripheral nerve block. One, using input from approximately 80 neurons, achieved sufficient precision to track a target that moved among three different levels (Pohlmeyer et al., 2009).

Fetz and colleagues adopted an approach inspired by his classic work, using the discharge of individual, voluntarily modulated neurons to control FES activation of individual muscles (Moritz et al., 2008). Their monkeys typically learned to control isometric wrist force in either the flexion or the extension direction after practice of only several tens of minutes.

11.3 METHODS

Most current BMIs control movement kinematics: a cursor on a screen or a robotic arm driven by a PD controller. However, signals recorded from the

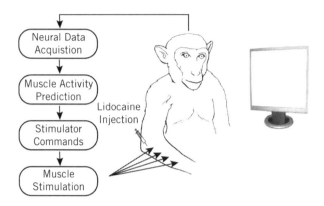

FIGURE 11.2

Cortically controlled FES for grasp implemented in an experimental monkey. The monkey's forearm muscles were temporarily paralyzed by injections of lidocaine to the motor nerves. Signals recorded from an intra-cortical microelectrode array implanted in primary motor cortex were used to predict the EMG activity during the monkey's attempted movements. These predictions were used to modulate the intensity of stimulation, much as with the Freehand neuroprosthesis in Figure 11.1. Our system allows the monkey voluntary control of the paralyzed muscles.

motor cortex should also be well suited to the control of movement dynamics, including the activity of muscles. We have begun experiments designed to harness the body's natural control signals as a means to control paralyzed muscles through FES. As a simple model of some of the deficits resulting from spinal cord injury, we temporarily paralyzed the forearm musculature with lidocaine injections (Figure 11.2). We recorded from the hand area of the primary motor cortex using chronically implanted 100-electrode arrays. We also implanted epimysial electrodes on a variety of muscles of the arm and hand that were used both to record EMG signals and to stimulate the muscles electrically. We used real-time predictions of EMG activity to control the input to a multi-channel FES stimulator that caused contractions of the paralyzed muscles. (All animal care, surgical, and research procedures were consistent with the Guide for the Care and Use of Laboratory Animals and were approved by the Institutional Animal Care and Use Committee of Northwestern University.)

11.3.1 Isometric Wrist Torque Tasks

The monkeys were initially trained to sit in a primate chair that faced a computer monitor. All of the behavioral tasks required them to interpret force feedback information that was represented by a moving cursor displayed on the monitor. Once the monkeys learned this basic association, they were trained on several specific tasks that were used for the FES experiments. One was an isometric wrist task in which the monkey was required to move the cursor from a central target at zero force to a peripheral target using various combinations of force generated along the flexion/extension and radial/ulnar deviation axes. The monkey's upper arm was restrained by a custom-fitted cast that maintained his elbow at a 90-degree angle. Force was measured by a 2-axis strain gauge mounted between casts on the monkey's hand and forearm. Cursor movement distance was proportional to the measured torque along each axis.

After holding in the center for 0.5 second, an outer target appeared. The monkey was required to move the cursor to the target within 5 seconds and to maintain that torque for 0.5 second in order to receive a juice reward. In some cases, cursor movement was constrained to the horizontal (flexion/ extension) axis. In other cases, the monkey was required to control torque along both axes. In the one-dimensional version of the task, as many as five targets were presented along the flexion/extension axis. The most distant targets required torques of approximately 35–50% of the monkey's maximum voluntary contraction (MVC). In the two-dimensional task, eight targets, separated by 45 degrees, were presented on the circumference of a circle, analogous to the center-out movement task adopted by many groups since its introduction (Georgopoulos et al., 1982). These tasks allowed us considerable control over the time course and direction of the torque exerted by the monkey.

11.3.2 Hand Grasp Tasks

Although the isometric task was well controlled and stereotyped, other tasks we studied were based on more functional movements. In the least constrained of these, the monkey was required to pick up a ball from a small tray and place it in a tube with a 60-mm opening. The balls were of a variety of diameters and weights: 40-mm, 130 g; 40-mm, 95 g; and 24-mm, 60 g. The monkey started each trial by placing his hand on a touchpad for 0.2 second. A "go" tone indicated the beginning of a 5-second reach time period during which the monkey attempted to pick up the ball from the tray. An infrared sensor in the device tray detected the monkey's initial grasping attempts. Removing the ball from the tray began another 5-second interval during which the monkey was required to place the ball in the tube. A successful ball return earned the monkey a juice reward. The next trial was initiated when the monkey returned his hand to the touchpad.

11.3.3 Surgical Methods

Following several months of training, when the monkeys reached a suitable behavioral criterion, a sequence of two surgical procedures was performed to implant the necessary recording and stimulation electrodes. (All surgery was conducted under aseptic conditions using Isoflurane inhalational anesthesia. Antibiotics (Amoxicillin or Cephazolin) and Dexamethasone were given pre- and postoperatively. Analgesics (Buprenex and Meloxicam) were also given postoperatively.)

11.3.3.1 *Electrode Array Implantation Surgery*

A single array composed of 100 silicon microelectrodes in a 10×10 grid (Blackrock Microimplantable Systems, Inc.) was chronically implanted in the hand area of the primary motor cortex (M1). A craniotomy was performed above M1, and the dura was incised and reflected. The electrode array was positioned on the crown of the right precentral gyrus, approximately in line with the superior ramus (medial edge) of the arcuate sulcus. In most cases, we used interoperative stimulation of the exposed cortical surface to determine the optimal implant site. A piece of artificial pericardium was applied above the array, the dura was closed using 4.0 Nurolon sutures, and another piece of pericardium was applied over it. The excised bone flap was replaced, and the skin was closed.

11.3.3.2 *Limb Implants*

In a separate procedure, electrodes (either epimysial or intramuscular) were implanted on a variety of muscles in the forearm and hand. The electrode leads were routed subcutaneously to a back connector using published methods (Miller et al., 1993). The electrodes were used for both bipolar recordings and monopolar stimulation. The full list of muscles included

- Those responsible for motion of the wrist: the extensor and flexor carpi ulnaris, the extensor and flexor carpi radialis (ECU, FCU, ECR, and FCR), and the palmaris longus (Pal).

- The common digit muscles: the extensor digitorum communis (EDC), the flexor digitorum superficialis (FDS), and the flexor digitorum profundus (FDP).

- The intrinsic hand muscles: the flexor pollicis brevis (FPB) and the first dorsal interosseous (1DI).

Both FDS and FDP were implanted with electrodes located in both the ulnar and radial compartments.

In the same procedure, custom-made nerve cuffs were implanted around the median, ulnar, and radial nerves at sites just proximal to the elbow. The cuffs were connected via cannulae to subdermal injection ports (Mentor Injection Domes) implanted in the upper arm. In this way, a block of all three nerves at the elbow caused paralysis of the wrist and hand muscles. To achieve the nerve block, we injected lidocaine or other local anesthetics into each port, providing a temporary means to paralyze the limb and mimic some of the effects of a C5-C6 spinal cord injury. The EMG activity was negligible after 5 to 20 minutes. (The details of this method have been published (Pohlmeyer et al., 2009)).

11.3.4 Data Collection

Data were simultaneously recorded from the EMG electrodes and the intra-cortical array while the monkey performed one or more of the motor tasks described earlier. A torque signal was also recorded during the isometric wrist task. The data were recorded using a 128-channel Cerebus system (Blackrock Microsystems, Inc.). Action potential waveforms and their corresponding timestamps were saved for later offline spike discrimination using Offline Sorter (Plexon Neurotechnology Research Systems, Inc.) or streamed to separate computers for real-time analysis and control. These methods are described in greater detail later.

The EMG signals were amplified, band-pass-filtered (4-pole, 50−500 Hz), and sampled at 2000 Hz. Subsequently, EMG was digitally rectified, low-pass-filtered (4-pole, 10-Hz Butterworth), and subsampled to 20 Hz. Subsequent analysis was carried out primarily using MATLAB (The Mathworks, Inc.) and is described in greater detail in a later section.

11.3.5 Linear Systems Identification

We used a linear decoder to predict EMG activity in each muscle using the set of available neural recordings. The selection of a linear decoder was based on previous studies demonstrating only small improvements in performance with more complex decoding algorithms for estimating both EMG signals (Pohlmeyer et al., 2007b) and limb kinematics (Gao et al., 2003; Lebedev and Nicolelis, 2006). The selection was also based on our motivating rationale that the use of a muscle-based decoder would further simplify the decoding of motor cortical recordings. It should be noted that most BMI applications, including ours, span a relatively small portion of the animal's natural motor behavior. As this range is expanded, nonlinear decoders are likely to become more important (Shoham et al., 2005).

The implemented decoder assumed that the EMG in each muscle could be described by Eq. (11.1), in which $x_k(t)$ is the N-point vector of neural recordings from electrode k; $h_k(t)$ is the M-point finite impulse response function (FIRF) relating $x_k(t)$ to the muscle EMG $z(t)$. $w(t)$ is a random process accounting for measurement noise, nonlinearities, and contributions from unmeasured inputs.

$$z(t) = \sum_{k=1}^{N} \sum_{\tau=0}^{M-1} h_k(\tau) x_k(t-\tau) + w(t) \tag{11.1}$$

The linear filters relating cortical recordings (input) to muscle activity (output) were efficiently estimated according to Eq. (11.2), in which $\hat{h}$ is the vector of estimated FIRFs relating each input to the output; Φ_{XX} is the matrix of correlation functions between all inputs; and $\vec{\Phi}_{Xz}$ is the vector of cross-correlations between each input and the output. Further details on the structure of these matrices can be found in Westwick et al. (2006).

$$\hat{h} = \Phi_{XX}^{-1}\vec{\Phi}_{Xz} \tag{11.2}$$

This identification process was repeated separately for each muscle in which EMG was recorded. We previously assessed the influence of FIRF length on EMG prediction accuracy (Pohlmeyer et al., 2007b) and found that 500-ms filters produced reliable and consistent predictions. That is the filter length that was used throughout this study. The FIRFs were estimated using 10–20 minutes of training data. When these numbers of data are available, we have found, Eq. (11.2) can be used directly and all available inputs can be incorporated into the identification. When fewer data are available, the identification process can be made more robust by restricting it to an optimal set of inputs (Sanchez et al., 2004; Westwick et al., 2006; Pohlmeyer et al., 2007b) and by using robust techniques to perform the matrix inversion required in Eq. (11.2) (Westwick et al., 2006).

One problem with using linear filters to predict EMG is the difficulty of accurately estimating periods of muscle quiescence. Low levels of noise in the estimates may substantially reduce the precision of control. This is true for both kinematic and EMG-based BMIs. However, it is a particularly important problem when using EMG predictions to drive FES because even low levels of continuous stimulation can increase the rate of muscle fatigue. We have demonstrated that nonlinear cascade models consisting of the linear FIRFs described earlier, followed by a static nonlinearity with characteristics similar to those of a threshold function, can enhance prediction of periods of muscle quiescence (Pohlmeyer et al., 2007b). Such models can be estimated from experimental data using published techniques (Bussgang et al., 1974; Hunter and Korenberg, 1986). In a closed-loop system as described in this chapter, the problem can be solved as well by employing a simple threshold function between the predicted muscle activations and the muscle stimulation commands. Such a threshold is commonly incorporated into FES systems to prevent small changes in the command signal to each muscle from triggering the onset of stimulation. Because our FES system incorporated such a threshold, we found it unnecessary to incorporate an additional static nonlinearity in our identification process; the use of two similar nonlinearities would have been redundant and would have led to unnecessary system complexity. The process of setting this threshold for FES stimulation is described below.

11.3.6 Nerve Block Effectiveness

It was important to assess the level of nerve block in these experiments, but compensating for changes in the monkey's level of motivation complicates an accurate assessment. However, we took advantage of our ability to predict the level of EMG activity to improve our estimates. We recorded 30-second segments of EMG activity together with the corresponding M1 discharge at frequent intervals throughout the experiment, both before and after nerve block. Since in most cases, the monkey was unable to complete the task under block, he was rewarded simply for attempting to do so. We computed the ratio between the predicted and the actual EMG activity as a measure of nerve block effectiveness. The percent block was expressed simply as

$$\text{Percent Block} = 100 \left(1 - \frac{EMG_{Actual}}{EMG_{Pr\,ed}} \right) \tag{11.3}$$

11.3.7 Stimulation

In order to use cortical recordings to control an FES system in real time, the basic EMG predictions need to be mapped into appropriate stimulator commands to cause the desired muscle activation. Our typical closed-loop FES experiments involved the stimulation and control of four to six muscles using a 16-channel stimulator with a common return electrode (FNS16, CWE). The stimulator output consisted of pulse width–modulated trains of biphasic pulses of fixed current and frequency. Stimulus frequency was common to all channels, typically 25 or 30 Hz, which was high enough to achieve nearly complete fusion of muscle twitches yet low enough to minimize fatigue effects. In a separate set of experiments for each monkey, we determined the current (typically 2–10 mA) necessary for each muscle such that pulse widths in the 50–100-μs range provided near-maximal muscle force. Using this stimulus frequency and current, we determined the threshold pulse width necessary to produce detectable force, as well as the pulse width at which significant nonlinear "spillover" effects were noted in the relation between pulse width and force. In any given session, we mapped the point at which the EMG prediction exceeded the noise floor by approximately two standard deviations to the stimulus threshold, and the peak EMG prediction to the spillover point or 200 μs, whichever was smaller.

Exceptions to these rules occurred when, during a closed-loop experiment, muscle fatigue significantly affected task performance. To compensate for the decreasing stimulation-evoked muscle force, we often increased the stimulation frequency to 30 Hz and, if needed, increased the currents by 25% or 50%.

11.3.8 Real-Time Control

Online sorting was performed to discriminate action potential waveforms based on the waveform clustering in principal component feature space. We made some effort to identify waveforms that appeared to result from single neurons, based on standard criteria: well-defined waveform clusters and a minimum inter-spike interval greater than 1.5 ms. However, the overlapping activity of well-modulated multiple units recorded from a single electrode were also included in our neural signal population despite failing the requirements for single-unit classification.

As soon as an action potential was discriminated, the Cerebus sent a UDP packet with the timestamp and neuron identification data encoded. A process running on a real-time Linux system received the packets and sorted the timestamps into different circular buffers for each neuron. The circular buffers were converted into the binned firing rates and passed to the linear filter that generates the predicted EMG activation. The EMG predictions were mapped to stimulus pulse width commands, as described above, and sent to the stimulator over a serial connection. The stimulator used these new values to update the stimulus pulses delivered on subsequent clock cycles.

11.4 RESULTS

The following subsections describe the results we achieved in our FES experiments.

11.4.1 Offline Signal Prediction

The data for our experiments were collected from two monkeys. One (EO) was trained to do both the 1D isometric wrist task and the ball grasp task. The other (OR) was trained to do the isometric wrist task under both 1D and 2D conditions. EMG predictions for the 1D task were based on seven sessions from monkey EO and 5 from monkey OR. EMG predictions for the 2D wrist task were based on five sessions with monkey OR. Grasp task results were from 17 sessions with monkey EO.

11.4.1.1 *Isometric Wrist Tasks*

Torque during the isometric wrist task was generated primarily by the flexor and extensor carpi radialis and ulnaris muscles. These muscles each pull roughly in the directions suggested by their names. *Radialis* and *ulnaris* refer to the radius and ulna bones of the forearm, situated on the thumb and little finger sides, respectively. These muscles were differentially modulated during the torques to the different targets. Both the neural signals and the EMG activity were well modulated during the 1D isometric wrist task, as can be seen by comparison with the torque signal (Figure 11.3). We were able

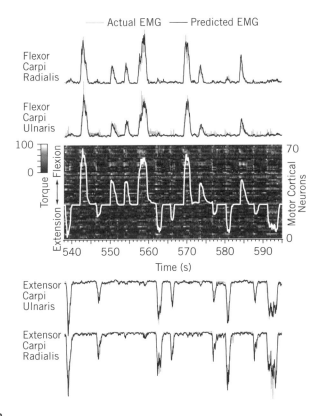

FIGURE 11.3

EMG predictions for four wrist muscles that generate isometric wrist torque. Predictions were based on the activity of 74 neurons. The discharge rate of each neuron was normalized to its maximum firing rate within this period. The corresponding flexion/extension torque signal is overlaid. Please see this figure in color at the companion web site: www.elsevierdirect.com/companions/9780123750273

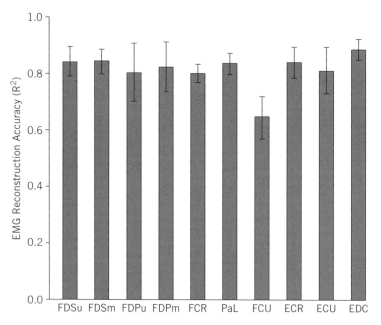

FIGURE 11.4

EMG prediction accuracy for each of the implanted muscles during the 1D wrist task. Bars show mean $+/-1SD$ (one standard deviation) of the multi-fold, cross-validation predictions calculated across 14 data sets.

to use information from the neural signals and simple linear filter models to predict the modulation of each muscle quite accurately. In this example, we used approximately 10 minutes of activity recorded from 77 neural signals to compute the filters. The predicted EMG signals shown here were computed from test data that were not used for generation of the decoder models. We quantified the accuracy of the predicted signals by computing the R^2 value between the actual and predicted EMG signals across the seven data sets collected from monkey EO. With a single exception (FCU), the average multifold cross-validation R^2 values for the task-relevant muscles were at or above 0.80 (Figure 11.4).

In a similar fashion, we used information from more than 100 neural signals recorded from monkey OR to predict EMG activity that occurred during the 2D task. In this task, the spatial patterns of muscle activity necessary to reach each of the targets were more varied than were those for the simple 1D task. The different combinations of EMG activity corresponding to the eight targets can be readily appreciated from Figure 11.5. In this example, prediction accuracy of the ulnaris muscles was somewhat lower than that of the examples shown

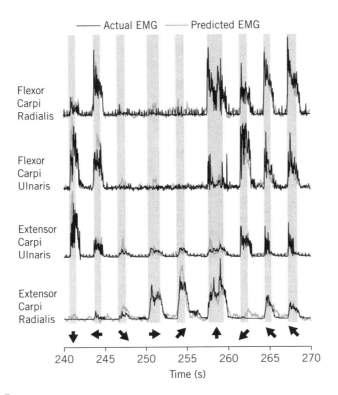

FIGURE 11.5

Predicted EMG activity for the 2D center-out isometric wrist task computed from more than 100 neural signals. Large arrows at the bottom of the figure indicate the direction of the target presented to the monkey during the period indicated by the gray shaded regions. Please see this figure in color at the companion web site: www.elsevierdirect.com/companions/9780123750273

for the 1D task. However, this did not appear to be a function of the higher dimensionality of the task. The predicted EMG signals were somewhat less accurate for both versions of the wrist task for data from monkey OR compared to monkey EO. There were, however, no consistent differences between the tasks for monkey OR (Figure 11.6).

11.4.1.2 *Grasp Task*

The grasp task differed from the isometric wrist task in several respects. It involved more complex muscle activation patterns, including co-activation, rather than reciprocal activation of antagonistic muscles of the wrist and fingers. A mix of brief, phasic bursts and longer tonic patterns of EMG activity typically appeared in a rapid sequence, making this task much less stereotypic than the isometric wrist task from trial to trial. Figure 11.7 is an example from

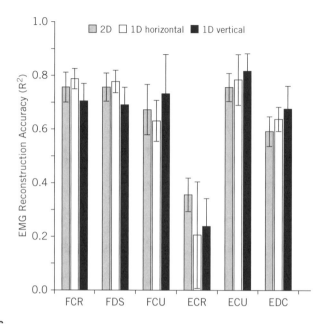

FIGURE 11.6

EMG prediction accuracy during the 2D wrist task for a single, representative day. There was no systematic difference in prediction accuracy between the 1D and 2D tasks.

monkey EO of three EMG signals recorded from finger and wrist flexors and their corresponding activity predicted from 106 neural signals. Flexor digitorum profundus (FDP), the main flexor muscle of the digits, was characterized by a relatively simple, several-hundred-ms burst that occurred during the time the ball was being grasped ("pick up"). The accuracy of this predicted signal was similar to that of the best signals during the isometric task. Flexor digitorum superficialis (FDS), on the other hand, was more complex, with briefer bursts that occurred somewhat less predictably, sometimes with repeated bursts within the grasp, and occasionally with short bursts between trials when the monkey was returning his hand to the touchpad. FDS was predicted with considerably lower accuracy than during the isometric task (Figure 11.3), but with values that were still close to those typically reported elsewhere for kinematic reaching signals. The wrist flexor FCR had perhaps the most complex pattern of activity and yet, in this example, it was predicted with an accuracy between that of FDP and FDS.

The results shown in Figure 11.6 are representative of the overall results from the grasp task. Predicted EMG signals typically had R^2 values somewhat lower than those of the isometric wrist task. The average multifold cross-validation R^2 values computed from 17 training data sets ranged

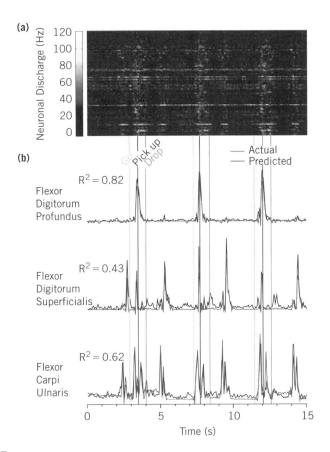

FIGURE 11.7

Grasp-related raw data: (a) the firing rates of 98 neuronal signals and (b) the corresponding actual and predicted EMG activity during a series of three normal grasps. The neuronal signals were strongly correlated with the bursts of muscle activity around the time the monkey began to reach toward ("Go") and grasp the ball ("Pick up"). The R^2 values calculated between the actual and predicted EMG signals for these muscles are reported on each graph. Please see this figure in color at the companion web site: www.elsevierdirect.com/companions/9780123750273

between 0.40 and 0.76 for six wrist and finger flexor muscles, with an average of 0.53.

11.4.2 Real-Time FES Control

The brain-controlled, real-time FES results from the isometric wrist task were based on seven experimental sessions with monkey EO. For the grasp task, the prediction results were from 17 model-building sessions with EO, and those for real-time FES control were from 20 experiments.

11.4.2.1 *Isometric Wrist Task*

As described above, the primary muscles involved in the isometric wrist task are the wrist flexors FCU and FCR and the extensors ECU and ECR. Because of the relatively weak forces generated by electrical stimulation and the substantial problem of fatigue (described later), we also stimulated the digit flexors FDS and FDP as well as the digit extensor EDC to increase the available force. Figure 11.8 illustrates the amount of wrist torque that could be generated by the monkey under blocked conditions and with BMI-controlled FES in the flexion and extension directions. While flexion torque reached 75% of normal, extension torque was limited to about 50%. However, the decreased magnitude of wrist torque was not the only control limitation imposed by FES activation. The stability of the torque was also less than normal, for at least two reasons.

For one, even with a stimulus frequency of 30 Hz, the muscle contractions were typically not completely fused since FES causes synchronous activation of motor units. This created a small ripple in the muscle force. In addition, there were lower-frequency fluctuations in the applied force, which may have

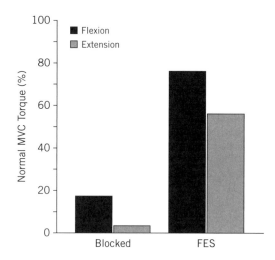

FIGURE 11.8

Maximum voluntary contraction (MVC) wrist torque under nerve block and FES conditions as a percentage of normal. The nerve blocks resulted in greatly diminished wrist strength. Cortically controlled FES allowed both monkeys to generate much greater torque voluntarily during the block.

been due to inadequacies in the predictive models of EMG or perhaps due to the loss of normal proprioceptive feedback that would also have been caused by the nerve block.

Furthermore, as noted later (Figure 11.10), FES-generated force was rather quickly reduced through fatigue—a consequence of abnormal motor unit recruitment, in this case because electrical stimulation preferentially recruited the large, fatigable fibers. Because fatigue was greater in the flexor muscles than in the extensors, the net result was that the practical operating torque range in both directions was limited to about 25% of normal. Consequently, the monkey's ability to reach targets at various levels was fundamentally limited by the precision with which the torque could be controlled and by the maximum torque that could be sustainably generated.

We chose the target size under FES control such that two nonoverlapping flexion targets could be presented within the monkey's assisted range. This target size was approximately twice that used under normal conditions. If the FES force range was divided into three different targets, the monkey was typically unable to control the torque level with sufficient precision. It was rarely possible to use more than one extension target. An example of three-target, cortically controlled FES performance is shown in Figure 11.9. To reduce the effects of fatigue, we used twice as many low-flexion as high-flexion targets. For all three targets, the monkey was able to generate appropriate torque. As under normal conditions, a successful trial required a 0.5-second hold time. In this one-minute segment of data, all of the FES-assisted trials were successful (light gray targets). However, catch trials were introduced in random trials, in which the stimulators were turned off unexpectedly at the time of the go cue. These trials served to demonstrate that the FES was necessary for the monkey to be able to complete the task. Two of these occurred within this series of trials (dark gray targets). During these trials, the monkey was essentially unable to produce any useful torque. However, although his attempts were unsuccessful, it is important to note that the neural discharge patterns could be used to indicate that he was still trying to complete the task. Note the discharge that occurred at the beginning of the two catch trials in Figure 11.9 at 547 and 573 seconds. Indeed, the activation during the second of the two catch trials was as strong as or stronger than that during any of the other FES trials. Similar results were obtained over the seven experimental sessions with monkey EO. Over these sessions, the monkey was unable to complete a single catch trial in 71 attempts. Under FES conditions, he performed at a 65% success rate compared to 95% under normal conditions.

Muscle fatigue is a serious problem in all FES applications, as electrical stimulation essentially reverses the normal motor unit recruitment order. Much of the force is produced by the activation of the largest, most

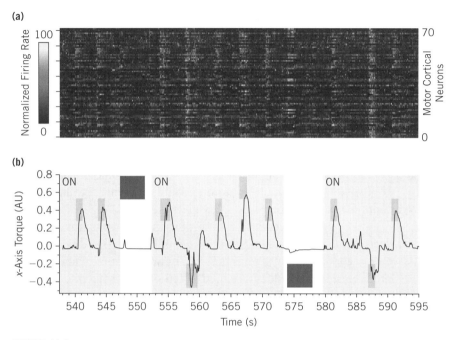

FIGURE 11.9

Cortically controlled 1D isometric wrist task during paralysis. (a) Firing rate plot of the neural activity that was used to control muscle stimulation. (b) Rectangles indicate the torque targets, including successful (medium gray) and failed (dark gray) trials. The cortically controlled FES system was operating only during the shaded periods. Note that the failures occurred during catch trials when the FES was turned off. However, the firing rate shows that the monkey was trying to do the task during these periods. Please see this figure in color at the companion web site: www.elsevierdirect.com/companions/9780123750273

fatigable muscle fibers. We found that as little as 5–10 minutes of trials under brain-controlled FES caused a significant force reduction. In addition to the direct effects on performance, fatigue also adversely affects the monkey's motivation. We quantified the amount of fatigue that occurred over a typical FES experimental session by stimulating with fixed trains (10 mA, 200 µs for 3 seconds) at the beginning and end of the session. The results for five such sessions are shown in Figure 11.10. The decrease in flexion torque was considerable, dropping to 33% of its initial value. The change in extension torque was much less pronounced, probably because there were fewer extension targets, of lower torque in the course of a typical FES session.

11.4.2.2 *Grasp Task*

Real-time FES experiments for grasp were carried out under median and ulnar nerve block, with the radial nerve intact. This paralyzed the wrist and finger flexors but left the extensors under the monkey's normal control. Thus, the

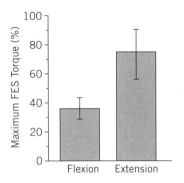

FIGURE 11.10

Fatigue resulting from FES at the end of the experimental session. There was a significant decrease in the torque generated by a standard stimulus train between the beginning and end of a typical FES session.

monkey was capable of opening his hand to prepare to grasp the ball or to release the ball without the assistance of the FES, but he needed FES assistance to close his hand to complete the grasp. Each experimental session began with a period of data collection under normal conditions in order to establish baseline performance and muscle activation levels. An example of one such block is shown in Figure 11.11. We injected lidocaine to block the median and ulnar nerves at time 0. After 15 minutes, the block was essentially complete, as determined by the loss of flexor muscle EMG activity (see Section 11.3) and the onset of significant motor deficits. The remainder of the session consisted of a series of 10-minute blocks in which the monkey attempted to complete the grasp task either with or without FES assistance. In most of the blocks, the monkey was not successful in any of the unassisted catch trials. He completed only 4 out of 57 catch trials, primarily by adopting adaptive strategies to scoop the ball in his hand using unblocked muscles and passive forces. In contrast, the average success rate in this session with FES assistance was 80%.

The success rates over 20 experimental sessions during normal-condition, closed-loop control and catch trials are summarized in Figure 11.12(a). FES assistance improved the rate by a factor of 7. We also quantified the monkey's performance in terms of the time required to complete a successful trial. We measured the time from the monkey's initial contact with the ball to the successful ball grasp for all three conditions. Under nerve block, the closed-loop FES system significantly improved the speed at which the monkey successfully completed trials (Figure 11.12(b)). It is important to note that the catch trial times represent only the successful trials, achieved mostly through an alternate motor strategy. The great majority of these trials were not successful.

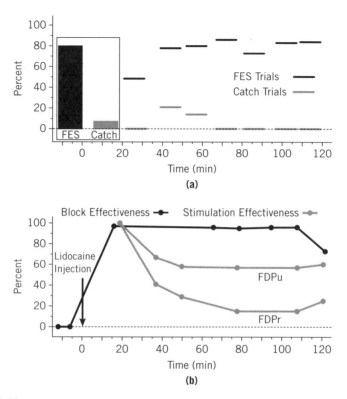

FIGURE 11.11

Summary of grasp task performance during median and ulnar nerve block. (a) Success rate for the seven experimental data sets for both FES trials and randomly inserted catch trials during which the stimulators were turned off. The monkey was essentially unable to pick up the ball during the catch trials, completing only 4 of 57 attempts. (b) Both nerve block effectiveness and stimulus effectiveness for FDPu and FDPr were tested at frequent intervals during the experiment. Please see this figure in color at the companion web site: www.elsevierdirect.com/companions/9780123750273

11.5 DISCUSSION

The following subsections discuss the successful outcomes of our FES experiments as well as the limitations of FES those experiments revealed.

11.5.1 Successful Proof of Concept

Our results demonstrate the feasibility of using a motor cortical BMI to control an FES system for restoring hand and wrist function following paralysis. Specifically, we demonstrated its utility for controlling the wrist in one and potentially two degrees of freedom, and for controlling simple hand grasp.

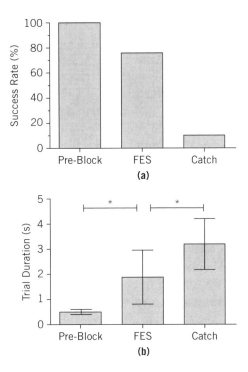

FIGURE 11.12

Summary of the results of 20 FES experimental sessions. (a) Success rates for normal, FES, and catch trial conditions comprising nearly 6000 trials. The monkey completed 77% of the trials with FES but only 11% when the stimulators were turned off. (b) Speed at which the successful trials were completed was significantly different for all three conditions ($p < 0.0001$, two-tailed Mann Whitney test). The great majority of catch trials failed as a result of the 5-second time-out.

In contrast to previous BMI studies, ours focused on decoding the intended muscle activation patterns rather than limb kinematics or torques. This greatly simplifies the use of a BMI for FES control, since the goal of an FES system is specifically to control the activation of paralyzed muscles. This approach allows us to harness the computational power of the brain for controlling the dynamics of movements; we do not have to rely on neuromechanical simulations to infer the most appropriate muscle activation patterns from a decoded movement trajectory.

11.5.2 Limitations in the Control of Complex Motor Tasks

The simple linear models used here yielded remarkably accurate EMG signal predictions, accounting for over 80% of the EMG variance during the isometric wrist task. The somewhat lower R^2 values in the ball grasp task may have arisen from the task's greater complexity. However, if this were the case, we would have expected to see a similar drop in the quality of predicted signals in the two-dimensional isometric wrist task compared to that in the one-dimensional task, was not the case.

In earlier work, we showed that decoders developed from training data collected in one task often did not generalize well to muscle activity recorded in a separate task, despite the fact that signals from either task could be accurately predicted independently (Pohlmeyer et al., 2007b). This phenomenon of limited task generalization may also have been at play when we used the cortical recordings to predict muscle activity during real-time control of hand grasp using FES. The EMGs recorded normally during the ball grasp task were transient, occurring only for brief periods associated with the ballistic nature of the movements used when unimpaired monkeys completed this task. However, the M1 and EMG activity used for decoder training data also included considerable sustained muscle activity resulting from the fact that the monkey often gripped a bar on the primate chair between successive ball grasps. The decoders constructed from this sustained activity appeared to have had somewhat poorer predictive power for the transient EMG recorded from FDS during the ball grasp task.

In addition to its dimensionality, the grasp task differs in another important way from the isometric wrist task. The latter was performed under FES control with patterns of predicted EMG that were very similar to patterns of normal EMG (Pohlmeyer et al., 2009). Although success rates were a bit lower and torque rise time a bit slower, behavior under FES was qualitatively normal. This was not the case with the grasp task. Although FES made it possible for the monkey to grasp the ball successfully despite paralysis, the movements were not normal. The overall posture of the hand was affected, and the strategies used differed noticeably from the norm. This was likely due to the fact that only a subset of the paralyzed muscles were activated by FES, requiring the monkey to adapt his motor strategies so that the grasping task could be completed. Hence, the real-time EMG predictions under FES conditions may have essentially involved generalization not unlike that between explicitly different motor tasks.

There are at least two important factors unrelated to the use of FES that may lead to problems with generalization. The first involves the high correlation of muscle activation patterns across many common tasks. Although the primate motor system is highly redundant with respect to the number of muscles available to control the skeletal degrees of freedom, activity within these muscles is highly correlated. These correlations lead to significant challenges when developing a robust neural decoder capable of generalizing to motor tasks other than those of the training data. Cortical neurons that contribute primarily to the activation of one muscle may nonetheless give excellent predictions of the activity of a second muscle, provided that both muscles remain well correlated. However, a decoder generated from such data would fail to generalize to novel tasks in which these two muscles were not used synergistically. This potential problem emphasizes the need for a robust training set that encompasses the range of muscle activation patterns relevant to the functional tasks of interest.

A second issue is that multiple neural pathways contribute to the activity of a given muscle. As one example, consider the role of peripheral feedback, which may alter muscle activity independent of descending commands. Proprioceptive feedback varies substantially with the environment a user interacts with. Unless the properties of this environment are also available as inputs to the decoder, generalization to interactions with different environments are not reliable.

A related challenge is that the role of motor cortical neurons can vary in a task-dependent manner, as demonstrated by Muir and Lemon (1983). These researchers identified motor cortical neurons whose relationship to the activity of their target muscle varied depending on whether a precision grip or a power grip was being performed. Similarly, Tanji and Evarts observed a differential role for motor cortical neurons in feedback control of distal arm muscles depending upon whether precision or ballistic movements were performed (Fromm and Evarts, 1981). Each of these findings illustrates the task-dependent role of the motor cortex in the generation of muscle activation and the corresponding challenges in developing a robust decoder for muscle activity based on recordings from a limited number of motor cortical neurons.

11.5.3 Limitations Related to the Use of FES for Control

There may be advantages to our approach as a result of the dynamical similarity between motor cortical discharge and muscle activity. However, the use of FES presents a number of formidable technical difficulties. Most are related to the efficacy with which muscles can be stimulated electrically and how well that stimulation can be used to replicate natural control.

One of the challenges encountered when using FES is fatigue, which can lead to a changing relationship between motor cortical activity and muscle activation. Electrical stimulation tends to invert the normal recruitment order, activating the fast fatigable fibers before the slow, fatigue-resistant fibers. In many of our experimental sessions, fatigue was evident within 5–10 minutes. Muscle fatigue is a complex process to which multiple mechanisms can contribute (Enoka and Stuart, 1992). These range from changes in the contractile properties of muscle to changes in descending motor cortical commands. Mechanisms with a peripheral origin, as in this case, alter the relationship between muscle stimulation and force, introducing an error between the intended motor output and that achieved by the FES system. The problem of fatigue may be mitigated somewhat in a chronic application because chronically stimulated muscles develop some level of fatigue resistance (Peckham et al., 1976; Kernell et al., 1987). Also, it may be true that most real-world applications would not require the intensive stimulation within a relatively short period of time that was typical of our experimental sessions.

FES also is limited in its ability to re-create the dexterity of movement observed in natural control. This limitation arises from the number of muscles that can be implanted in a given subject and the efficacy with which those muscles can be stimulated. At most, we stimulated six muscles in these experiments, which was sufficient to restore a rudimentary form of palmar grasping but much less than what would be needed for more dexterous control. While stimulators with more channels exist, they may not be sufficient to restore natural control because of the many challenges associated with controlling muscle force electrically.

The first of these challenges is associated with the difficulty of fully recruiting the fibers within a given muscle. We used electrodes to activate each of the FES target muscles in our experiments. These have the advantage of being highly selective, at least at low stimulation levels. However, it is difficult to activate an implanted muscle fully, since the high-stimulation currents needed to achieve full activation often spread to neighboring muscles, resulting in "spillover" stimulation that degrades the fidelity of individual muscle control.

Cuff electrodes, placed around the nerves innervating each target muscle, can be used to generate much more complete muscle activation with substantially lower currents. However the recruitment order problem remains. This technique works well for larger muscles, but is not feasible for smaller ones, many which have restricted access to their isolated motor nerves. Furthermore, beyond the multiple heads that are recognized in some muscles, many (in particular the extrinsic digit muscles) have multiple compartments that preferentially affect individual digits.

An alternative is to use multicontact electrodes placed more proximally on peripheral nerves innervating multiple muscles of interest. We have begun to test this approach using both multicontact flat interface nerve electrodes (FINE) (Leventhal and Durand, 2003) and the multi-electrode Utah "slant" array (Branner et al., 2004). These technologies hold the promise of effective, selective activation of muscles or muscle compartments, including the intrinsic hand muscles. However, their utility for dexterous control of the primate hand during functional tasks has yet to be demonstrated.

11.6 **FUTURE DIRECTIONS**

Although our present results demonstrate the feasibility and promise of a muscle-based decoder for FES-controlled hand grasp, there are a number of areas in which the current system might be expanded or improved. Each presents interesting applications and challenges in the area of signal processing.

11.6.1 Control of Higher-Dimensional Movement Using Natural Muscle Synergies

Control of more complex, higher-dimensional grasping movements through FES will certainly require refinements in our ability to activate many muscles. However, generating the necessary control signals that can be expected to remain robust across a variety of movements will also present considerable challenges. There are at least 30 muscles with insertions distal to the elbow that are involved in the control of hand movements. Despite this fact, the kinematics of even rather complex manipulation tasks can be predicted with only six or seven components (Santello and Soechting, 2000; Todorov and Ghahramani, 2004). A number of studies have reported that several different groups of muscles appear to be activated synergistically across a broad range of movements. This inferred organization is based on spinal cord stimulation as well as recordings from groups of muscles during behavior (Kargo and Giszter, 2000; d'Avella et al., 2003; Ting and Macpherson, 2005; Tresch and Jarc, 2009). The groups appear to act as a basis set for the construction of more complex movements.

This organization also seems to be reflected in the central nervous system, where it has been shown that that individual neurons in M1 nearly always project to more than one muscle (Fetz and Cheney, 1980; Schieber, 1996). It is possible to describe the "muscle space" preferred directions of individual M1 neurons in terms of the muscle or muscles with which a given neuron's activity is correlated. Neurons tend to cluster within restricted regions of this muscle space, in areas that correspond to groups of synergistically activated muscles (Holdefer and Miller, 2002; Morrow et al., 2007). The appeal of these results is that they suggest dimensionality reduction may occur within the nervous system, perhaps at the spinal level, to reduce the complexity of control at the central level.

Although there remains some debate as to the role or even existence of muscle synergies in the control of natural movement, such an approach may provide a useful tool for controlling movements through FES. Importantly, the use of synergies that exploit the natural dynamics of the musculoskeletal system may be used to simplify control without substantially sacrificing motor performance or flexibility (Berniker et al., 2009). By specifically decoding the activation patterns for groups of muscle synergies rather than individual muscles, it may also be possible to increase the robustness of an FES controller. This increase would arise from the reduced chance of erroneously predicting the activation of an individual muscle, since muscles with a relatively small weighting in a given synergy would have the benefit of being associated with the commands to muscles with a deeper modulation and therefore an increased signal-to-noise ratio. A number of computational techniques have been used to estimate muscle synergies, including principal component analysis, independent component analysis, and non-negative matrix factorization (see the comparison of algorithms in Tresch etal. (2006)). The last has the benefit of restricting muscle activations to non-negative values, which can be directly mapped to muscle stimulation commands.

11.6.2 **Adaptation**

While we stress the utility of simple linear decoders, it is possible that further performance gains may be made with adaptive decoders. Such decoders are likely to be beneficial for the implementation of practical BMIs, in which training data is generally not available (Hochberg et al., 2006; Kim et al., 2008). Just as the learning of complex motor tasks requires substantial adaptation at the cortical, subcortical, and peripheral levels of the neuromuscular system, it is likely that the challenges just identified will be partially met by a system in which the user has the ability to adapt neural commands as needed to obtain the desired motor output. Indeed, in our experiments we did not find a strong correlation between our ability to predict EMG signals offline and the online, closed-loop performance of the same decoder. This suggests that the monkeys were able to adapt their motor commands to compensate for changes in the relationship between their intended movements and those actually produced by the FES.

A co-adaptive BMI controller that benefited from both motor learning and algorithm adaptation was described several years ago (Taylor et al., 2002). In that case, the decoder was modified after trials containing errors so that the pattern of recorded neural discharge would generate the desired movement. More recently, investigators tested the ability of human subjects to learn to control a two-dimensional cursor using a 19-dimensional digit movement signal recorded from a data glove. They discovered that small changes in the transform between digit movements and cursor movements that only partially corrected errors led to a more rapid increase in performance than did either a fixed decoder or large changes that fully corrected the errors (Danziger et al., 2008). A similar approach applied to higher-dimensional FES control signals that lacks fixed kinematic targets will present new challenges. It may be beneficial to combine this approach with the muscle synergy-based control outlined earlier.

Several more extreme examples of adaptation have been reported, in which monkey subjects were able to learn to adapt even to decoders that had been intentionally scrambled (Jarosiewicz et al., 2008; Ganguly and Carmena, 2009). In one case, monkeys were able to learn, over a number of days, two different decoders and to switch rapidly between them. Their performance on a fixed decoder was actually better after several days than their performance when a decoder was recomputed from new training data at the beginning of each session (Ganguly and Carmena, 2009). The investigators suggested that optimal performance would be achieved by allowing the monkeys simply to learn the properties of a fixed decoder. However, they did not attempt to modify the decoder optimally.

These results, demonstrating the importance of neural adaptation, are in accordance with our general philosophy of using relatively simple decoders and relying on the computational power of the brain to use them to their fullest potential. It remains to be seen if more complex, adaptive decoders will exhibit a substantial improvement in performance during tasks involving more degrees of freedom than those used in the present study. If so, they are likely to require adaptive algorithms tailored to the behavior and learning rates of each subject.

Acknowledgments

This work was supported in part by grant NS053603 from the National Institute of Neurological Disorders and Stroke to L.E. Miller, and by a postdoctoral fellowship from the Fonds de la Recherche en Santé du Québec to C. Ethier. Additional funding was supplied by the Chicago Community Trust through the Searle Program for Neurological Restoration. We also wish to acknowledge the surgical assistance of Dr. Sonya Paisley Agnew, Division of Plastic and Reconstructive Surgery, Northwestern University Feinberg School of Medicine.

References

Alon, G., McBride, K., 2003. Persons with C5 or C6 tetraplegia achieve selected functional gains using a neuroprosthesis. Arch. Phys. Med. Rehabil. 84, 119–124.

Anderson, K.D., 2004. Targeting recovery: priorities of the spinal cord-injured population. J. Neurotrauma 21, 1371–1383.

Batista, A.P., Buneo, C.A., Snyder, L.H., Andersen, R.A., 1999. Reach plans in eye-centered coordinates. Science 285, 257–260.

Berniker, M., Jarc, A., Bizzi, E., Tresch, M.C., 2009. Simplified and effective motor control based on muscle synergies to exploit musculoskeletal dynamics. Proc. Natl. Acad. Sci. U. S. A. 106, 7601–7606.

Branner, A., Stein, R.B., Fernandez, E., Aoyagi, Y., Normann, R.A., 2004. Long-term stimulation and recording with a penetrating microelectrode array in cat sciatic nerve. IEEE Trans. Biomed. Eng. 51, 146–157.

Bussgang, J.J., Ehrman, L., Graham, J.W., 1974. Analysis of nonlinear systems with multiple inputs. Proc. IEEE 62, 1088–1119.

Carmena, J.M., Lebedev, M.A., Crist, R.E., O'Doherty, J.E., Santucci, D.M., Dimitrov, D., et al., 2003. Learning to control a brain-machine interface for reaching and grasping by primates. PLoS Biol. 1, 193–208.

Cheney, P.D., Fetz, E.E., 1980. Functional classes of primate corticomotorneuronal cells and their relation to active force. J. Neurophysiol. 44, 773–791.

Creasey, G., Elefteriades, J., DiMarco, A., Talonen, P., Bijak, M., Girsch, W., et al., 1996. Electrical stimulation to restore respiration. J. Rehabil. Res. Dev. 33, 123–132.

Danziger, Z., Fishbach, A., Mussa-Ivaldi, F.A., 2008. Adapting human-machine interfaces to user performance. Conf. Proc. IEEE Eng. Med. Biol. Soc. 1, 4486–4490.

d'Avella, A., Saltiel, P., Bizzi, E., 2003. Combinations of muscle synergies in the construction of a natural motor behavior. Nat. Neurosci. 6, 300–308.

Enoka, R.M., Stuart, D.G., 1992. Neurobiology of muscle fatigue. J. Appl. Physiol. 72, 1631–1648.

Evarts, E.V., 1968. Relation of pyramidal tract activity to force exerted during voluntary movement. J. Neurophysiol. 31, 14–27.

Fagg, A.H., Ojakangas, G.W., Miller L.E., Hatsopoulos. N.G., 2009. Kinetic trajectory decoding using motor cortical ensembles. IEEE Trans. Neural Syst. Rehabil. Eng. 17, 487–496.

Fetz, E.E., Cheney, P.D., 1980. Postspike facilitation of forelimb muscle activity by primate corticomotoneuronal cells. J. Neurophysiol. 44, 751–772.

Fetz, E.E., Finocchio, D.V., 1975. Correlations between activity of motor cortex cells and arm muscles during operantly conditioned response patterns. Exp. Brain Res. 23, 217–240.

Fromm, C., Evarts, E.E., 1981. Relation of size and activity of motor cortex pyramidal tract neurons during skilled movements in the monkey. J. Neurosci. 1, 453–460.

Galvani, L., 1791. De viribus electricitatis in motu musculari commentarius. Bon. Sci. Art. Inst. Acad. Comm. 7, 363–418.

Ganguly, K., Carmena, J.M., 2009. Emergence of a stable cortical map for neuroprosthetic control. PLoS Biol. 7, e1000153.

Gao, Y., Black, M.J., Bienenstock, E., Wu, W., Donoghue, J.P., 2003. A quantitative comparison of linear and non-linear models of motor cortical activity for the encoding and decoding of arm motions. In: 1st International IEEE/EMBS Conference on Neural Engineering, vol. 1. Capri, Italy, pp. 189–192.

Georgopoulos, A.P., Ashe, J., Smyrnis, N., Taira, M., 1992. The motor cortex and the coding of force. Science 256, 1692–1695.

Georgopoulos, A.P., Kalaska, J.F., Caminiti, R., Massey, J.T., 1982. On the relations between the direction of two-dimensional arm movements and cell discharge in primate motor cortex. J. Neurosci. 2, 1527–1537.

Graupe, D., Kohn, K.H., 1998. Functional neuromuscular stimulator for short-distance ambulation by certain thoracic-level spinal-cord-injured paraplegics. Surg. Neurol. 50, 202–207.

Handa, Y., Hoshimiya, N., 1987. Functional electrical stimulation for the control of the upper extremities. Med. Prog. Technol. 12, 51–63.

Hart, R.L., Kilgore, K.L., Peckham, P.H., 1998. A comparison between control methods for implanted FES hand-grasp systems. IEEE Trans. Rehabil. Eng. 6, 208–218.

Hatsopoulos, N., Joshi, J., O'Leary, J.G., 2004. Decoding continuous and discrete motor behaviors using motor and premotor cortical ensembles. J. Neurophysiol. 92, 1165–1174.

Hepp-Reymond, M.C., Wyss, U.R., Anner, R., 1978. Neuronal coding of static force in the primate motor cortex. J. Physiol. Paris 74, 287–291.

Hochberg, L.R., Serruya, M.D., Friehs, G.M., Mukand, J.A., Saleh, M., Caplan, A.H., et al., 2006. Neuronal ensemble control of prosthetic devices by a human with tetraplegia. Nature 442, 164–171.

Holdefer, R.N., Miller, L.E., 2002. Primary motor cortical neurons encode functional muscle synergies. Exp. Brain Res. 146, 233–243.

Hoshimiya, N., Naito, A., Yajima, M., Handa, Y., 1989. A multichannel FES system for the restoration of motor functions in high spinal cord injury patients: a respiration-controlled system for multijoint upper extremity. IEEE Trans. Biomed. Eng. 36, 754–760.

Humphrey, D.R., Schmidt, E.M., Thompson, W.D., 1970. Predicting measures of motor performance from multiple cortical spike trains. Science 170, 758–761.

Hunter, I.W., Korenberg, M.J., 1986. The identification of nonlinear biological systems: Wiener and Hammerstein cascade models. Biol. Cybern. 55, 135–144.

Jarosiewicz, B., Chase, S.M., Fraser, G.W., Velliste, M., Kass, R.E., Schwartz, A.B., 2008. Functional network reorganization during learning in a brain-computer interface paradigm. Proc. Natl. Acad. Sci. U. S. A. 105, 19486–19491.

Johnston, T.E., Betz, R.R., Smith, B.T., Benda, B.J., Mulcahey, M.J., Davis, R., et al., 2005. Implantable FES system for upright mobility and bladder and bowel function for individuals with spinal cord injury. Spinal Cord 43, 713–723.

Kalaska, J.F., Hyde, M.L., 1985. Area 4 and area 5: differences between the load direction-dependent discharge variability of cells during active postural fixation. Exp. Brain Res. 59, 197–202.

Kargo, W.J., Giszter, S.F., 2000. Rapid correction of aimed movements by summation of force-field primitives. J. Neurosci. 20, 409–426.

Keith, M.W., Peckham, P.H., Thrope, G.B., Stroh, K.C., Smith, B., Buckett, J.R., et al., 1989. Implantable functional neuromuscular stimulation in the tetraplegic hand. J. Hand Surg. Am. 14, 524–530.

Kennedy, P.R., Bakay, R.A., 1998. Restoration of neural output from a paralyzed patient by a direct brain connection. Neuroreport 9, 1707–1711.

Kernell, D., Donselaar, Y., Eerbeek, O., 1987. Effects of physiological amounts of high- and low-rate chronic stimulation on fast-twitch muscle of the cat hindlimb. II. Endurance-related properties. J. Neurophysiol. 58, 614–627.

Kim, S.P., Simeral, J.D., Hochberg, L.R., Donoghue, J.P., Black, M.J., 2008. Neural control of computer cursor velocity by decoding motor cortical spiking activity in humans with tetraplegia. J. Neural Eng. 5, 455–476.

Lauer, R.T., Peckham, P.H., Kilgore, K.L., 1999. EEG-based control of a hand grasp neuroprosthesis. Neuroreport 10, 1767–1771.

Lebedev, M.A., Nicolelis, M.A., 2006. Brain-machine interfaces: past, present and future. Trends Neurosci. 29, 536–546.

Leventhal, D.K., Durand, D.M., 2003. Subfascicle stimulation selectivity with the flat interface nerve electrode. Ann. Biomed. Eng. 31, 643–652.

Liberson, W.T., Holmquest, H.J., Scot, D., Dow, M., 1961. Functional electrotherapy: stimulation of the peroneal nerve synchronized with the swing phase of the gait of hemiplegic patients. Arch. Phys. Med. Rehabil. 42, 101–105.

Miller, L.E., van Kan, P.L.E., Sinkjaer, T., Andersen, T., Harris, G.D., Houk, J.C., 1993. Correlation of primate red nucleus discharge with muscle activity during free-form arm movements. J. Physiol. London 469, 213–243.

Moritz, C.T., Perlmutter, S.I., Fetz, E.E., 2008. Direct control of paralysed muscles by cortical neurons. Nature 456, 639–642.

Morrow, M.M., Jordan, L.R., Miller, L.E., 2007. Direct comparison of the task-dependent discharge of M1 in hand space and muscle space. J. Neurophysiol. 97, 1786–1798.

Muir, R.B., Lemon, R.N., 1983. Corticospinal neurons with a special role in precision grip. Brain Res. 261, 312–316.

Musallam, S., Corneil, B.D., Greger, B., Scherberger, H., Andersen, R.A., 2004. Cognitive control signals for neural prosthetics. Science 305, 258–262.

Peckham, P.H., Keith, M.W., Kilgore, K.L., Grill, J.H., Wuolle, K.S., Thrope, G.B., et al., 2001. Efficacy of an implanted neuroprosthesis for restoring hand grasp in tetraplegia: a multicenter study. Arch. Phys. Med. Rehabil. 82, 1380–1388.

Peckham, P.H., Knutson, J.S., 2005. Functional electrical stimulation for neuromuscular applications. Annu. Rev. Biomed. Eng. 7, 327–360.

Peckham, P.H., Marsolais, E.B., Mortimer, J.T., 1980. Restoration of key grip and release in the C6 tetraplegic patient through functional electrical stimulation. J. Hand Surg. Am. 5, 462–469.

Peckham, P.H., Mortimer, J.T., Marsolais, E.B., 1976. Alteration in the force and fatigability of skeletal muscle in quadriplegic humans following exercise induced by chronic electrical stimulation. Clin. Orthop. Relat. Res. 114, 326–333.

Pfurtscheller, G., Muller, G.R., Pfurtscheller, J., Gerner, H.J., Rupp, R., 2003. 'Thought'–control of functional electrical stimulation to restore hand grasp in a patient with tetraplegia. Neurosci. Lett. 351, 33–36.

Phillips, C.G., 1975. Laying the ghost of 'muscles versus movements'. Can. J. Neurol. Sci. 66, 209–218.

Piccolino, M., 1998. Animal electricity and the birth of electrophysiology: the legacy of Luigi Galvani. Brain Res. Bull. 46, 381–407.

Pohlmeyer, E.A., Oby, E.R., Perreault, E.J., Solla, S.A., Kilgore, K.L., Kirsch, R.F., et al., 2009. Toward the restoration of hand use to a paralyzed monkey: brain-controlled functional electrical stimulation of forearm muscles. PLoS ONE 4, e5924.

Pohlmeyer, E.A., Perreault, E.J., Slutzky, M.W., Kilgore, K.L., Kirsch, R.F., Taylor, D.M., Miller, L.E., 2007a. Real-time control of the hand by intracortically controlled functional neuromuscular stimulation. In: IEEE 10th International Conference on Rehab Robotics, vol. 10. Noordwijk, The Netherlands, pp. 454–458.

Pohlmeyer, E.A., Solla, S.A., Perreault, E.J., Miller, L.E., 2007b. Prediction of upper limb muscle activity from motor cortical discharge during reaching. J. Neural Eng. 4, 369–379.

Popovic, D., Stojanovic, A., Pjanovic, A., Radosavljevic, S., Popovic, M., Jovic, S., et al., 1999. Clinical evaluation of the bionic glove. Arch. Phys. Med. Rehabil. 80, 299–304.

Popovic, D., Tomovic, R., Schwirtlich, L., 1989. Hybrid assistive system–the motor neuroprosthesis. IEEE Trans. Biomed. Eng. 36, 729–737.

Popovic, M.B., 2003. Control of neural prostheses for grasping and reaching. Med. Eng. Phys. 25, 41–50.

Popovic, M.R., Popovic, D.B., Keller, T., 2002. Neuroprostheses for grasping. Neurol. Res. 24, 443–452.

Prochazka, A., Gauthier, M., Wieler, M., Kenwell, Z., 1997. The bionic glove: an electrical stimulator garment that provides controlled grasp and hand opening in quadriplegia. Arch. Phys. Med. Rehabil. 78, 608–614.

Riehle, A., Requin, J., 1995. Neuronal correlates of the specification of movement direction and force in four cortical areas of the monkey. Behav. Brain Res. 70, 1–13.

Sanchez, J., Principe, J., Carmena, J., Lebedev, M., Nicolelis, M.A., 2004. Simultaneus prediction of four kinematic variables for a brain-machine interface using a single recurrent neural network. Conf. Proc. IEEE Eng. Med. Biol. Soc. 7, 5321–5324.

Santello, M., Flanders, M., Soechting, J.F., 1998. Postural hand synergies for tool use. J. Neurosci. 18, 10105–10115.

Santello, M., Soechting, J.F., 2000. Force synergies for multifingered grasping. Exp. Brain Res. 133, 457–467.

Santhanam, G., Ryu, S.I., Yu, B.M., Afshar, A., Shenoy, K.V., 2006. A high-performance brain-computer interface. Nature 442, 195–198.

Schieber, M.H., 1996. Individuated finger movements: rejecting the labeled-line hypothesis. In: Wing, A.M., Haggard, P., Flanagan, J.R. (Eds.). Hand and Brian. The Neurophysiology and Psychology of Hand Movements. Academic Press, San Diego, pp. 81–97.

Scott, S.H., Kalaska, J.F., 1997. Reaching movements with similar hand paths but different arm orientations. I. Activity of individual cells in motor cortex. J. Neurophysiol. 77, 826–852.

Scott, T.R., Peckham, P.H., Kilgore, K.L., 1996. Tri-state myoelectric control of bilateral upper extremity neuroprostheses for tetraplegic individuals. IEEE Trans. Rehabil. Eng. 4, 251–263.

Sergio, L.E., Hamel-Paquet, C., Kalaska, J.F., 2005. Motor cortex neural correlates of output kinematics and kinetics during isometric-force and arm-reaching tasks. J. Neurophysiol. 94, 2353–2378.

Serruya, M.D., Hatsopoulos, N.G., Paninski, L., Fellows, M.R., Donoghue, J.P., 2002. Instant neural control of a movement signal. Nature 416, 141–142.

Shoham, S., Paninski, L.M., Fellows, M.R., Hatsopoulos, N.G., Donoghue, J.P., Normann, R.A., 2005. Statistical encoding model for a primary motor cortical brain-machine interface. IEEE Trans. Biomed. Eng. 52, 1312–1322.

Snoek, G.J., IJzerman, M.J., in 't Groen, F.A., Stoffers, T.S., Zilvold, G., 2000. Use of the NESS handmaster to restore handfunction in tetraplegia: clinical experiences in ten patients. Spinal Cord 38, 244–249.

Solomonow, M., Aguilar, E., Reisin, E., Baratta, R.V., Best, R., Coetzee, T., et al., 1997. Reciprocating gait orthosis powered with electrical muscle stimulation (RGO II). Part I: Performance evaluation of 70 paraplegic patients. Orthopedics 20, 315–324.

Taylor, D.M., Tillery, S.I., Schwartz, A.B., 2002. Direct cortical control of 3D neuroprosthetic devices. Science 296, 1829–1832.

Thach, W.T., 1978. Correlation of neural discharge with pattern and force of muscular activity, joint position, and direction of next movement in motor cortex and cerebellum. J. Neurophysiol. 41, 654–676.

Ting, L.H., Macpherson, J.M., 2005. A limited set of muscle synergies for force control during a postural task. J. Neurophysiol. 93, 609–613.

Todorov, E., Ghahramani, Z., 2004. Analysis of the synergies underlying complex hand manipulation. Conf. Proc. IEEE Eng. Med. Biol. Soc. 6, 4637–4640.

Tresch, M.C., Cheung, V.C., d'Avella, A., 2006. Matrix factorization algorithms for the identification of muscle synergies: evaluation on simulated and experimental data sets. J. Neurophysiol. 95, 2199–2212.

Tresch, M.C., Jarc, A., 2009. The case for and against muscle synergies. Curr. Opin. Neurobiol. 19, 601–607.

Vodovnik, L., Kralj, A., Stanic, U., Acimovic, R., Gros, N., 1978. Recent applications of functional electrical stimulation to stroke patients in Ljubljana. Clin. Orthop. Relat. Res. 131, 64–70.

Weber, D.J., Stein, R.B., Chan, K.M., Loeb, G.E., Richmond, F.J., Rolf, R., et al., 2004. Functional electrical stimulation using microstimulators to correct foot drop: a case study. Can. J. Physiol. Pharmacol. 82, 784–792.

Wessberg, J., Stambaugh, C.R., Kralik, J.D., Beck, P.D., Laubach, M., Chapin, J.K., et al., 2000. Real-time prediction of hand trajectory by ensembles of cortical neurons in primates. Nature 408, 361–365.

Westwick, D.T., Pohlmeyer, E.A., Solla, S.A., Miller, L.E., Perreault, E.J., 2006. Identification of multiple-input systems with highly coupled inputs: application to EMG prediction from multiple intracortical electrodes. Neural Comput. 18, 329–355.

Wuolle, K.S., Van Doren, C.L., Thrope, G.B., Keith, M.W., Peckham, P.H., 1994. Development of a quantitative hand grasp and release test for patients with tetraplegia using a hand neuroprosthesis. J Hand Surg. Am. 19, 209–218.

Index